Designing Switch/Routers

This book focuses on the design goals (i.e., key features), architectures, and practical applications of switch/routers in IP networks. The discussion includes some practical design examples to illustrate how switch/routers are designed and how the key features are implemented. *Designing Switch/Routers: Architectures and Applications* explains the design and architectural considerations as well as the typical processes and steps used to build practical switch/routers.

The author describes the components of a switch/router that are used to configure, manage, and monitor it. This book discusses the advantages of using Ethernet in today's networks and why Ethernet continues to play a large role in Local Area Network (LAN), Metropolitan Area Network (MAN), and Wide Area Network (WAN) design. The author also explains typical networking applications of switch/routers, particularly in enterprise and internet service provider (ISP) networks.

This book provides a discussion of the design of switch/routers and is written to appeal to undergraduate and graduate students, engineers, and researchers in the networking and telecom industry as well as academics and other industry professionals. The material and discussion are structured to serve as standalone teaching material for networking and telecom courses and/or supplementary material for such courses.

Designing Switch/Routers

Architectures and Applications

James Aweya

CRC Press
Taylor & Francis Group
Boca Raton London New York

CRC Press is an imprint of the
Taylor & Francis Group, an **informa** business

First edition published 2023
by CRC Press
6000 Broken Sound Parkway NW, Suite 300, Boca Raton, FL 33487-2742

and by CRC Press
4 Park Square, Milton Park, Abingdon, Oxon, OX14 4RN

CRC Press is an imprint of Taylor & Francis Group, LLC

ISBN: 978-1-032-31770-0 (hbk)
ISBN: 978-1-032-31583-6 (pbk)
ISBN: 978-1-003-31125-6 (ebk)

DOI: 10.1201/9781003311256

Typeset in Times
by SPi Technologies India Pvt Ltd (Straive)

Contents

Preface

Volume 1 of this two-part book discussed in greater detail the fundamental concepts underlying the design and operation of IP devices and networks, and the various methods for designing, specifically, switch/routers. **Volume 1** described in particular, the basic ideas that are very important and essential for understanding the role of switch/routers in IP networks, and the procedures and tools used for designing them. **Volume 2** focuses more on the design goals (i.e., a switch/router's key features), architectures, and practical applications of switch/routers in IP networks. The discussion includes some practical design examples to illustrate how switch/routers are designed as well as the steps involved in implementing the key features.

Chapter 1 and 2 of this volume discuss the design goals and features switch/router manufacturers consider when designing their products. The discussion in **Chapter 1** covers the basic features typically seen in the typical high-end, high-performance switch/routers. **Chapter 1** discusses the requirements and benefits of non-blocking architectures, device and port/interface scalability, QoS control, network security and device access control, multicast forwarding, and node redundancy and resiliency. The discussion on node redundancy and resiliency covers the mechanisms that allow a routing device to continue forwarding packets while the control plane recovers from a failure, and the control plane software restart mechanisms that allow the software components/processes of the control plane to be restarted to recover state information when a fault occurs without disrupting the data or forwarding plane.

Chapter 2 discusses the advanced and value-added features seen in switch/routers. The discussion includes device management (using Simple Network Management Protocol (SNMP)), device access (using Terminal Access Controller Access-Control System Plus (TACACS+), Remote Authentication Dial-In User Service (RADIUS), IEEE 802.1X, and Secure Shell (SSH)), traffic monitoring (using sFlow, NetFlow, and Remote network monitoring (RMON)), value-added features (such as Dynamic Host Configuration Protocol (DHCP) and Domain Name System (DNS)), and the factors to consider when designing for energy efficiency.

Chapter 3 discusses the design and architectural considerations, as well as the typical processes and steps used to build practical switch/routers. The discussion covers how a switch/router can be built by coalescing all the concepts and methods discussed in **Volume 1** and the previous chapters of **this volume**. This chapter also discusses the components of a switch/router that are used in configuring, managing, and monitoring it.

The aim of this book is to provide the reader with a deeper understanding of the basic concepts underlining switch/routers, and the various methods and tools available for designing them. The discussion in **Volume 1** considers the switch/router as a generic Layer 2 and Layer 3 forwarding device without placing emphasis on any particular manufacturer's device. The underlining concepts and design methods are positioned to be applicable not only to this generic switch/router but also to the typical switch/router seen in the industry. **Chapters 4 and 5** of this volume discuss real-world switch/router architectures that employ the concepts and design methods described in the previous chapters. This is to provide the reader with example archi-tectures that illustrate how real-world switch/routers are designed.

Once considered a networking technology mainly for data services, Ethernet has evolved over the years to be widely accepted as a network convergence technology for delivering the full array of packet-based data, voice, and video services. **Chapter 6** discusses the advantages of using Ethernet in today's networks and why Ethernet continues to play a bigger role in Local Area Network (LAN), Metropolitan Area Network (MAN), and Wide Area Network (WAN) design. The discussion includes the benefits of using Ethernet, industry trends related to the use of Ethernet, and the role of Ethernet in network infrastructure virtualization.

Chapter 7 covers the typical networking applications of switch/routers, particu-larly, in enterprise and Internet service provider (ISP) networks. A switch/router uses Ethernet at Layer 1 and 2, and IP at Layer 3. Thus, given that Ethernet has evolved to become the dominant and widely used technology for building LANs, MANs, and WANs, this evolution has also influenced how switch/routers are used in enterprise and ISP networks. The discussion in this chapter includes some industry trends and some important applications of switch/routers such as in data centers and application hosting, high-performance computing, and enterprise infrastructures.

Author

James Aweya, PhD, is a Chief Research Scientist at the Etisalat British Telecom Innovation Center (EBTIC), Khalifa University, Abu Dhabi, UAE. He was a technical lead and senior systems architect with Nortel, Ottawa, Canada, from 1996 to 2009. He was awarded the 2007 Nortel Technology Award of Excellence (TAE) for his pioneering and innovative research on timing and synchronization across packet and TDM Networks. He has been granted 70 US patents and has published over 54 journal papers, 40 conference papers, and 43 technical reports. He has authored six books including this book.

Author

James Aryan, PhD, is a Chief Research Scientist at the Etisalat British Telecom Innovation Center (EBTIC), Khalifa University, Abu Dhabi, UAE. He was a technical lead and senior systems architect with Nortel, Ottawa, Canada, from 1996 to 2009. He was awarded the 2007 Nortel Technology Award of Excellence (TAE) for his pioneering and innovative research on failure and synchronization across packet and TDM Networks. He has been granted 20 US patents and has published over 58 journal papers, 40 conference papers, and 13 technical reports. He has authored six books including this book.

1 High-Performance Switch-Routers
Part 1: Basic Features

1.1 INTRODUCTION

Current and emerging switch/routers (as well as switches and routers) are designed to meet the varying needs and challenges of modern networks which are continuously evolving. Real-time transactions and services (e.g., high-definition broadcast video, Virtual Reality/Augmented Reality, massive Internet-of-Things (IoTs), remote surgery, factory automation, power system control, real-time traffic information service, autonomous driving), data centers, cluster computing, and Grid computing applications are some examples that are most definitely more dependent on network performance than other applications. In most cases, these applications require the network to deliver consistently high-performance and non-stop availability over long time periods. Therefore, designing networks for time- and resource-critical applications involves selecting network devices that have the performance and robustness features that can provide the optimum environment for successful running of these applications.

This chapter and the next, describe the features that current and next-generation switch/routers must possess if they are to successfully support time- and resource-critical applications as well as emerging business models and service demands. Switch/routers are typically based on Ethernet and IP because of the evolution and proven effectiveness and practicality of these networking technologies for converged networks (that handle a wide range of applications and services). Ethernet and IP have become the networking technology of choice for building scalable, high-reliability, service-aware LANs, MANs, and WANs.

In order to deliver predictable service levels and meet required Service Level Agreements (SLAs), network designers need to employ network devices with well-balanced designs characterized by a wide range of features and attributes. To meet these challenges, current high-capacity switch/routers are designed with a number of technical goals in mind. Most switch/router designs strive to meet most if not all of these requirements. This chapter and the next, discuss the key features (some of which are advanced or optional (value-added)) that switch/routers support.

A high-availability network has become increasingly important for business operations. At the same time, it is equally important for a network designer to understand the total cost of ownership (TCO) for network infrastructure. It is now customary for enterprises to compare and contrast network equipment vendors' technical capabilities and the various performance and feature capabilities of competing platforms, as

DOI: 10.1201/9781003311256-1

well as the operational issues that impact TCO and the ability to meet day-to-day network performance requirements.

The current generation of market-leading switch/routers has been designed from the outset to meet the most demanding requirements for increased levels of aggregated traffic in data centers, as well as the access, aggregation, and core tiers of networks. Switch/routers with additional advanced features beyond the basic ones provide the enabling features for modern networks that are rapidly employing consolidation and virtualization of networking, computing, and storage resources. Some of these advanced features (some of which can be considered optional when considering customer requirements) are discussed in these chapters.

This chapter discusses the requirements and benefits of non-blocking architectures, device and port/interface scalability, quality of service (QoS), network security and device access control, multicast forwarding, and network node redundancy and resiliency. The discussion on node redundancy and resiliency covers the mechanisms that allow a routing device to continue forwarding packets while the control plane recovers from a failure, in addition to the control plane software restart mechanisms that allow the software components/processes of the control plane to be restarted in order to recover state information when a fault occurs, without disrupting the data or forwarding plane.

1.2 NON-BLOCKING ARCHITECTURE

A switch/router might serve as a point where all network user traffic converges, as a result, it must hold enough bandwidth capacity for all the expected connected users. The switch/router must fulfill this requirement with a switch fabric capable of accommodating the heaviest traffic loads. The modules within the switch/router including the network interface modules are typically built with high-performance, high-bandwidth components (e.g., crossbar switch fabrics, forwarding engines, packet classifiers) capable of serving traffic to and from all the network interfaces on the module without any sacrifice in performance, or any packet being blocked by another packet due to contention for limited internal processing resource. The use of virtual output queuing (VOQ) eliminates head-of-line (HOL) blocking [TAMIRY88], making the switch fabric more available for data transfer.

HOL blocking occurs when a single queue on each input port on the switch fabric is used to forward traffic to the output ports [KAROLHM87]. Let us assume that the packet at the head of the input queue (i.e., HOL packet) is destined for a congested destination port while the packets following it are destined for non-congested destination ports. The input queue is prevented from sending packets to all other non-congested destination ports because the packet destined to the congested destination port is holding up (or blocking) all other packets in the queue destined for the non-congested destinations. Only when the congested destination port is clear can the HOL packet and all the other packets following can be forwarded. Older crossbar switch fabric architectures with single queuing at input ports suffer from this problem because they do not have multiple, parallel input queuing paths (VOQs) to allow packets to non-congested output ports to bypass the HOL packet.

Another type of HOL blocking occurs when low-priority packets whose destination is congested block high-priority packets needing to traverse the switch fabric. Older crossbar switch architectures also suffer from this type of problem because they do not have multiple priority input queues.

The use of an input queue per each output port on each input port in conjunction with appropriate traffic scheduling algorithms effectively avoids both of these HOL blocking conditions [DAIPRAB00] [MCKEOWN96] [MCKEOWN98]. In this case, when an input port has data for multiple destination ports, the crossbar switch fabric provides simultaneous parallel paths from the input port to the destination ports. One congested destination port does not affect any of the other non-congested destination ports. In the event that multiple input ports have data for the same destination port, the traffic scheduling algorithm ensures that all input ports are able to deliver their packets to the destination port without locking out any particular input port. No single input port is allowed to dominate any other input port.

The combination of wire-speed performance and non-blocking architecture makes it easier for the network designers to provide bandwidth guarantees for the wide range of applications on the network. A non-blocking architecture, also, avoids packet losses and makes packet delay and packet delay variation (PDV) more predictable. Low latency is a key attribute of high-performance network-based computing and is also important for delay-sensitive applications (e.g., real-time streaming voice and video) especially in large-scale networks where network node delay and PDV could significantly build up.

Some desirable attributes of non-blocking switch/router architectures can be summarized as follows:

- The switch/router should support full line-rate, non-blocking packet forwarding performance even when the chassis is fully populated with the densest interface cards available.
- Non-blocking architecture provides the most consistent and predictable performance possible (delay, PDV, and packet loss), irrespective of traffic patterns.
- Packet forwarding performance should not be compromised (should be line-rate and non-blocking) even when all other Layer 3 services enabled, such as the full range of traffic management/control and QoS services supported on the device.

1.3 HIGH SCALABILITY

Network planning and deployment can occur over an extended period, and budget constraints may require a phased deployment. The scalability of a switch/router is determined by its ability to support application/service requirements in a rapidly expanding network environment. Network designers need to deploy network infrastructures that address today's and tomorrow's requirements with a scalable architecture designed to support growth and evolution. A well-designed switch/router makes it easy to deploy a solution today that can be upgraded later to support more ports, higher capacity ports, or switch stacking or clustering as needed. Scalability can also be achieved by having the switch/router support upgradeable line cards, power supplies, etc.

1.3.1 STACKABLE SWITCHES

Using *switch stacking* technology, multiple switches are interconnected to create a single logical unit, using stack interconnect cables. The *stack* (which refers to the set of interconnected switches) behaves as a single unit that is managed by a member switch (the master switch) that is elected as the master switch of the stack. Using stacking, two or more physically separate switches (up to 9) may be interconnected through stack cables so that they form one logical switch (Figure 1.1). Switch members of a stack can be categorized as *master* or *slave*.

A stack is managed as one unit and has a single IP address for management. Stacking allows the backplane of the individual switches to be connected, providing full backplane speed connectivity between the switches. Stacking is easily scalable but has a limited number of stack members and is generally limited to a single wiring

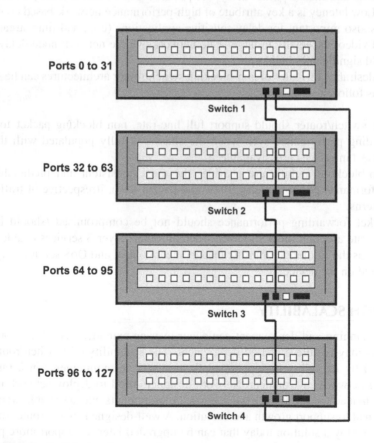

FIGURE 1.1 Switch stacking example.

closet. Although stacking offers low entry price and high performance, it is difficult to add or delete stack members and perform detailed troubleshooting in the stack.

For cost-effective and rapid scaling at the network edge, a network designer may deploy switches that support stacking on a virtual chassis. The stacking technology could support stacked configurations in which as many as up to eight switches, for example, can be interconnected while maintaining the operational simplicity of a single switch. In addition, the stacking technology could support higher switching capacity between stacking units, providing a high-capacity interconnect across the stack.

A stacked configuration might support multi-gigabits of stacking bandwidth per unit. Stacking could be done over copper and fiber cables. For example, a stacked configuration might be built using 10 Gigabit Ethernet copper or fiber connections. When fiber connections are used, the stacked configuration can be extended between racks, floors, and buildings with fiber lengths up to several hundred meters. This allows flexible stacked configurations in which stacked units can be separated by longer distances and not necessarily co-located.

The stacked system (configuration) typically operates as a single logical chassis (with a single IP management address) and supports cross-member trunking, mirroring, switching, static routing, NetFlow or sFlow [RFC3176], multicast snooping, and other switch functions across the stack. The stacked system might have a single configuration file and support remote console access from any stack member. Support for active-standby controller failover, stack link failover, and hot insertion and removal of stack members delivers the resilience that is typical of higher end modular switches.

1.3.2 STACKABLE VERSUS MODULAR SWITCHES

Generally, switches come in two form factors that allow for scalability: stackable versus modular. Table 1.1 presents the relative advantages of stackable versus modular switch/routers. The two form factors are quite complementary, as evidenced by the fact that most medium to large networks employ a mix of modular and stackable products in both wiring closets and data centers [FOR10FPWC08].

TABLE 1.1
Relative Advantages of Stackable versus Modular Switch/Routers

Stackable Switches	Modular Switches
• Better low-end scalability (fewer standalone desktop/closet switches) • Lower entry price • Lower price per port • Less rack space consumed per user port	• Greater resiliency/redundancy features • Better high-end scalability • Better control of subscription ratio • More bandwidth for local switching • Better power efficiency (Gbps/Watt) • Greater flexibility and upgrade ability • Better ability to assimilate new technologies • Longer service life

Stackable switches offer advantages for smaller wiring closets, where the growth rate in the number of user ports required is fairly low, and where there may be a considerable degree of sensitivity to the capital equipment component of TCO. Because of these advantages, fixed configuration and stackable switches tend to be preferred by the majority of small- to medium-sized enterprises.

On the other hand, modular switches are generally preferred for larger wiring closets where the emphasis is on future-proofing the wiring closet investment. Modular devices have the advantage of superior flexibility to accommodate rapid growth, assimilate new technologies or industry standards, and minimize TCO measured over a longer service lifetime. The service lifetime of the modular switch can often be extended a number of times through a series of backward-compatible upgrades to the various subsystems, including the switch fabric, route processors, line cards, and power and cooling systems.

1.3.3 SWITCH CLUSTERING

In today's converged networks, reliability goes beyond individual node performance. A high number of networks now demand the support of switch clustering technology to extend sub-second failover and full session load-sharing across the network infrastructure. Unlike switch stacking, *switch clustering* is a logical group of independent switches that extends beyond the wiring closet and are not interconnected using stack cables (Figure 1.2). Members of the cluster are not necessarily co-located and have

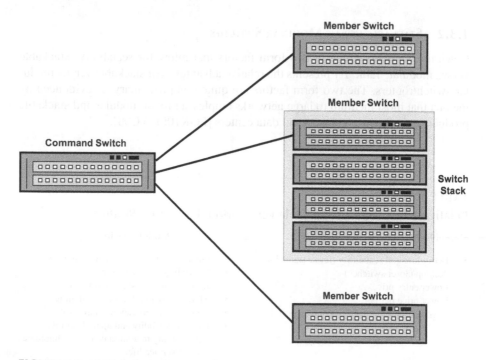

FIGURE 1.2 Switch clustering example.

no particular physical boundary. The cluster has a single point of authentication and management.

Switch clustering technology enables up to a maximum of N switches to be interconnected (N equal to 16 in the Cisco Switch Clustering technology [CISC2950DSS] [CISC3850SSW] [CISCCACCFCS]) to form a single managed IP addressed network system (managed simultaneously through a single IP address). *Switch Clustering* is essentially a method to manage a group of switches without the need to assign an IP address to every switch. The cluster is provided a single virtual IP address for full cluster management which can be Web-based management. In [CISC2950DSS] [CISC3850SSW] [CISCCACCFCS], switches within a cluster have one of these roles:

- **Command Switch**: Each cluster (using Cisco Switch Clustering) has a master switch called a *command switch*. A switch cluster must have one switch designated as the command switch and up to 15 switches can be *member switches*. The total number of switches in a cluster cannot exceed 16 switches. The command switch provides the primary management interface for managing the entire switch cluster. The command switch is the single point of access for configuring, managing, and monitoring the member switches. The command switch is typically the only switch configured with an IP address within the switch cluster. All communication with the switch cluster is through the command switch's IP address. The command switch receives each management request and then redirects it to the appropriate member switch. For redundancy, the network manager can configure a *backup* or *standby command switch*. The standby command switch must be the same model as the command switch [CISC2950DSS] [CISC3850SSW] [CISCCACCFCS]. One or more switches can be designated as standby command switches to avoid loss of contact with cluster members. A *cluster standby group* is a group of standby command switches. For redundancy, a second switch (the standby switch) can be assigned an IP address and then the overall cluster can be managed through a single virtual IP address. If the primary command switch fails, the standby command switch seamlessly takes over the management of the switch cluster, while the network administrator can still access the switch cluster via the virtual IP address.
- **Member Switches**: All other switches (apart from the command switch) serve as *member switches*. Member switches are switches that have actually been added to a switch cluster. An IP address is typically not configured on a member switch. It receives management commands that the command switch has redirected to it.
- **Candidate Switches**: A *candidate switch* is a switch that the network administrator can add to the switch cluster as a member switch. Basically, candidate switches are cluster-capable switches that have not yet been added to a cluster.

The benefits of clustering switches include the management of switches regardless of their interconnection media and their physical locations. The switches can be in the same location, or they can be distributed across a Layer 2 network. Cluster members are connected through the management VLAN of the command switch [CISC2950DSS] [CISC3850SSW] [CISCCACCFCS].

Switch clustering provides protection against any individual component, link, or node failure. This solution provides for sub-second recovery combined with user session-based load-balancing, all leveraging standards-based dynamic link aggregation at the edge of the network, both at a user and server.

1.3.4 GENERAL SCALABILITY FEATURES

Scalability enables the network designer to size the network to meet current needs and still have room for future growth. The goal is a network that is flexible enough to meet the demands of today but can scale and adapt as the organization grows. As network traffic increases, network operators can easily upgrade to, for example, 10 and higher multi-gigabit Ethernet to provide high-capacity connectivity to the network backbone and/or high-performance data servers.

A scalable forwarding performance requires a distributed non-blocking switch architecture with line card implementations that maintain full line-rate forwarding performance (possibly, full distributed forwarding) and predictable latency irrespective of the traffic load, QoS features, access control lists (ACLs), traffic monitoring, or other services that have been configured for traffic control or security. Scalable port density is achieved through the combination of line cards with scalable forwarding performance and a scalable switch fabric which has ample switching capacity to deliver non-blocking full line-rate performance for large numbers of high-density line cards.

A switch/router may also employ specialized ASICs, advanced network processors, and high-speed memory on each line card to provide a scalable high-performance architecture. In this architecture, the forwarding information base (FIB) is downloaded to the hardware-based forwarding engine on each line card. In this case, each line card is pre-populated with an FIB for wire-speed packet forwarding performance.

The control plane of the switch/router comprises the software processes used to construct and maintain the FIB, handle Layer 2/Layer 3 forwarding functions, and perform the necessary network management tasks. A scalable control plane allows the switch/router to support large FIBs and also provide the performance capacity to prevent contention among control functions from degrading the system's response time to network topology changes or other changing conditions in the network.

In addition to scalable forwarding and control planes, overall system scalability has a number of additional dimensions including the number of peers/neighbors supported, the number of routing and forwarding table entries, the number of MAC addresses and VLANs supported, the type and depth of VLAN stacking supported, the number of standard and extended ACLs that can be configured, the amount of packet buffering allocated to each network interface, the aggregate bandwidth supported by the switch fabric, and the packet forwarding performance/capacity provided by each line card.

1.4 HIGH PORT AND INTERFACE DENSITY

In most network environments, space in the wiring closet and networking room is premium and scarce. This makes high interface density another important requirement for high-end switch/routers, especially those deployed in large data centers and large enterprise and service provider networks. In these applications, switch/routers should be capable of supporting large numbers of Gigabit Ethernet and multi-gigabit Ethernet interfaces in compact enclosures. Therefore, the port density (measured in terms of ports per *rack unit* (RU)) and the maximum number of ports per device are important metrics that should be considered when making growth projections for the wiring closet.

High-density switching simplifies network design, makes efficient use of scarce rack space, and minimizes the cost and complexity of interconnecting and managing multiple switch chassis in the data center or across enterprise and service provider networks. High-density switch/routers, capable of performing Layer 2 and Layer 3 forwarding, provide an effective design approach for networks, allowing even an entire design of a network segment to fit within a single platform within each wiring closet, resulting in savings in equipment space and operating costs.

Thus, high port density has become an integral component of switch/router scalability. Scalable switch/routers are designed to accommodate subsequent generations of network technologies without radical architectural changes or render obsolete earlier generation systems. Highly scalable, high port count switch/routers can reduce the total number of devices required in the traditional two-tier data center network based on access and distribution/aggregation tiers. Even greater reduction and consolidation can be achieved by using high-density switch/routers to collapse the two tiers of data center switches into a single access/aggregation tier.

High density of ports in a rack space enables organizations to design highly flexible and cost-effective networks. In addition, organizations can utilize various combinations of short-range and long-range transceivers to provide a variety of user connectivity options. High port density is a requirement regardless of the type of connectivity. The switch/router should provide enough ports to support some of the most complex networking environments. In addition to being suitable for the traditional enterprise environment, a high-density switch/router is an ideal solution for high-performance computing environments and Internet Exchanges and Internet Service Providers (IXPs and ISPs) where non-blocking, high-density switches are needed.

1.5 IMPROVED MODULARITY AND FLEXIBILITY – MODULES WITH MIXED 10/100/1000 MB/S, 10 GB/S, AND HIGHER GIGABIT SPEED INTERFACES

For flexibility, the switch/router could offer a wide range of interfaces and uplink technologies. The switch/router/router could offer high-density 10/100/1000 Mb/s, 10 Gb/s, 25 Gb/s, 40 Gb/s, and even higher multi-gigabit Ethernet modules that can readily co-exist in the same system. To ensure compatibility with existing infrastructures and provide long-term investment protection, the switch/router architecture could support interfaces ranging from basic 10/100/1000 Mb/s Ethernet to higher multi-gigabit interfaces.

To support flexible network design, the switch/router could be built (ordered or field-upgraded) with a wide range of low- to high-capacity interfaces, allowing for a full breadth of networking interconnectivity. With some interfaces acting as uplinks, network designers can easily build redundancy into their networks and also take advantage of low-cost optical fiber connectivity.

The high-density optical interfaces enable network operators to design flexible, cost-effective networks that can grow with application requirements. Designers can mix and match various interface combinations enabling the switch/router to offer both short-range "fiber at the desktop" and "fiber-to-the-home" connectivity. The switch/router can be deployed to deliver connectivity to the desktop, high-density aggregation within the distribution layer, and connectivity for high-performance computing, grid computing, and network-attached storage. Support for Ethernet jumbo frames of up to 9,126 bytes may be included to ensure faster file transfer between high-end servers within the data center and reduce server CPU load.

Multi-gigabit interfaces provide large enterprise customers non-disruptive migration to higher speeds and performance levels with a lower TCO. Such multi-gigabit interfaces significantly increase the core network capacity to support high-bandwidth, mission-critical business applications. These high-capacity interfaces can also be deployed to create high-speed enterprise networks across significantly distant campuses, or within existing data centers where over-provisioning of critical links is needed to avoid degradation of network performance. The multi-gigabit uplinks supported by the switch/router ensure that data center can be effectively connected to other switches, enabling concurrent support for low-latency and mission-critical applications, and high-volume network traffic.

1.6 ADVANCED QoS FEATURES FOR GUARANTEED DELIVERY OF TIME-CRITICAL, DELAY-SENSITIVE TRAFFIC

In both the enterprise and service provider networks, the underlying issue for managing QoS is satisfying the wide range of end-user application requirements. Using traffic control along with admission control mechanisms provides a more promising path toward supporting guaranteed service levels. QoS mechanisms are a set of mechanisms that protect selected traffic from the effects of network congestion when the network resources on a link, port, or switch fabric are oversubscribed. These mechanisms may be applied to avoid excessive delay and delay variation for real-time applications. Other mechanisms may also be applied to avoid packet loss for critical eBusiness data applications.

By incorporating comprehensive QoS capabilities in network devices, network operators would be in a better position to protect mission-critical applications and delay-sensitive traffic (Figure 1.3). Latency, delay variations and data loss that would be minor issues in traditional data networks become major concerns when dealing with real-time traffic like streaming voice and video. In modern multiservice networks, incorporating QoS control mechanisms has become an essential requirement. These QoS features have to be fully compatible with industry standards to allow

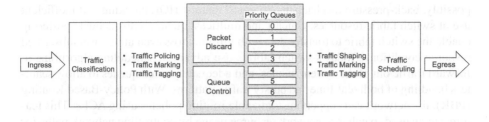

FIGURE 1.3   Key QoS elements in Layer 2/3 forwarding.

device interoperability and ensure customer confidence in the long-term viability of provided solutions.

More diverse and robust QoS capabilities like IETF (Internet Engineering Task Force) Differentiated Services (DiffServ), IP Precedence marking, IEEE 802.1p, QoS mapping, sophisticated classification and low latency queuing, traffic management, among others, have to be supported. These QoS features enable more efficient use of network bandwidth and optimizing of existing network resources and capabilities. Typically, networks provide packet classification and marking at the edge in order to simplify QoS control at the core. By classifying, prioritizing, policing, and marking traffic, networks are in a better position to deliver the right service levels for mission-critical and quality-sensitive applications.

Network devices that support superior QoS features enable network operators to provide and ensure high-quality services throughout the network and end to end. A network device may identify, classify, re-classify, police, and mark traffic prior to delivery based on a number criteria and network operator policy. The criteria can include network device port, source/destination MAC address, IEEE 802.1p priority field, source/destination IP address, IP Precedence or Differentiated Services Control Point (DSCP) fields, and TCP/UDP port numbers. Traffic can be classified by network to differentiate various traffic flows and enforce network resource allocation policies based on above criteria.

Once classified, traffic is queued and scheduled for delivery – the network operator can exercise control over how the system services the queues. Weighted Round Robin (WRR) queuing ensures that all packets can be delivered and lower-priority packets are not starved for bandwidth. Strict Priority (SP) queuing ensures highest-priority traffic always gets serviced first, ahead of all other traffic. Combining SP with WRR guarantees highest priority traffic delivery and servicing of lower priority queues.

A switch/router with advanced QoS features should be capable of performing rate-limiting, giving the network the control needed to regulate bandwidth utilization. On the ingress, extended ACLs, in combination with rate-limiting traffic policies, can be used to balance, fine-tune, and control bandwidth consumption. On the egress, outbound rate limiting can be used to control bandwidth per interface and per priority queue. Using these advanced QoS features, services can be delivered without applications suffering reduced performance.

QoS also has to be built into the switch fabric of network devices. Both ingress and egress buffering can be used, including the required traffic scheduling and possibly, back-pressure mechanisms, to guard against HOL blocking and inefficient use of switch fabric resources. Unicast and multicast queues with adequate buffering enable the switch fabric to transfer data with minimal loss even under oversubscribed resource conditions. Today's switch fabrics use advanced scheduling algorithms to forward traffic out of the system queues, and adequately sized queues to allow seamless handling of both real-time and bursty traffic patterns. With Policy-Based Routing (PBR), the network can support customizable routing policies using ACLs. This feature can be used to enhance network resource usage by controlling network paths for different traffic flows.

Considering the QoS requirements described above, perhaps the biggest problem with today's switch/routers is the performance degradation they suffer when classification and QoS features are turned on. As opposed to the traditional software-based architectures, the architectures with hardware-based classification and packet forwarding provide line-rate forwarding performance. Most high-performance architectures use custom ASICs that operate at line-rate and support packet classification, filtering, statistics collection, traffic policing, and shaping. These ASICs also support Layer 2 and Layer 3 hardware-based protocol-specific processing at line-rate. With hardware-based and service-aware QoS capabilities, high-performance switch/routers enable network operators to honor end-user traffic with varying QoS requirements.

1.7 ENHANCED NETWORK SECURITY AND ACCESS CONTROL

Today's enterprises require cost-effective, flexible, and secure solutions for communication. Lapses or failures in network security can have a costly impact on the profitability of companies. Security of networks has become a major concern for companies due to the proliferation of viruses and other threats in network which can be spread via infected or unauthorized systems. Malicious software can spread behind firewalls and intrusion-detection systems (IDSs) that are supposed to protect against network intrusions.

Securing the control plane of routing devices has also become critical to allow proper operation of the devices and the network as a whole. If the control plane is compromised, it is almost impossible to guarantee the correct operating state of the network. A compromised control plane can lead to unintended routing, and service disruption where packets are not forwarded and delivered to their intended destinations.

A key element of proper network design is the support of comprehensive security services for all aspects of network operation. A multi-level approach to security is also required to protect the network, data centers, network devices, services, and applications from external or even internal attacks. Layer 2 and Layer 3 ACLs and filtering capabilities in addition to advanced security features in switch/routers, for example, allow network operators to build firewalls that prevent unauthorized network access. Permit/deny filters can be created based on Layer 2 and Layer 3 packet information (e.g., source and destination MAC and IP addresses, UDP/TCP port numbers, higher-layer protocol information).

Modern switch/routers are now required to offer advanced security features for Layer 2 and Layer 3, network access control (NAC), control plane protection, and

DoS protection. Multiple levels of network device access, user authentication and authorization, and protection for Distributed Denial of Service (DDoS) attacks should be integral elements of the switch/router architecture. The security features can include protection against TCP SYN and Internet Control Message Protocol (ICMP) DoS attacks, Spanning Tree Root Guard and Bridge Protocol Data Unit (BPDU) Guard to protect Spanning Tree Protocol (STP) operation, and broadcast and multicast packet rate limiting. In some configurations, the switch/router offers additional security features including dynamic Address Resolution Protocol (ARP) inspection and Dynamic Host Configuration Protocol (DHCP) snooping to protect against address spoofing and man-in-the-middle attacks.

To protect the network against DoS attacks, the network operator can disable the forwarding of ICMP messages and also enable an option to rate limit ICMP and TCP SYN packets. The switch/router monitors, throttles, and locks out ICMP and TCP SYN traffic both to the management address of the switch/router and traffic transiting the system. Enabling this feature secures and protects the network from user-generated DoS attacks or aiding them. To prevent "user identity theft" (spoofing), the switch/routers may support DHCP snooping, Dynamic ARP inspection, and IP source guard. These three features work together to deny spoofing attempts and to defeat man-in-the-middle attacks.

Network operators can rely on security features such as port authentication and IEEE 802.1X authentication [IEEE 802.1X]. The operator can use dynamic policy assignment to control network access and perform targeted authorization on a per-user level. IEEE 802.1X (MAC Port Security) can be used to control the MAC addresses allowed per port. Additionally, the switch/router may support enhanced MAC policies with the ability to deny traffic to and from a MAC address on a per-VLAN basis. This powerful tool allows network administrators to control access policies per endpoint device.

Standards-based NAC enables network operators to deploy inter-operable and future-proof NAC solutions for authenticating network users and validating the security posture of a connecting device. Support for policy-controlled MAC-based VLANs provides additional control of network access, allowing policy-controlled assignments of devices to Layer 2 VLANs.

Once a network port is operational, the network administrator can use both regular and extended ACLs to control access to and through the network. Network operators can enable access control policies that can permit or deny traffic based on a wide variety of identification characteristics, such as source/destination MAC addresses, source/destination IP addresses, and TCP/UDP port numbers – further protecting and restricting network access from malicious users. Some high-end switch/routers implement ACL lookups in hardware, ensuring that security and protection for the network does not adversely affect routing and packet forwarding performance.

The switch/router may support configurable user-selectable security such as MAC address lockdown. The network administrator can assign a single MAC address or a group of addresses to an individual port in order to prevent unauthorized users from accessing the network. Using Remote Authentication Dial-In User Service (RADIUS) authentication servers [RFC2865] [RFC2866] [RFC3576], the network administrator can enable IEEE 802.1X port-based authentication – ensuring that the switch/

router first authenticates a user before allowing the port to transmit data onto the network. This also grants users secure mobility, while maintaining the integrity and security of the network against unwarranted breaches. In addition, enhanced Spanning Tree features such as Root Guard and BPDU Guard prevent rouge hijacking of the Spanning Tree root and maintain a contention- and loop-free environment especially during dynamic network deployments.

Organizations may need to set up lawful traffic intercept due to today's heightened security environment. For example, in the United States, the Communications Assistance for Law Enforcement Act (CALEA) requires businesses be able to intercept and replicate data traffic directed to a particular user, subnet, port, etc. This capability is particularly essential in networks implementing IP phones. A switch/router may provide the capability necessary to support this requirement through ACL-based mirroring, MAC filter-based mirroring, and VLAN-based mirroring.

Network operators may apply a "mirror ACL" on a port to mirror a particular traffic stream based on IP source/destination address, TCP/UDP source/destination ports, and IP protocols such as ICMP, and IGMP, etc. A MAC filter may be applied on a port to mirror a traffic stream based on a source/destination MAC address. VLAN-based mirroring is another option for CALEA compliance. Many enterprises have service-specific VLANs, such as voice VLANs. With VLAN mirroring, all traffic on an entire VLAN on a switch can be mirrored to a remote server.

Support for embedded, hardware-based NetFlow or sFlow traffic sampling [RFC3176] could allow extending the switch/router's security shield to the network edge. The switch/router may use NetFlow or sFlow to provide real-time network visibility so that organizations can more effectively manage the transactions flowing throughout the network. NetFlow and sFlow use traffic sampling to achieve scalable network monitoring on high-speed gigabit rate networks. This allows network devices to implement high-speed threat mitigation solutions that use intrusion detection systems based on inspecting NetFlow or sFlow traffic samples for possible network attacks.

In response to a detected attack, the switch/router can apply a security policy to the compromised port. Automated NetFlow or sFlow plus threat detection and mitigation stops network attacks in real time, without human intervention. This is one example of advanced security capabilities that can provide a network-wide security umbrella without the cost and added complexity of supplementary security sensors.

The consolidation of data centers simplifies the deployment of virtualization technologies but also places increasing emphasis on reliable and effective data center security. In the virtual data center, security could be provided by:

- Application and control VLANs to provide traffic segregation
- Wire-speed switch/router ACLs applied to intra-data center and inter-data center traffic
- Stateful virtual firewall capabilities that can be customized to specific application requirements within the data center
- Security-aware functionality for load balancing and other traffic management and acceleration functions

- Intrusion Detection System/Intrusion Prevention System (IDS/IPS) functionality at full wire-speed for real-time protection of critical data center resources from attacks
- Authentication, authorization, and accounting (AAA) for controlled user access to the network and network devices, enforcing policies defining user authentication and authorization profiles

In addition to application VLANs, control VLANs can be configured to isolate control traffic (exchanged by the network devices) from application traffic. For example, control VLANs may be defined to carry routing updates between switch/routers. In addition, a redundant pair of load balancers or stateful firewalls may share a control VLAN to permit traffic flows to failover from a primary to secondary appliance without loss of state or session continuity. In a typical network design, trunk links may be configured to carry a combination of traffic for application VLANs and link-specific control VLANs. Where application VLANs are extended beyond the data center, the trunks between the data center core switches will carry a combination of Layer 2 and Layer 3 traffic.

1.8 EFFICIENT MULTICAST SUPPORT

The use of broadcast video, video conferencing, online multi-user gaming, financial trading, and other one-to-many applications requires networks that support scalable multicast services. The use of video conferencing applications in the workplace, for example, requires support for scalable multicast services from the network edge to the core. Service providers that want to offer high-end services such as IPTV or Video-On-Demand services benefit a lot from the multicast features supported in network devices.

To support business-critical multicast applications, high-performance switch/routers typically support efficient multicast capabilities using hardware-based multicasting that efficiently forwarding traffic on high-speed interfaces. A good switch/router design replicates multicast packets only once at distribution points, sending them to all destination ports (with multicast receivers) simultaneously. It forwards a single copy of a multicast transmission only to ports having multicast receivers. This approach results in high data transmission efficiency and low latency.

Switch/routers typically support a rich set of Layer 2 multicast snooping features that facilitate advanced multicast services delivery. Internet Group Management Protocol (IGMP) snooping for IGMP versions 1, 2, and 3 (IGMPv1, IGMPv2, IGMPv3) is typically supported. Support for IGMPv3 source-based multicast snooping improves bandwidth utilization and security for multicast services. To enable multicast service delivery in IPv6 networks, the switch/router has to support Multicast Listener Discovery (MLD) version 1 and 2 (MLDv1, MLDv2), which are the multicast protocols used in IPv6 environments.

Relevant multicast protocols and capabilities like Protocol Independent Multicast Sparse Mode (PIM-SM) and PIM Dense Mode (PIM-DM), PIM Source Specific Multicast (PIM-SSM), Multicast Source Discovery Protocol (MSDP), multicast

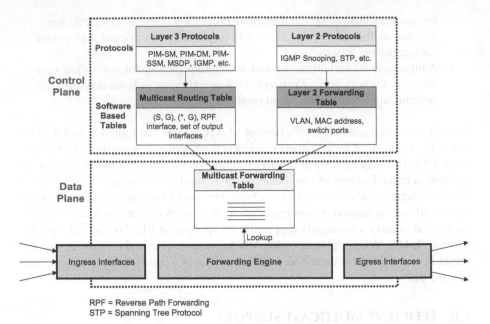

RPF = Reverse Path Forwarding
STP = Spanning Tree Protocol

FIGURE 1.4   IP multicast routing protocols, and Layer 2 and Layer 3 forwarding components.

Rendezvous Points, and PIM snooping may also be supported (Figure 1.4). These protocols provide various ways of "pruning" multicast traffic distribution to reduce unnecessary and wasteful bandwidth consumption, thus, improving overall network performance while conserving bandwidth.

In Layer 3 networks, support for IGMP (v1, v2, and v3), IGMP Proxy, PIM-SM, PIM-SSM, and PIM-DM optimizes multicast routing and network utilization. A switch/router may support storm control features to contain and intelligently forward rather than broadcast multicast traffic. IGMP and PIM snooping improves bandwidth utilization in Layer 2 networks by restricting multicast flows to only those switch ports that have multicast receivers.

PIM and IGMP Snooping offer efficient handling of multicast traffic in Layer 2 topologies by identifying ports that have active multicast receivers and forwarding the multicast stream only on these ports. This dramatically improves the performance of multicast routing, allowing many more streams to transit the network. Routers and switch/routers use the PIM Snooping feature to acquire multicast routes, enabling them to intelligently forward multicast traffic rather than blindly broadcast multicast traffic in a Layer 2 domain.

1.9   NODE REDUNDANCY AND RESILIENCY

We start our discussion by defining some key terms:

- **Mean Time Between Failures (MTBF)**: The MTBF is the average time between system failures (i.e., the average system uptime).

- **Mean Time To Repair (MTTR)**: The MTTR is the average time to restore or repair a system from the failed state (i.e., average system downtime).
- **Availability**: Availability is the probability that a system can be used or will operate satisfactorily when needed at a given point in time. The benchmark availability of carrier-grade network devices is "five-nines" (i.e., 0.99999), which is equivalent to saying the device is available for service 99.999% of the time (downtime per year is equal to 5.26 minutes). Availability translates into the fraction of time a system can be used or operate satisfactorily when needed. Availability can be calculated as the ratio of the average uptime (MTBF) of a system to the sum of the average uptime and average downtime (MTTR). The sum (MTBF + MTTR) is the total amount of time or the observation window.
- **Reliability**: Reliability is the probability that a system will operate as intended (i.e., perform its intended functions) without interruptions over a given period of time.
- **Redundancy**: Redundancy is the duplication or inclusion of extra functions or critical components in a system with the goal of increasing the reliability or improving the actual performance of the system. This is typically done by providing backup functions or components in the system. The following are some common redundancy configurations:
  - **1:1 Redundancy (One-for-One or One-to-One Redundancy)**: There is one passive standby component for an active component (active/passive redundancy).
  - **1+1 Redundancy (One-plus-One Redundancy)**: There is one active standby component for an active component (active/active redundancy).
  - **N+1 Redundancy**: There is one active standby component for N active components.
  - **N+M Redundancy**: There are M active standby components for N active components.
- **Fault Tolerance**: Fault tolerance is the ability of a system to continue operating uninterrupted (i.e., with its intended operations) even when there is a failure on one or more of its components or functions. In network devices (switches, routers, switch/routers, etc.), fault tolerance is provided via hardware, software, or a combination of the two.
- **High-Availability**: High-availability is a characteristic of a system and is the ability to avoid interruption of its intended operations (and loss of service) by managing or reducing failures and minimizing unplanned system downtime. High-availability systems are designed to be dependable enough to operate continuously without unplanned interruptions or failures.

The above discussion shows that increasing the reliability of a system's components (hardware and software components) improves the availability of the system. Consequently, improving the reliability of a network's components (switches, routers, switch/routers, links, etc.) improves the availability of the network. Also, the availability of a system (a network or its elements) is influenced by the fault tolerance of the system's components. A fault tolerant system is designed to have features such that, in the event of the failure of a system component, a backup or standby component immediately takes over, preventing loss of service.

Introducing redundant components in a system significantly reduces its downtime. Redundant components improve the MTBF of the system and consequently its availability. Other than preventing system component failures from causing service outages, redundancy provides a means for planned or scheduled system maintenance, as well as system hardware/software upgrades. The effectiveness of a redundancy scheme employing passive standby components depends on the speed of the switchover from active to passive component.

Because of the costs associated with providing redundancy, enterprise networks and service provider networks typically implement different levels of redundancy and fault tolerance at the network edge and core. A network outage can be as a result of the loss of network elements such as switches, routers, switch/routers, and links. The loss of a network element in turn can be due to a partial or complete failure of system hardware and software, system configuration errors and other operational errors, scheduled hardware and software upgrades, and so on.

Many modern-day businesses are conducted via the Internet and service interruptions can have a serious impact on customer satisfaction and business revenues. Enterprises have become more reliant on their networks for day-to-day business operations making system downtime less and less tolerable. High-availability network designs are common in businesses where access to information is very time-sensitive, such as financial services, telecommunications, healthcare, and emergency services.

For example, online stock brokerages generally demand that information requests and transaction demands of thousands of customers are handled with minimal delays. Orders that are not processed in a timely manner in many cases have significant, direct financial repercussions as well as long-term implications for customer satisfaction and loyalty.

The availability of services increasingly depends on highly resilient networks and devices that can continue operation in spite of network component failures, software errors or upgrades, or malicious attempts to disrupt service.

1.9.1 General Observations – Device and Network Level Reliability

As organizations build distributed data centers and more business-critical information assets are centralized in a number of data center locations, the issue of non-stop accessibility to business-critical data and computing has become even more important. With data center and server farm consolidation and virtualization, resiliency and reliability have become even more critical aspects of the network design. The requirements of business continuity call for data centers with highly redundant architectures designed to eliminate as many single points of failure as possible. This is because the impact of a failed network resource is now more likely to extend to multiple applications and larger numbers of user flows. Therefore, the virtual data center (representing the consolidated and virtualized data centers) requires the combination of high resiliency devices, and end-to-end network design that takes maximum advantage of active-active or active-passive redundancy configurations, with rapid failover to standby resources.

Furthermore, in order to guarantee business continuity even if an entire data center is lost due to a catastrophic event, critical data must be replicated in remote data centers that have the computing resources available to continue to support key business operations. Network-level redundancy and data center replication/redundancy are even more effective when the servers and networking devices are designed with the robustness and high-availability features that minimize the number of device failures (that require failovers to redundant backup devices) while simultaneously minimizing the impact of any failover that does occur.

Redundant and resilient network designs ensure high-availability operation for businesses that are less tolerant of or very sensitive to service interruptions. Redundancy provides a pathway to high-availability by preventing equipment failures from causing service outages, as well as providing a means for hitless system maintenance and upgrade. Redundancy alone does not guarantee high-availability, which depends more on the levels of redundancy, elimination of single points of failure, detection of failures as they occur, and fast switchover to redundant components.

As the design of network devices continues to keep up with advances in electronics, semiconductor, and optical technologies, switch/routers also continue to scale in terms of link speed, forwarding performance and port density. This trend has made device-level redundancy and resiliency an indispensable system attribute. For example, high degrees of traffic aggregation at the network access and core create the situation where even short periods of interrupted system operation can disrupt a large number of traffic flows and users, potentially violating numerous SLAs.

A high-availability network design always involves a degree of redundancy at the network device level. This means designing a network for very high levels of availability requires highly resilient network devices. Thus, high-availability becomes essential at two levels: at the individual device level (e.g., switch, router, switch/router, servers, etc.) and the overall network.

Although much attention has been given to the internal workings of the switch, switch/router, router, data center, server and storage components, and facilities, organizations must not neglect the role of the network in enabling non-blocking, wire-speed connectivity between elements within the network itself. The impact of a suboptimal network design can have grave consequences and negate all other investments the organization may have made in the design and deployment of high-availability devices.

The best approach to maximizing the reliability, resiliency, and stability of network devices involves the application of sound design principles:

- Eliminating single points of failure in as many system components as possible, including both hardware and software
- Constraining any failures that do occur to only one system component or subsystem
- Ensuring that, when a subsystem requires updating, becomes compromised, or fails, its recovery can be accomplished quickly without disrupting the operation of the overall system

The resiliency for a switch/router device implies its ability to continue network operations in the face of the full range of fault conditions, including the failure of hardware and software components, software faults and restarts, link failures, protocol restarts, and attempts by intruders to disrupt normal traffic flow. Non-stop hardware performance and failover technologies are crucial to maintaining network reliability.

1.9.2 Device-Level Hardware Resiliency Features

Network devices used by enterprises and service providers are expected to offer maximum uptime to enable them meet business demands and stay competitive. Business networks are expected to stay operational to provide service to end-user mission-critical traffic, meet strict SLAs made to customers, and attract and retain customer loyalty. Thus, network availability and reliability have become indispensable for most organizations. Although high-availability is usually built into the core and distribution layers through redundant systems, making a single network device more resilient becomes most important in the wiring closet, data center access, and network access. This is because there are a number of single points of failure in the traditional architectures used in these application areas.

1.9.2.1 Examples of Device Component Redundancy Configurations

As a stand-alone device, a high-performance network device must provide an extremely robust platform allowing for resilient networking. For example, the robustness of the control plane design is critical because a high percentage of system failures and service interruptions are directly traceable to control plane deficiencies. A typical high-performance system would support redundant switch fabric modules (N+1 redundancy), route processors (1+1 or 1:1 redundancy), AC or DC power supplies (N+M redundancy), and hot-swappable modules such as line cards, power supplies, and cooling fan trays.

Network solutions serving mission-critical traffic require high-availability modules, especially, the power supplies and cooling fan systems. A network device could fulfill this requirement with dual, hot-swappable AC or DC power supplies. The power modules could be load-sharing supplies providing 1+1 redundancy. The line cards and all of the redundant subsystems could support Online Insertion and Removal (OIR) and could be also fully monitored for fault conditions or out-of-range environmental parameters.

Redundant, hot-swappable components provide non-stop service delivery as explained by the following examples:

- **Route Processor Module**: The route processor (or control engine) is responsible for running the control plane functions in the system which comprises the tasks performed by the routing protocols (e.g., RIP, OSPF, IS-IS, BGP) and the management protocols (e.g., SNMP, ICMP, IGMP), and the tools used for system configuration and troubleshooting. The IP routing protocols are used for advertising network topologies, exchanging routing information, and computing routes to network destinations (intra-network and inter-network routes). A switch/router may be configured with

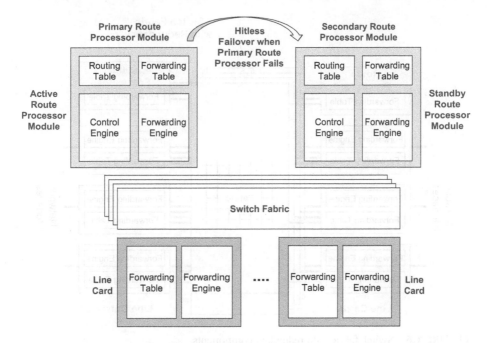

FIGURE 1.5 Dual route processors with hitless failover.

dual CPU/route processor modules supporting sub-second detection and failover (Figure 1.5).

- **Switch Fabric Element Redundancy**: The switch fabric is responsible for transferring packets between the line cards and between the line cards and the route processor. A switch/router may be configured with one or more redundant switch fabric modules supporting millisecond or lower failover performance (Figure 1.6).
- **Hitless Route Processor Failover**: Stateful failover ensures that the forwarding engines on the line modules (in a distributed forwarding architecture) are not impacted by a route processor failover (Figure 1.5). This capability enables non-stop packet forwarding in the event of a route processor module failure.
- **Redundant Power Supplies**: A switch/router may be designed to support N+M power supply module redundancy for AC and DC power configurations.
- **Redundant Chassis Cooling Fans**: Traditional computing devices and network devices use cooling fans to accomplish a number of functions: draw cooler air into the chassis from the outside, expel warmer air from within the chassis to the outside, and blow cooler air across internal processors and heat sinks, and power supplies. This is done to maintain a cooler ambient temperature and improve system performance. A switch/router may support N+M cooling fan modules to allow continuous operation irrespective of system traffic loading and outside temperature conditions.

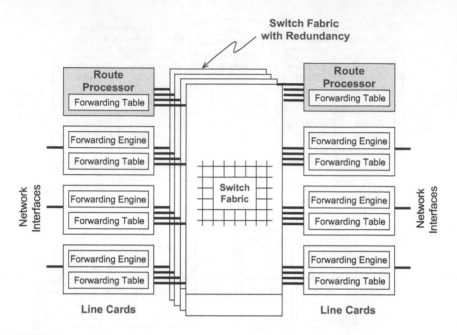

FIGURE 1.6   Switch fabric with redundant components.

Additional design features include intake and exhaust temperature sensors, and cooling fan spin detection to aid in rapid identification of abnormal or failed operating conditions to help minimize MTTR.

1.9.2.2   Carrier-Grade Equipment Requirements

A carrier-grade network device (switch, router, and switch/router) is typically expected to meet the following requirements [HUSSFAULT04]:

- The system should provide no worse than "five-nines" (0.99999) availability, that is, system downtime should not be greater than 5.26 minutes.
- The route processor, switch fabric, line card, power supplies, and cooling fan modules should have redundancy with appropriate functions to monitor system status and trigger switchover to standby modules.
- The route processor module (i.e., the hardware and software running the control plane functions) should not be a single point of failure, and the forwarding engine (i.e., the hardware and software running the data (or forwarding) plane functions) operations should not be disrupted due to failure of the control plane.
- The system should be capable of service recovery in the event of link and node failures.
- No single hardware fault in the system should result in a loss of system control plane and management functions, or a degradation or loss of end-user traffic.

As discussed above, an effective way of improving system availability and reducing unplanned system downtime is to provide system component redundancy, that is, redundant route processors, switch fabrics, line cards, power supplies, cooling fans, and so on. Generally, network-level fault tolerance is achieved using both node and network-level redundancy techniques. The reliability of the device's hardware and software (including software stability) are crucial for building reliable and high-availability networks. Also, the use of effective Operation and Maintenance (OAM) mechanisms is indispensable to carrier-grade service delivery. Service providers use a variety of OAM tools to identify and isolate network faults, leading to reduced service downtime and high level of network availability.

Planned outages such as performing system software upgrades are a big contributor to system and network downtime. This means planned and unplanned outages have to be reduced if carrier-grade availability is to be achieved. The downtime due to planned outages such as software upgrades can be reduced using In-Service Software Upgrade (ISSU) which allows the software on a network device to be updated without taking the device offline and thereby disrupting service.

1.9.2.3 Implementing Control Plane Redundancy

We discussed in Chapter 5 of Volume 1 of this two-part book ("Control Plane and Data Plane in the Router or Switch/Router" section) that the control plane and data plane of a routing device are interrelated components that can still be implemented as independent components in the device. The control plane includes the routing protocols that supply the routing information used to construct the routing tables.

Although most high-performance routers and switch/routers are designed to support redundant hardware components, the control plane software is still very vulnerable to failure and is a major cause of most system failures. Control plane tasks are generally less time critical when compared to data or forwarding plane tasks, and as a result they are run on a centralized or dedicated route processor module (see Chapter 5 of Volume 1). The protocols that constitute the control plane and run on the centralized processor serve as the overall intelligence of the system, that is, the router's or switch/router's brains. The control plane functions are pivotal to the operations of the system. For this reason, designers typically protect the route processor against failure by providing route processor redundancy (1:1 redundancy).

The data forwarding functions and forwarding tables collectively form the data (or forwarding) plane. The operations of the data plane (i.e., the packet forwarding operations) are time-critical in nature and can be implemented on multiple distributed forwarding engine modules or online cards to improve the packet forwarding performance of the device. The routing protocols, forming part of the control plane and running in the route processor, exchange network topology information (i.e., routing information) with other routing devices to build the routing table, from which the forwarding table is generated. The forwarding table is then used to determine the next-hop nodes and outgoing interfaces of packets transiting the routing device.

1.9.2.3.1 Traditional Dual-Route Processor Redundancy Techniques

In this section, we described some of the traditional methods for route processor redundancy.

1.9.2.3.1.1 Cold Restart Redundancy (or Cold Standby Redundancy)

In the most basic dual-route processor redundancy scheme called the *cold restart redundancy* scheme (also referred to as *cold standby redundancy*), when one route processor fails, the other route processor (i.e., the standby processor) reboots the routing device [LEELIMONG06]. When this happens, the system is unavailable for the time it takes to reboot the routing device. When a switchover to the standby processor takes place, the system operation is interrupted for some time.

The standby processor has its own software image and configuration file, allowing it to operate as an independent processor. In the cold restart redundancy configuration, when the standby processor takes over and reboots the system, the system stops forwarding traffic, routing information is lost, and all protocol sessions and connections are lost. In this redundancy scheme, the following events take place from the time a failure occurs to the first time the system forwards the first packet [LEELIMONG06]:

1. A failure is detected.
2. Software is loaded on the standby processor and it is booted.
3. New configuration is loaded on the standby processor.
4. Line cards are reset and reloaded.
5. New configuration is loaded on the line cards.
6. System exchanges routing information with neighbors, learns routes, sends keepalive messages, and forwards user traffic.

In the cold restart redundancy scheme, the entire system stops functioning for the duration it takes to retore the system. The system stops forwarding traffic entirely during this period. The main benefit of this scheme is that the system will restart without manual intervention through the rebooting of the standby processor (which takes control of the system).

1.9.2.3.1.2 Warm Restart Redundancy (or Warm Standby Redundancy)

In a slight variant and improvement of the cold restart redundancy scheme called *warm restart redundancy* (or *warm standby redundancy*), the standby processor loads a software image when it boots up and initializes (i.e., places) itself in the standby mode [LEELIMONG06]. In the event of a fatal hardware or software fault on the active processor, the system performs a switchover to the standby processor, which reinitializes itself as the active processor, reloads all the line cards, and restarts the system.

The warm standby scheme improves upon the cold restart redundancy scheme by eliminating Step 2 ("software is loaded on the standby processor and it is booted") and Step 3 ("new configuration is loaded on the standby processor") in the cold restart scheme. The elimination of these steps reduces the failure recovery time because the standby processor has already started the boot-up process, allowing it to take control of the system faster.

1.9.2.3.2 Limitations of the Traditional Dual-Route Processor Redundancy Techniques

In the traditional (simplistic) design with dual primary (active) and secondary (standby) route processors, control plane software restart and switchover are disruptive and can produce periods of packet forwarding inactivity (interruptions in packet forwarding) [HUSSFAULT04]. When a hardware/software failure is detected on the active route processor, the system switches over to the standby route processor, resets this processor's control plane software, and resets the forwarding engines in the line cards. This process results in a disruption of the packet forwarding operations plus service outage [HUSSFAULT04].

For example, when a control plane routing protocol such as OSPF or IS-IS restarts on the router's standby route processor, the neighbor routers of this router will detect the restart, and then originate OSPF Link State Advertisements (LSAs) (or IS-IS Link State Packets (LSPs)) to find ways of routing around (i.e., bypassing) the failing/restarting router. Routers that receive the new LSAs (or LSPs) will compute new routes to avoid the restarting router. This form of control plane switchover causes undesirable interruptions of packet forwarding, generates extra and unnecessary routing information traffic, and causes neighbor routers to initiate OSPF (or IS-IS) Shortest Path First (SPF) recomputations [HUSSFAULT04].

Techniques that minimize packet forwarding disruptions and service outages are more desirable. Newer more improved route processor redundancy schemes, further eliminate Step 4 ("line cards are reset and reloaded") and Step 5 ("new configuration is loaded on the line cards") in the cold restart redundancy scheme. Eliminating these steps allows the routing device to keep the line cards functioning during the switchover to the standby processor. The system does not reload or reinitialized the line cards, allowing them to continue forwarding traffic. It is noted in [LEELIMONG06] that the newer redundancy schemes reduce the switchover time by 90 percent compared to the warm restart redundancy scheme.

1.9.2.3.3   Enhanced Dual-Route Processor Redundancy Schemes

This section describes newer methods that address the limitations of the traditional dual-route processor redundancy schemes. These enhanced methods incorporate mechanisms that lessen disruptions of the forwarding plane operations than systems using the traditional route processor redundancy methods discussed above. The following are some of the newer control plane software redundancy schemes that are used to reduce system downtime when unplanned control plane failures occur [HUSSFAULT04]:

- In the first approach, the active and the standby route processors run identical copies of the control plane software, and the two software instances are executed independently without any communication between the two instances. Identical control packets are sent to and received from both software instances. Control packets received by the router are duplicated and sent to both software instances and are processed at the same time. When sending control packets, packets from the active processor are sent while those from the standby processor are discarded. In this approach, the router does not need to save the state from the active processor on the standby processor.
  - o The main advantage of this approach is that control plane failure in the active processor and switchover to the standby processor is not visible to neighbor routers, and therefore, does not require changes to the control plane protocols.

- o One main disadvantage here is the system requires extra processing to duplicate control packets (extra processing load) and both software instances are required to be started at the same time (extra mechanism needed for a simultaneous start).
- o Another disadvantage is that this approach precludes the ability to perform software upgrades or downgrades since both route processors are required to execute at the same time.
- In the second approach, the active and the standby route processors run identical copies of the control plane software, and the two software instances are executed in complete lock step using inter-process (or inter-instance) communication. The receive and send processes that handle control packets are identical to the first approach described above.
  - o The main advantage of this approach is that it provides fast system recovery times and does not require changes to the control plane protocols.
  - o The main disadvantages of this approach are: i) there is some design complexity involved in synchronizing route processor state and ii) it introduces a requirement that the two software instances must be executed synchronously.
  - o The requirement of synchronous execution of the two software instances precludes software upgrades and downgrades as in the first approach.
- In the third approach, the active and the standby route processors are instantiated with copies of the control plane software, but the active processor executes its copy while the standby processor does not. However, the standby processor receives information for state synchronization from the active processor and maintains a partial state of the active software instance. The active processor runs the routing protocols and establishes protocol sessions, exchanges routing information with neighbors, and constructs and maintains the routing and forwarding tables. The inactive standby processor, on the other hand, does not exchange any routing information with neighbors. When switchover occurs, the standby processor takes over, reestablishes protocol sessions with neighbor routers, and resynchronizes its state information.
  - o This approach allows the router after a restart to continue forwarding packets as the control plane recovers (see next point). This approach also allows control plane software upgrades, a feature that contributes to high-availability and carrier-grade system performance.
  - o The main disadvantage of this approach is that it requires the standby processor to reestablish protocol sessions and to recover control plane state information upon a restart. To provide this capability, extensions to the control plane protocols are required, plus support from neighbor routers in maintaining forwarding state information, while the router enters full service after a restart.
  - o A variant of the third approach is to allow the standby processor to maintain complete state information so that switchover can take place without the standby processor having to reestablish protocol sessions with

neighbor routers. The main disadvantage of this latter approach is that it is not scalable because of the requirement of maintaining complete state information of the active processor.

Typically, high-performance routers adopt the third approach and its variants to provide control plane redundancy and fault tolerance.

1.9.2.4 Stateful Control Plane Switchover

We start our discussion by defining the following important terms [HUSSFAULT04]:

- **Control Plane State Recovery Period**: A routing device is in the recovery period when it is in the process of regaining possession of its control plane state and updating its existing forwarding table entries. Note that after a control plane failure has occurred, the control plane will restart and reestablish sessions with neighbor routing devices to recover control plane state.
- **Stateless Control Plane Component**: This refers to a control plane component that is only capable of recovering its state completely from neighbor routing devices after a restart.
  - In the first and second redundancy methods above, the active processor does not need to save state on the standby processor; instead, it recovers the state completely from neighbor routing devices after a restart.
- **Stateful Control Plane Component**: This refers to a control plane component that is capable of saving some state on a standby processor after a restart.
  - In the third approach above, the active processor saves partial state on the standby processor and recovers missing state information from neighbor routing devices after a restart.

Stateful Switchover (SSO) is a control plane redundancy feature that allows automatic nondisruptive switchover to the control plane software of the standby processor upon detection of hardware/software failures in the active processor.

By using a combination of control plane and data plane separation, redundant route processors, and fault-tolerant control plane software, routers are in a better position to perform nondisruptive stateful control plane switchover (to a standby processor) when hardware and software failures are detected in the active processor.

The SSO feature reduces the time spent in Step 6 ("system exchanges routing information with neighbors, learns routes, send keepalive messages, and forwards user traffic") in the cold restart redundancy scheme described above [LEELIMONG06] (see "Traditional Dual Route Processor Redundancy Techniques" section above). The SSO feature allows the active processor to communicate key routing and interface state information to the standby processor, thereby, reducing the time the standby processor takes to learn routes and for the network to converge.

When using the SSO feature, both active and standby processors typically run the same software and have the same configuration so that the standby processor will always be ready to assume control when the active processor fails. The system synchronizes the configuration information from the active processor to the standby processor at system startup and whenever changes to the active processor configuration occur.

Following the initial synchronization between the two route processors, when SSO takes place, the state information is already maintained between the two processors including forwarding state information. During switchover, the execution of the routing protocols and system control is transferred from the active processor to the standby processor. The time required to switch over from the active processor to the standby processor is implementation-dependent. Some systems can perform the switchover very quickly with zero packet drops.

1.9.2.5 Non-stop Packet Forwarding

Normally, when a router restarts, all of its peer routers will detect that the router went down and came back up, a transition that can result in a routing flap. The routing flap can propagate across multiple routing domains, causing network instabilities and routing loops. Although the routing device may be forwarding packets, the routing flaps caused by the control plane switchover can cause network instabilities which can degrade the performance of the overall network. Non-stop Forwarding (NSF) helps to suppress routing flaps in networks with routers supporting SSO, thereby reducing network instabilities.

NSF refers to a high-availability feature that allows the forwarding engine(s) in a routing device to continue forwarding packets while the control plane recovers from a fault. When a control plane fails, the forwarding engine continues to forward packets using its existing forwarding table entries. To implement NSF, the control plane and data (forwarding) plane functions in the routing device must be separated; plane separation is a requirement (see the "Control Plane and Data Plane in the Router or Switch/Router" section in Chapter 5 of Volume 1).

The decoupling of the control plane and the data plane provides the ability to implement SSO and NSF in routing devices. It is noted in [HUSSFAULT04] that regardless of whether a stateless or stateful control plane is used, a prerequisite for NSF is the ability of the system to preserve packet forwarding state as the switchover from active to standby processor takes place. A forwarding state represents the information in the FIB entries such as the IP destination address/prefix, next-hop IP address, outbound interface, and Layer 2 rewrite information.

To implement NSF, the following architectural components are required [HUSSFAULT04]:

- For a control plane that uses stateful components, the SSO capability is required.
- The forwarding engine(s), for example, in the line cards, must remain unaffected by the control plane SSO and must be capable of continuing to forward packets using existing or pre-SSO forwarding information. The router must be capable of maintaining forwarding state information as the switchover takes place.
- After control plane switchover, the routing protocols running in the control plane must be capable of restarting, recovering routing information from

neighbor routing devices or locally, and updating existing forwarding state for the forwarding engines.

- The router must be capable of recovering control plane state information without causing undesirable effects such as disruption of the packet forwarding operations and routing loops.
- The router performing control plane switchover must have available to it (be able to receive) cooperation and support from neighbor routing devices.

NSF allows a routing device to continue forwarding traffic on known routes while the system restores routing protocol information following a control plane switchover. The line cards continue to forward traffic while the standby processor assumes control from the failed active processor during the control plane switchover. The line cards remain up through the switchover, and keep and use current forwarding state provided by the active route processor. With NSF, the peers of a routing device operate in such a way that the network does not experience routing flaps.

Thus, for NSF to work, routing protocols such as OSPF, IS-IS, and BGP have been enhanced with features such as Graceful Restart [HUSSFAULT04] [LEELIMONG06]. These enhancements have made these routing protocols NSF-aware and capable, allowing routing devices running these protocols to detect a switchover, and take the necessary actions to continue forwarding traffic, while at the same time, recover routing information from peer routing devices.

1.9.2.6 Packet Forwarding in a Routing Device with SSO and NSF

We summarize in this section, the packet forwarding operations in a routing device that supports SSO to a standby processor, and NSF in the forwarding engines. We assume the active and standby processors each run one instance of the same routing protocol in the control plane (e.g., RIP, EIGRP, OSPF, IS-IS, etc.), and the standby processor can receive control plane state information via inter-process communications:

1. The routing protocol running in the active processor establishes protocol sessions with neighbor routing devices, starts exchanging routing information with them, and constructs the RIB.
2. The active processor takes the RIB contents and generates the main FIB entries. Initially, the active processor distributes the entire FIB to the line cards and the standby processor. Thereafter, it distributes only incrementally changes to these components as routing information changes.
3. Each line card receives the main FIB entries and populates its local FIB. The local FIB is then used by the line card's forwarding engine for packet forwarding.
4. Let us assume the active processor experiences a hardware or software failure that causes the system to trigger an automatic switchover to the standby processor.
5. When the line cards learn about the switchover to the standby processor, via, for example, detecting the loss of IPC connectivity with the active processor (or any other means), they will mark the existing forwarding state as requiring refreshing, and will start a *stale timer* [HUSSFAULT04].

a. However, the line card's forwarding engine will continue to use the existing forwarding state to forward arriving packets.

b. If the line card determines that the existing forwarding state has not been refreshed before the stale timer expires, it will delete the forwarding state information.

6. The protocol running on the standby processor, which has now assumed the active role, will reestablish protocol sessions with neighbor routers, exchange routing information, and reconstruct the RIB and FIB.

   References [HUSSFAULT04] [LEELIMONG06] describe in detail the procedures that routing protocols use to reestablish protocol sessions and relearn routing information gracefully without causing undesirable network effects such as routing loops.

7. The FIB client in each line card connects with the new active processor, which then copies its latest main FIB contents and updates all distributed line card FIBs and the FIB of the new standby processor.

8. Each line card's forwarding engine then uses its updated FIB for local packet forwarding. Note that the line card, upon stale timer expiration, will delete all FIB entries still marked as stale (i.e., entries that have not been refreshed after switchover).

In Step 4, the active and standby route processors may send keepalive messages to each other. If the standby route processor fails to receive keepalives for say 20 seconds (using a non-configurable timer), it enters a message in a local message log file. If after 300 seconds (using a configurable failover timer), the standby route processor attempts to assume the active role for the router. When it succeeds, the router sends an alarm generated to notify the network administrator that the active route processor has failed.

Using the above steps, the packet forwarding plane operations are not disrupted during and after control plane switchover. As noted earlier, the use of NSF requires the preservation of forwarding state as control plane switchover takes place. Unlike the FIB which is used directly in the data (or forwarding) plane for packet forwarding, the RIB has no direct role in the forwarding plane, and as a result does not need to be preserved as switchover takes place. Instead, the RIB can be rebuilt after the switchover takes place.

Protocols such as OSPF, EIGRP, IS-IS, BGP, and MPLS have been enhanced with newer features (such as Graceful Restart capabilities) to allow them to restart and reconstruct the RIB without causing detrimental network effects such as routing loops [HUSSFAULT04] [LEELIMONG06]. The traditional routing protocols designs are not capable of reconstructing the RIB without disrupting the packet forwarding operations and causing undesirable network effects such as routing loops.

1.9.2.7 Control Plane Component Modularity and the Issue of Component Restartability

A *restartable* software component is one that can recover from fatal runtime errors. In a system with control plane redundancy, employing software components that are restartable enables the system to correctly recover from failures, thereby, making switchover unnecessary. However, when a restartable software component fails to restart correctly when a fatal error occurs, this failure will cause the system to switchover to a standby processor in order to recover from the failure [HUSSFAULT04]. Component restartability is not meant to address all forms of failures; some fatal errors cannot be recovered from (this discussion is beyond the scope of this book).

The control plane redundancy schemes and SSO feature discussed above assume that the control plane software of the routing device executes as a "monolithic unit" with one or more inseparable components (or processes) that share critical system resources and data structures. Thus, given that the control plane software is composed of inseparable components and is incapable of being restarted individually, a failure of any one component will result in the failure of the whole software (i.e., all components), thereby necessitating the system to perform control plane switchover [HUSSFAULT04]. This also means systems that use control plane software with components/processes that are not restartable, will require stateful control plane redundancy schemes and switchover to recover from failures.

The above discussion shows that restartability of software components/processes when used together with SSO, provides further fault tolerance to a system. In a system with restartable software components, upon system initialization, a system management module is responsible for instantiating all components of the control plane software, plus monitoring their health status. As soon as the system management module detects a control plane software component failure, it will restart the failed component without disrupting the operations of other components, or triggering control plane switchover. Upon successful restart, the failed component/process will first recover its preserved state information and then resume normal operation [HUSSFAULT04].

It is important to note that, unlike SSO, which is used with systems with control plane software redundancy, component or process restartability can be used to improve the fault tolerance and availability of systems with a single control plane processor as well as systems with redundant control plane processors [HUSSFAULT04]. In a system with a single processor, component restartability allows the system to recover from unplanned control plane software failures, leading to improved system availability. In a system with control plane redundancy, SSO and component restartability, when used together, improves the fault tolerance, availability, and reliability of the control plane and the device as a whole. In the latter case, the system can recover when a restartable component experiences a minor fault without performing a switchover to the standby processor; it can also perform a switchover when it experiences a major hardware or software fault. Ideally, using a design that allows complete switchover to a standby processor and software components that do not have shared data structures, offers a higher level of fault tolerance.

1.9.3 DEVICE-LEVEL SOFTWARE RESILIENCY FEATURES

Some high-end, high-performance systems employ highly modularized operating system software that provides system resiliency at the software level [FOR10HAL08] [FOR10TSR06]. Software modularity minimizes downtime and boosts operational efficiency. Enabling modular software subsystems to run as independent processes provides a number of benefits. This enhancement provides the following benefits:

- Minimizes unplanned downtime through process self-healing
- Simplifies software updates through subsystem In-Service Software Upgrades (ISSU)

Other benefits include better fault containment, process restartability, process check-pointing, more efficient use of memory, etc.

In modular software, all major functions are implemented as separate processes, with each process supported by its own protected memory space. This modularity prevents a fault in one process from affecting or corrupting other processes. A higher degree of modularity could even allow individual processes to be started or stopped independently, and individual software components to be upgraded without stopping or disrupting the remaining processes.

High-end, high-performance switch/routers also support a number of mechanisms for resiliency at the manageability/serviceability level. These mechanisms enable system upgrades, reconfiguration, fault correction, and component repair without taking the device out of service. Other manageability features help to reduce the time to diagnose and repair faults, which have an impact on system availability. A switch/router may support hitless Layer 2/Layer 3 software upgrades, and graceful restart routing (allowing fast convergence in the event of a route processor failure).

1.9.4 UNDERSTANDING ONLINE INSERTION AND REMOVAL OR HOT SWAPPING

OIR, also referred to *hot swapping*, allows a faulty part on a hardware device (e.g., a routing device) to be removed and replaced with a functioning one without affecting the system operation. For example, when a replacement line card is inserted into a slot in a routing device, it will detect the power supply available to it, and will initialize itself to start operating in the system.

A system that supports hot swapping of line cards, for example, will detect when a change in the system's line card configuration occurs and will reallocate system resources to allow all line cards and their attached network interfaces to function properly. Hop swapping allows the interfaces of a swapped line card to be reconfigured while the existing line cards and their interfaces on the system remain unchanged.

A system that supports hot swapping has software functions (a subsystem) that handle the tasks involved removing and inserting a line card. When the system detects a hardware change, it sends a hardware interrupt to the hot swapping software subsystem, and the subsystem reconfigures the system accordingly [LEELIMONG06]. When a line card is inserted in a slot, the system analyzes and initializes it in such a way that, it is available/ready for the network administrator to configure it as would be done when the overall device is powered on.

The initialization routines used by the system during hot swapping of a line card are the same as those used when the system is powered on. The system allocates resources to the new line card and interfaces so that they can start functioning properly. Also, when a line card is removed from a slot, the system either frees or alters the resources associated with the empty slot and indicates a change in the slot's status.

If a system does not support hot swapping of line cards and a line card is removed from a slot, this action may disrupt the traffic being processed by the system, and in the worse situation, the system might reboot. It is therefore important to know if a particular system supports hot swapping because, removing a card from a system that does not support such capabilities can cause permanent damage to the card, or worse, the whole system.

1.9.5 SINGLE LINE CARD RELOAD

On older routing devices, the only way to correct a hardware failure or software error on a particular line card is to execute a process (for example, the Cbus Complex process on Cisco 7500 Series Routers) that reloads every line card on the system backplane [LEELIMONG06]. The main disadvantage of this process is that, during the long time it takes for this process to complete, the system does not route any network traffic.

The *single line card reload* feature was introduced to enable a system to correct a line card failure by allowing the system to automatically reload the microcode on the failed line card [LEELIMONG06]. As the single line card reload process takes place, all network interfaces and routing protocols on the other line cards connected to the system backplane remain active (and continue to forward traffic). The reload process does not affect the correct functioning of the system.

The key benefit here is the single line card reload improves the availability of the system when compared to the older reload process. The single line card reload feature reloads only the line card experiencing hardware or software failure rather than reloading all line cards connected to the system backplane, allowing the active line cards to continue to forward traffic.

1.9.6 LAYER 1/LAYER 2 PROTOCOL RESILIENCY FEATURES

Software features such as protected link groups, IEEE 802.3ad Link Aggregation (now moved to the standalone standard, IEEE 802.1AX-2008 – Link aggregation) [IEEE802.1AX], and trunk groups provide alternate paths for traffic in the event of a link failure. IEEE 802.3ad Link Aggregation (e.g., aggregating up to eight links) provides scalable, and cross-module trunking for resilient high-capacity connections between switches. A Link Aggregation Group (LAG) based on the IEEE 802.3ad specification bundles multiple physical Ethernet links of the same speed into a higher bandwidth logical link (Figure 1.7).

A major benefit of LAG is that it provides economical scaling of bandwidth for inter-switch links within the distribution/aggregation and core layers of a Layer 2 switched network, as well as for links connecting access routers to distribution switches. For example, a LAG comprising multiple Gigabit Ethernet links offers a granular bandwidth expansion path between 1 Gigabit Ethernet and 10 Gigabit Ethernet. For example, as a deployed 10 Gigabit Ethernet link in a network eventually becomes saturated, the use of a 10 Gigabit Ethernet LAG will allow bandwidth

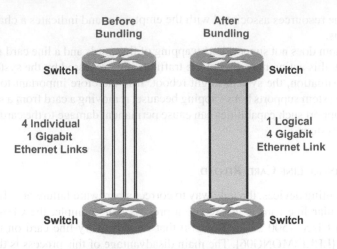

FIGURE 1.7 Bundling of multiple physical Ethernet links.

scaling until the next multi-gigabit Ethernet (e.g., 40 and 100 Gigabit Ethernet) is ready for deployment. LAGs also offer the benefit of greater resiliency, especially when the links in the group are distributed across multiple line cards. Links can be added or deleted from the LAG without disrupting traffic or rebooting the system.

To provide self-healing topologies in Layer 2 network configurations, switch/routers typically support industry-standard Ethernet protocols including the most recent versions of the Spanning Tree Protocol (STP) [IEEE802.1D98] and link aggregation, in addition to optic-, link-, and switch-level fault detection and correction features. The Layer 2 features include the Rapid Spanning Tree Protocol (RSTP), originally specified as IEEE 802.1w, [IEEE802.1D04], and Multiple Spanning Tree Protocol (MSTP), originally specified as IEEE 802.1s [IEEE802.1Q05].

IEEE 802.1D STP [IEEE802.1D98] was widely used in Layer 2 networks for failure recovery and loop avoidance. However, STP was conservatively designed for large diameter networks of arbitrary topology and reacts slowly to failures even in relatively simple networks, taking tens of seconds to provide recovery from link and node failures.

Switches now support IEEE 802.1w RSTP that provides fast convergence in case of link or Spanning Tree root failure. RSTP allows switches to maintain knowledge of multiple paths to the root. When the primary link fails, the system fails over to the secondary link in a matter of milliseconds, placing the secondary link in the forwarding state immediately without going through listening and learning states in the older STP.

In the event of root bridge failure, RSTP accelerates the aging of protocol information, allowing rapid failure detection. With RSTP, the switch can achieve very fast convergence in simple 2-tier configurations, in the range of tens or hundreds of milliseconds. RSTP dramatically improves the spanning tree convergence time to subsecond by automatically renegotiating port roles in case of a link failure without relying on timers.

MSTP was developed to provide an efficient means of supporting multiple instances of Spanning Tree (ST) as would be required for deploying numerous

VLANs in a switched Ethernet LAN. Instead of a separate instance of ST for each VLAN, MSTP allows a group of VLANs to share a common instance of ST. MSTP allows the formation of Multiple Spanning Tree (MST) regions that can run Multiple MST Instances (MSTI). Multiple regions and other STP bridges are interconnected using one single Common Spanning Tree (CST). Enhanced Spanning Tree features such as Root Guard and Bridge Protocol Data Unit (BPDU) Guard prevent rogue hijacking of a Spanning Tree root and maintain a contention and loop-free environment especially during dynamic network deployments.

A switch/router may support logical fault detection through software features such as Link Fault Signaling (LFS), Remote Fault Notification (RFN), Protected Link Groups, and Unidirectional Link Detection (UDLD). Sub-second fault detection utilizing LFS and RFN ensures rapid fault detection and recovery:

- Link Fault Signaling (LFS) is a Physical Layer protocol that ensures bi-directional communication on a link between two Ethernet switches, thereby allowing switches on both sides to disable the link.
- Remote Fault Notification (RFN) enabled on Ethernet ports notifies the remote port whenever the fiber cable is either physically disconnected or has failed. When this occurs, the link is disabled by the switches on both sides.
- Protected Link Groups minimize disruption to the network by protecting critical links from loss of data and power. In a protected link group, one port in the group acts as the primary or active link, and the other ports act as secondary or standby links. The active link carries the traffic. If the active link goes down, one of the standby links takes over.
- UDLD monitors a link between two switches and brings the ports on both ends of the link down if the link fails at any point between the two devices.

Some switch/routers also support stability features such as Port Flap Dampening. Port Flap Dampening increases the resilience and availability of the network by limiting the number of port state transitions on an interface. For example, if a port transitions from an Up to a Down state and back, more than three times in five seconds, the port is disabled. After 10 seconds, it is automatically re-enabled. This reduces the protocol overhead and network inefficiencies caused by frequent state transitions occurring on misbehaving ports.

1.9.7 LAYER 3 PROTOCOL RESILIENCY FEATURES

Resilient Layer 2 and Layer 3 protocols provide fast service restoration in the event of link or device failures:

- RSTP (IEEE 802.1w) and MSTP (IEEE 802.1s) for general Layer 2 topologies
- Virtual Router Redundancy Protocol (VRRP) for redundant router configurations
- Equal Cost Multi-Path (ECMP) routing for routed backbones

- BGP-Guard complements MD5 security for BGP sessions to protect against session disruption by restricting the number of hops the BGP session can traverse

VRRP [RFC5798] provides high network availability by routing IP traffic from hosts on Layer 2 (Ethernet) networks without relying on the availability of any single router. VRRP enables one or more routers to act as backup routers to other routers on a network as shown in Figure 1.8. In the event of a router failure in the virtual router group (VRG), one of the backup routers will automatically and seamlessly perform the tasks of the failed router.

With VRRP, a virtual IP address is specified as the default gateway IP address. The virtual IP address is shared among the routers in the VRG, one being set as the master router, the rest as backup routers. When the master router is unavailable, one of the backups becomes the master. When VRRP detects that the designated active router (master) has failed, the selected backup router assumes control of the VRP's MAC and IP addresses.

A number of enhancements have been proposed within the Internet Engineering Task Force (IETF) to improve the resiliency of IP. Bidirectional Forwarding Detection (BFD) [RFC5880] [RFC5881], is a protocol for detecting faults in the bi-directional path between two forwarding engines, including faults on physical interfaces, sub-interfaces, data links and, to the extent possible, the forwarding engines

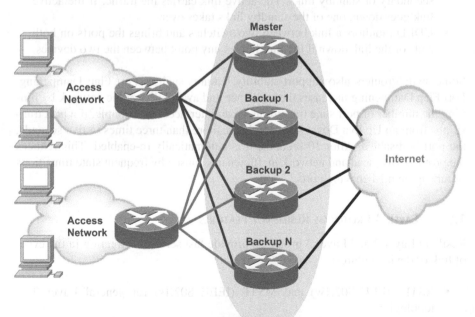

FIGURE 1.8   VRRP in an N-Router configuration.

themselves. It is intended to be independent of media, data protocols and routing protocols, and to operate with very low latency. BFD provides low-overhead detection of faults even on physical media that do not support failure detection of any kind, such as static routes, virtual circuits, tunnels and Multiprotocol Label Switching (MPLS) Label Switched Paths (LSPs). For example, BFD enables Ethernet to detect media faults, fault detection capabilities that Ethernet was not originally designed to support.

The BFD protocol is a simple Hello mechanism that detects failures in a network. BFD works with a wide variety of network environments and topologies. Two routing devices may be configured to exchange BFD packets. Hello packets are sent at a specified, regular interval. A neighbor failure is declared when any one of the routers stops receiving a reply for a specified time interval. The BFD failure detection timers have shorter time limits than the standard route failure detection mechanisms used in routers, allowing faster failure detection.

IP Fast Reroute (FRR) mechanisms [RFC5286] [RFC5714] are other enhancements in IP routing. IP FRR refers to the set of technologies that provide fast rerouting capability using standard IP forwarding and routing techniques. IP FRR is designed to protect against link or router failure by having a router use locally determined repair paths. Analogous to the technique employed in MPLS Fast Reroute, these IP-oriented mechanisms are applicable to networks using conventional IP routing and forwarding. IP FRR allows a router to compute backup routes to repair a network failure locally, without the immediate need to inform other routers about the failure. Such mechanisms prevent packet loss, for example, caused by the "microloops" that form as a result of temporary inconsistencies between the routing tables in the routers.

1.9.8 FOCUS – LOAD BALANCING FEATURES

In addition to providing robustness for large network designs, IP routing allows optimum load balancing using ECMP routing of traffic over the redundant paths that are prevalent in highly meshed IP networks (which are designed for both high-performance and maximum availability). ECMP routing is a load balancing technology that optimizes flows across multiple IP paths between any two IP subnets or networks.

Both RIP and OSPF support only ECMP routing, while EIGRP supports ECMP and Unequal Cost Multi-Path (UCMP) routing (see [AWEYA2BK21V1] [AWEYA2BK21V2]). For example, when RIP or OSPF, determines multiple routes to the same network destination with equal routing metric values, it can install these routes in the routing table for the router to perform ECMP routing. The specifics of UCMP routing in EIGRP are described in [AWEYA2BK21V2]. ECMP can apply load balancing to TCP and UDP packets on a per-flow basis, when needed.

Load balancing is the ability of a network device like a switch/router or router to distribute traffic over two or more of its network ports that lead to the same network destination address. Load balancing increases the utilization of network segments, thus increasing network data throughput. A switch/router may offer, for example, the following two load balancing schemes for IP traffic: per-packet load balancing and per-destination load balancing.

Per-packet load balancing sends data packets over multiple paths without regard to the individual end-host or user session. Though packets can be evenly distributed in this scheme, they can also arrive out of order at the destination end-system because per-packet load balancing typically uses round-robin packet scheduling on the multiple paths (which may have unequal path delays). *Per-destination load balancing*, on the other hand, performs load sharing by ensuring that packets to a given destination address always take the same path. Note that the path utilization of this scheme might not be as efficient as per-packet load balancing, but the packets are received out of order.

One of the shortcomings of per-destination load balancing is that it does not take into account the source of the packet. If multiple sources (servers) send packets to the same destination, these packets use the same path even if there are other underutilized paths available. A switch/router may instead implement the more efficient scheme of *source and destination-based load balancing*. Essentially, this method takes certain bits from the source and destination IP addresses and maps this addressing information into a specific path; allowing the different servers to send traffic on different paths.

REVIEW QUESTIONS

1. What is head-of-line (HOL) blocking and how is it avoided in network devices?
2. Explain the main difference between switch stacking and switch clustering.
3. What is the difference between the availability and reliability of a system?
4. What is the difference between system redundancy and fault tolerance?
5. What is the difference between 1:1 redundancy and 1+1 redundancy?
6. Explain the main difference between cold restart redundancy and warm restart redundancy.
7. What is the difference between a stateless control plane component and a stateful control plane component?
8. What is Stateful Switchover (SSO)?
9. What is Non-stop Forwarding (NSF)?
10. What is a restartable software component and what are its benefits in a routing system?
11. What is Online Insertion and Removal (OIR) or Hot Swapping?
12. What is single line card reload in a routing system?
13. What is the purpose of the Virtual Router Redundancy Protocol (VRRP)?
14. What is Equal Cost Multi-Path (ECMP) routing?
15. What is per-packet load balancing and what are its main disadvantages?
16. What is per-destination load balancing and what are its main disadvantages?

REFERENCES

[AWEYA2BK21V1]. James Aweya, *IP Routing Protocols: Fundamentals and Distance Vector Routing Protocols*, CRC Press, Taylor & Francis Group, ISBN 9780367710415, 2021.

[AWEYA2BK21V2]. James Aweya, *IP Routing Protocols: Link-State and Path-Vector Routing Protocols*, CRC Press, Taylor & Francis Group, ISBN 9780367710361, 2021.

[CISC2950DSS]. Cisco Systems, *Catalyst 2950 Desktop Switch Software Configuration Guide*, Cisco IOS Release 12.1(6)EA2b, March, 2002.

[CISC3850SSW]. Cisco Systems, "Cisco Catalyst 3850 Series Switches StackWise-480 Architecture", *White Paper*, 2015.

[CISCCACCFCS]. Cisco Systems, "Configuring and Analyzing Clustering on Catalyst Fixed Configuration Switches", *Document ID: 4085*, November 16, 2007.

[DAIPRAB00]. J. Dai and B. Prabhakar, "The throughput of data switches with and without speedup", *IEEE INFOCOM 2000*, Tel Aviv, Israel, March 2000, pp. 556–564.

[FOR10FPWC08]. Force10 Networks, "Future-proofing the Wiring Closet with Resilient and Scalable Modular Switch/Routers", *White Paper*, 2008.

[FOR10HAL08]. Force10 Networks, "The Hardware Abstraction Layer: Enabling FTOS to Span the Switching and Routing Infrastructure with a Consistent Feature Set and Unified Management", *White Paper*, 2008.

[FOR10TSR06]. Force10 Networks, Next Generation Terabit Switch/Routers: Transforming Network Architectures, White Paper, 2006.

[HUSSFAULT04]. I. Hussain, *Fault-Tolerant IP and MPLS Networks*, Cisco Press, 2004.

[IEEE802.1D98]. Media Access Control (MAC) Bridges, 1998.

[IEEE802.1D04]. IEEE Std 802.1D – 2004 - Media Access Control (MAC) Bridges.

[IEEE802.1Q05]. IEEE Std. 802.1Q-2005, Virtual Bridged Local Area Networks.

[IEEE 802.1X]. IEEE Std 802.1X-2004 - Port-Based Network Access Control.

[IEEE802.1AX]. IEEE Std 802.1AX-2008 - IEEE Standard for Local and Metropolitan Area Networks — Link Aggregation.

[KAROLHM87]. M. Karol, M. Hluchyj, S. Morgan: "Input versus Output Queueing on a Space-Division Packet Switch", *IEEE Transactions on Communications*, Vol. COM-35, No. 12, December 1987, pp. 1347–1356.

[LEELIMONG06]. Kok-Keong Lee, Fung Lim, and Beng-Hui Ong, *Building Resilient IP Networks*, Cisco Press, 2006.

[MCKEOWN96]. N. McKeown, V. Anantharam and J. Walrand, "Achieving 100% Throughput in an Input-Queued Switch," *IEEE INFOCOM 96*, pp. 296–302, 1996.

[MCKEOWN98]. N. McKeown and A. Mekkittikul, "A Practical Scheduling Algorithm to Achieve 100% Throughput in Input-Queued Switches", *IEEE INFOCOM 98*, pp. 792–799, 1998.

[RFC2865]. C. Rigney, S. Willens, A. Rubens, W. Simpson "Remote Authentication Dial In User Service (RADIUS)", *IETF RFC 2865*, June 2000.

[RFC2866]. C. Rigney, "RADIUS Accounting", *IETF RFC 2866*, June 2000.

[RFC3176]. InMon Corporation's, "sFlow: A Method for Monitoring Traffic in Switched and Routed Networks", *IETF RFC 3176*, September 2001.

[RFC3576]. M. Chiba, M. Eklund, D. Mitton, and B. Aboba, "Dynamic Authorization Extensions to Remote Authentication Dial In User Service (RADIUS)", *IETF RFC 3576*, July 2003.

[RFC5286]. A. Atlas and A. Zinin, "Basic Specification for IP Fast Reroute: Loop-Free Alternates", *IETF RFC 5286*, September 2008.

[RFC5714]. M. Shand and S. Bryant, "IP Fast Reroute Framework", *IETF RFC 5714*, January 2010.

[RFC5798]. S. Nadas, "Virtual Router Redundancy Protocol (VRRP) Version 3 for IPv4 and IPv6", *IETF RFC 5798*, March 2010.

[RFC5880]. D. Katz and D. Ward, "Bidirectional Forwarding Detection (BFD)", *IETF RFC 5880*, June 2010.

[RFC5881]. D. Katz and D. Ward, "Bidirectional Forwarding Detection (BFD) for IPv4 and IPv6 (Single Hop)", *IETF RFC 5881*, June 2010.

[TAMIRY88]. Y. Tamir and G. Frazier, "High-Performance Multi-Queue Buffers for VLSI Communication Switches", *Proceedings of the 15th International Symposium on Computer Architecture*, ACM SIGARCH, Vol. 16, No. 2, May 1988, pp. 343–354.

2 High-Performance Switch/Routers

Part 2: Advanced and Value-Added Features

2.1 IMPROVED MANAGEABILITY AND LOWER TOTAL COST OF OWNERSHIP

Operational optimization is one of the most important components of running an organization. Operational costs in many cases are typically a much larger portion of a business's total budget than capital costs. Being able to use the network to lower the operational costs justifies the initial hardware investment. There are at least three areas of investigation here: energy costs, management costs, and total cost of ownership (TCO) over time.

In addition to providing QoS, traffic control, and high-availability features, networks should be made as simple, robust, and easy to manage as possible in order to control/minimize TCO. As a result, there is the requirement to provide network administrators with comprehensive tools for configuring, managing, monitoring, diagnosing/debugging, and securing the network. Switch/routers can be designed to offer intelligent network management solutions that reduce the complexity of updating, monitoring, and managing network-wide features such as access control lists (ACLs) for QoS and security control, rate-limiting policies, VLANs, software and configuration updates, and network alarms and events.

The discussion in this chapter covers the advanced and value-added features typically seen in high-performance switch/routers. The discussion includes device management (using Simple Network Management Protocol (SNMP)), device access (using Terminal Access Controller Access-Control System Plus (TACACS+), Remote Authentication Dial-In User Service (RADIUS), IEEE 802.1X, and Secure Shell (SSH)), traffic monitoring (using sFlow, NetFlow, and Remote network monitoring (RMON)), value-added features (such as Dynamic Host Configuration Protocol (DHCP) and Domain Name System (DNS)), and the key factors to consider when designing for energy efficiency.

2.2 GENERAL SWITCH MANAGEMENT

The management capabilities of network devices contribute greatly to the ease of deployment and maintenance of a network. A network device could be equipped with a variety of management tools that aim to provide effective network management

DOI: 10.1201/9781003311256-2

and the operational flexibility demanded by an organization's business requirements. These tools can include Command Line Interfaces (CLIs), Web-based device management tools, SNMP-based management tools (SNMPv1, SNMPv2, and SNMPv3), network traffic profiling and statistics collection tools (sFlow, NetFlow, and RMON), ACLs, device access control tools (TACACS+, RADIUS, SSH), per VLAN statistics tools, etc.

A wide variety of network management tools and applications could be provided to enable FCAPS (fault, configuration, performance, and security) management, and other functions. The management features in a network device could provide the following capabilities:

- Simplify network administration tasks using SNMP-based platforms, allowing the network to be managed from any SNMP-based management station.
- Allow more flexibility for network administration using remote (out-of-band) management through SNMP or SSH connections initiated on any management interface (e.g., console, auxiliary, or Ethernet ports), allowing the network to be managed from anywhere in the network.
- Provide centralized network management using Web-based, graphical interface tools (in addition to the CLI), empowering network operators to seamlessly control software and configuration updates (e.g., easy-to-use Graphical User Interfaces (GUIs) for system configuration from standard Web browsers).
- Reduce the costs of administering software upgrades and allow remote downloads of new revisions of operating systems and other software without hardware changes (using, for example, Trivial File Transfer Protocol (TFTP) or more secure tools offering similar data transfer functions).
- Simplify network device diagnosis and troubleshooting using status light-emitting diodes (LEDs) that allow the user to visually monitor operation of power supplies, cooling fans, and interfaces. This could be done, for example, via Ethernet port LEDs (RX and TX LEDs, link state, Ethernet indicators), and the route processor LEDs (status LEDs, fan and power supply LEDs, operational LEDs, portable external storage card LEDs, Ethernet RX and TX LEDs, link-state LEDs, etc.).
- Provide protection against unauthorized device and network configuration changes by requesting users requiring local access or remote access (via SSH, for example) to provide verifiable credentials.

Most network equipment vendors provide a feature-rich unified network management platform (mostly Web-based) for their devices. Such management systems greatly simplify network operations (provisioning, troubleshooting, and alarm reporting). Typically, these systems offer multilevel access security on console and secure Web management interfaces that enable network administrators implement the needed mechanisms to prevent unauthorized users from accessing or changing device and network configurations.

Such feature-rich management systems may employ Java-based network configuration and management tools that displays, in graphical detail, network and application-level traffic information. This allows network operators to accurately monitor overall network operations, zero in on hot spots, and quickly diagnose and troubleshoot difficulties before they develop into widespread network problems.

We describe in three sections below, the following important general management functions and their requirements in network devices (including switch/routers):

- Management with SNMP
- Secure Access
- Traffic Monitoring

2.3   MANAGEMENT WITH SNMP

The various versions of SNMP standardized by the IETF (Internet Engineering Task Force) are Application Layer protocols that define message formats and procedures for monitoring, collecting, organizing, and exchanging information about managed devices in an IP network. SNMP is part of the TCP/IP protocol suite (see Chapter 3 of Volume 1 of this two-part book) and is presently the most widely used protocol for monitoring and managing network devices. It is the de-facto industry standard for managing network devices in communication networks. SNMP supports a number of functions that can be used to send information to a network device in order to modify or change its behavior.

2.3.1   SNMP COMPONENTS

A typical SNMP-managed network has the following three components (Figure 2.1): SNMP manager (or network management system (NMS)); SNMP agent; and Management Information Base (MIB) (also called Management Information Database). A *managed device* is a network element (e.g., switch, router, switch/router, cable modems, server, workstation, printer, IP telephone, IP video camera, etc.) on which some form of monitoring and management is performed. Most network devices are designed and bundled with SNMP functions. This allows the device to communicate with an NMS located somewhere in the network.

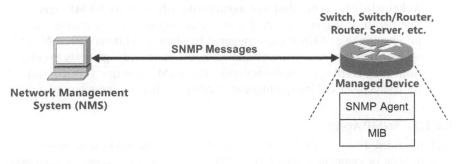

FIGURE 2.1   SNMP components.

A managed device implements some SNMP functions and an SNMP interface that allows bidirectional (read and write) access or unidirectional (read-only) access to both network-related and device-specific information. A managed device exchanges the information gathered in it, which may be only node-specific, with the NMS. The other SNMP components shown in Figure 2.1 are described below.

2.3.1.1   Network Management System

In a typical network using SNMP, one or more administrative processing nodes called *SNMP managers* have delegated the task of monitoring and managing a group of devices in the network. The SNMP manager is typically implemented as part of a broader NMS. The SNMP manager is typically a standalone device (e.g., a work-station) but may be implemented as part of a shared computing system. The NMS supports applications that send/receive SNMP messages when it is monitoring and controlling the managed devices in a network. To unburden the managed devices, the NMS generally supports the bulk of the memory and processing resources required for network management. A managed network may support one or more NMSs.

An SNMP manager supports a set of management applications that are designed specifically for data analysis, fault recovery, etc. The SNMP manager maintains a database of management information that is extracted from the MIBs of all the SNMP agents (associated with the managed devices) in the network. The SNMP manager also includes a user interface through which a network administrator can monitor and control the network and devices. The following are key functions of the SNMP manager:

- **Sends queries to SNMP agents**: The SNMP manager may send requests to an SNMP agent for information. The SNMP manager polls the managed devices in the network for information about activities, events, and network connectivity.
- **Gets responses from SNMP agents**: The agent responds to requests sent by the SNMP manager for information, and performs actions as directed by the SNMP manager.
- **Sets variables in SNMP agents**: The SNMP manager translates the network administrator's requirements and instructions into a format that is actually used to remotely monitor and control a managed device in the network. The SNMP manager can request, store, or change values of MIB variables associated with the SNMP agent.
- **Acknowledges events that are asynchronously sent by SNMP agents**: The SNMP manager may receive important but unsolicited management information (e.g., SNMP traps triggered by alarm conditions) from SNMP agents. These are sent as asynchronous notifications (of significant events) from any SNMP agent in the network. The SNMP manager is responsible for intercepting and interpreting each SNMP notification message.

2.3.1.2   SNMP Agent

Each managed device in an SNMP network that runs a network-management software module or component called an *SNMP agent* which exchanges management information with the NMS using SNMP messages. An SNMP agent has local knowledge of the management information that is collected in the managed device, and is also able to translate that information into a format (SNMP-specific form) that is readable by the NMS via SNMP. The agent is also able to translate management instructions received from the NMS (via SNMP) into a format that is understood by the managed device.

The following summarize the key functions of the SNMP agent:

- Monitors and gathers management information about its local environment.
- Stores and retrieves management information in its local MIB(s).
- Sends notification signals to the SNMP manager about events.
- May act as a proxy for network nodes that are non–SNMP manageable.

The SNMP agent responds to requests for management information and performs actions as directed by the SNMP manager. The SNMP agent controls access to its local MIB, including the collection of management objects that can be read or modified by the SNMP manager.

A managed device may support a *master SNMP agent* and a number of *SNMP subagents* that reside on different modules on the same device such as line cards, route processor, switch fabric, power supplies, etc. In this architecture, the master SNMP agent could delegate the processing of SNMP requests to the subagents. Each subagent will then be responsible for handling a specific set of MIBs in the system.

2.3.1.3 Management Information Base

SNMP presents and organizes management data about managed systems in the form of variables (or managed device parameters) that are stored in an *MIB*. Typically, an MIB contains a well-defined set of statistical and control variables for a managed device in a network. The MIB variables (accessible via SNMP) describe the system status and configuration and can be remotely queried (and, in some cases, modified/ changed) by management applications. SNMP allows active management tasks to be carried out on a managed device such as performing configuration changes through remote modification of variables in the MIB.

MIB variables are organized in hierarchies, but SNMP does not define or specify which variables a managed device should support. Instead, an extensible design approach is adopted in SNMP which allows systems/applications to define their own variables and hierarchies, and also describe these in data structures called MIBs. An MIB describes the management data structure of a system or subsystem in the managed device. MIBs use a hierarchical namespace that contains data items referred to as *object identifiers* (OID).

An MIB consists of managed objects, each identified by an OID. Every OID in an MIB is organized hierarchically. An MIB hierarchy can be represented in a tree structure defining grouping of MIB objects into related sets, while also indicating each individual variable OID. A "*leaf*" in the tree structure represents the actual managed object instance, for example, representing an event, activity, or resource in the managed device.

Each OID in the MIB identifies a specific variable (or object) that can be set or read using SNMP. Particularly, each OID is a variable that represents an identifiable aspect of the managed device; a collection of OIDs makes up an MIB. Each OID is unique and represents specific characteristics of a managed device. When an MIB is queried, the value returned for each OID could be different, for example, it may be a text, number, counter, etc.

A management station performs the monitoring function by retrieving the value of MIB objects. An SNMP manager may perform a task at an SNMP agent (i.e., effect an action) or change the configuration settings in it, by modifying the value of specific variables. Using SNMP, the SNMP manager may retrieve and/or modify the value of objects, and receive notifications from the SNMP agent of significant events. The notation used in MIBs is defined by IETF standard RFC 2578 [RFC2578] (Structure of Management Information Version 2.0 (SMIv2), which is a subset of Abstract Syntax Notation One (ASN.1)).

MIBs are developed as either *enterprise-specific* (i.e., proprietary) or as *standard*. The standard MIBs are published in documents referred to as Requests For Comments (RFCs) by the IETF. Each equipment manufacturer typically develops and supports its own enterprise-specific MIBs. Any organization can design and implement its own proprietary MIB as long as the MIB is defined under an enterprise-specific OID. Among the standard or proposed standard MIBs, the most basic and popular MIB is the group of managed objects known as MIB-II [RFC1213]. Some of the components of MIB-II are considered optional for most systems, while some have already been deprecated by newer RFCs. SNMP agents typically implement only a subset of MIB-II.

2.3.2   SNMP VERSIONS

Currently, there are three major versions of SNMP that have been developed. The first standardized version is SNMPv1. The newer versions, SNMPv2c and SNMPv3, were designed to include features that provide improvements in system performance, flexibility in use and deployment, and security.

- **SNMPv1 [RFC1155], [RFC1157], [RFC1213]**: SNMPv1 was the first standardized version of SNMP and is still the most widely used version. It has become the de-facto standard for network management and is supported by a majority of vendors, allowing in many cases, multi-vendor network management. SNMPv1 has poor security features, as a result, many subsequent developments over the years led to the current security features seen in SNMPv3. SNMPv1 provides only *community-based security*, where *community strings* (which are transmitted in cleartext) are viewed as actual passwords. With this, SNMPv1 provides only a simple authentication service that allows all SNMP messages to be identified and checked to see if they are authentic SNMP messages. SNMPv1 sends the community string over a network in an unencrypted form.
- **SNMPv2 [RFC1441], [RFC3584]**: In addition to introducing the *GetBulkRequest* and *InformRequest* messages, SNMPv2 was designed to

provide improvements in system performance, manager-to-manager communications, and security. Also, it provided a new *party-based security* feature which was later viewed by the industry to be overly complex, resulting in this version not being widely implemented or adopted.

o **SNMPv2c [RFC1901], [RFC2578], [RFC3416], [RFC3417]**: SNMPv2c (or Community-Based SNMP version 2) retains many aspects of SNMPv2 without its new party-based security system which was viewed as unnecessarily complex. SNMPv2c was designed to provide greater efficiency and more functionality than SNMPv1 (by introducing the *GetBulkRequest* message, simplified Trap message format, and expanding on error identifications). SNMPv2c provides some enhancements to SNMPv1 and SNMPv2 but uses instead the simple community-based security mechanism employed in SNMPv1. The improvements provided by SNMPv2c eventually led to it being widely considered as the de facto SNMPv2 standard, and many of its features were later added as part of SNMPv3. The changes made in SNMPv2c ended up making it not compatible with SNMPv1, mainly in the areas of message formats and protocol operations. The message header and protocol data unit (PDU) formats in SNMPv2c are different from those used in SNMPv1 messages. Also, SNMPv2c supports two protocol operations that are not used in SNMPv1. Thus, to overcome the incompatibilities in SNMPv1 and SNMPv2c, two SNMPv1/v2c coexistence strategies were defined in [RFC3584], which provide various definitions for SNMP proxy agents and bilingual network-management systems.

o **SNMPv2u [RFC1909], [RFC1910]**: SNMPv2u (or User-Based SNMP version 2) was developed as a compromise, that is, an SNMP version that promises or offers greater security than SNMPv1, while excluding the high-complexity security feature of SNMPv2. The *user-based security* mechanism of SNMPv2u was eventually adopted as one of SNMPv3's two security frameworks.

• **SNMPv3 [RFC3411], [RFC3412], [RFC3414], [RFC3415] [RFC3416], [RFC3417], [RFC3418]**: Aside from the introduction of a significant number of features and enhancements from cryptographic security perspective, SNMPv3 did not make changes to the protocol, particularly, in regards to the messages types used. However, SNMPv3 comes across as much different from the older versions because of the addition of new terminology, concepts, and textual conventions. The earlier versions were seen as lacking practical security features and SNMPv3 tried to remedy this. So, the most prominent and visible changes made in SNMPv3 were to define features and behaviors (security and remote configuration enhancements) that made it more secured for many network tasks such as configuration, accounting, and fault management. The development of SNMPv3 focused strongly on the two main aspects of security and administration. Security was addressed by including both strong authentication and encryption of SNMP messages. SNMPv3 was designed to offer authentication, message integrity, and encryption. For administration, SNMPv3 also addressed two aspects related to notification originators and proxy forwarders. The new changes offered by SNMPv3 were to facilitate secure remote configuration and administration of SNMP entities in a network and address more effectively the issues related to deployment, performance management, fault management, and accounting in larger networks. SNMPv3 also defines a number of security models that cater to a broader range of application scenarios: User-based Security Model (USM), View-based Access Control Model (VACM), and a Transport Security Model (TSM) that provides support for SNMPv3 over Secure Shell (SSH), SNMPv3 over Transport Layer Security (TLS), and Datagram Transport Layer Security (DTLS). The TSM provides a method for SNMP message authentication and encryption over external security channels.

Until the development of SNMPv3, security was considered to be one of the biggest weaknesses of SNMP. Authentication in SNMPv1 and SNMPv2 was basically achieved using a password (community string) transmitted in cleartext between an SNMP manager and agent. The IETF recognizes SNMPv3 as the current standard version of SNMP ([RFC3411] to [RFC3418], also known as STD0062), and considers the earlier versions (SNMPv1, SNMPv2, SNMPv2c, SNMPv2u) to be historic or obsolete. However, in practice, most SNMP implementations do not support a single SNMP version but instead support multiple versions, with the typical combination comprising SNMPv1, SNMPv2c, and SNMPv3, but with SNMPv1 more commonly featured.

2.3.3   SNMPv1/SNMPv2c INTEROPERABILITY

Proxy agents and bilingual network-management systems have been defined in [RFC3584] to overcome incompatibilities between SNMPv1 and SNMPv2c.

2.3.3.1   SNMPv2 Proxy Agents

The capabilities defined in [RFC3584] allow an SNMPv2 agent to serve as a proxy agent for managed devices that are based on the older but still widely used SNMPv1:

- An SNMPv2 NMS sends an SNMP message that is targeted at an SNMPv1 agent.
- The SNMP message is sent to an SNMPv2 proxy agent somewhere in the network.
- The SNMPv2 proxy agent receives and forwards all *GetRequest*, *GetNextRequest*, and *SetRequest* messages unchanged to the intended SNMPv1 agent.
- However, the proxy agent converts *GetBulkRequest* messages to *GetNextRequest* messages, and then forwards them to the SNMPv1 agent.

The proxy agent receives SNMPv1 trap messages from the SNMPv1 agent, maps them to corresponding SNMPv2 trap messages, and forwards these messages to the NMS.

2.3.3.2 Bilingual Network Management Systems

SNMP has gone through significant changes and upgrades since its inception. However, SNMPv1 and SNMPv2c are the most widely implemented versions. The use of SNMPv3 is growing (because of its enhanced security features) but still has not caught up and reached considerable use compared to the two older versions.

A *bilingual SNMPv2 NMS* supports both SNMPv1 and SNMPv2 operations. A dual-SNMP environment is provided by adding a management application in the bilingual NMS that is responsible for contacting agents. The NMS first examines a local database information to determine whether a particular agent supports SNMPv1 or SNMPv2. The information stored in the database specifies the SNMP language the agent speaks. The NMS then exchanges information with the agent using the appropriate version of SNMP.

2.3.4 SNMP Message Types

As described in Chapter 3 of Volume 1, SNMP is one of the protocols that operate in the Application Layer of the TCP/IP protocol suite. SNMP messages are carried in User Datagram Protocol (UDP) datagrams, and an SNMP agent (in a managed device) receives SNMP requests on UDP port 161. An SNMP manager may use any available UDP port (i.e., source port) to send SNMP requests to UDP port 161 on the SNMP agent. The SNMP agent may generate a response to be sent back to the SNMP manager using the originating source port specified in the request message.

The SNMP manager receives notifications from managed devices (SNMP *Trap* and *InformRequest* messages) on UDP port 162. However, the SNMP agent may generate SNMP notification messages from any available UDP port. When SNMP is used with Transport Layer Security (TLS) or Datagram Transport Layer Security (DTLS), the SNMP agent receives SNMP requests on UDP port 10161 and sends notifications to UDP port 10162 [RFC6353]. Both TLS and DTLS are primarily used to provide privacy and data integrity (i.e., prevent SNMP messages from being eavesdropped, tampered with, or forged) when two or more applications are communicating.

SNMPv3 categorizes SNMP protocol data units (PDUs) according to the following classes:

- **Read-Only**: These are messages (SNMP *GetRequest*, *GetNextRequest*, and *GetBulkRequest*) sent via a polling mechanism to read management information in a managed device.
- **Read-Write**: These are messages (SNMP *SetRequest*) sent to change management information in a managed device (i.e., to change the device's operation/behavior).
- **Response**: These are messages (SNMP *Response*) sent in response to a received request.
- **Notification**: These are messages (SNMP *Trap* and *InformRequest*) sent by a managed device to notify an SNMP manager of the occurrence of an event in the device.

In the following, we describe the SNMP messages that are specifically assigned to each of these SNMPv3 classes.

SNMPv1 defines five PDU types:

- *GetRequest*: This is a (manager-to-agent) message sent by an SNMP manager to a managed device to retrieve the value of a variable or list of variables from the device. The managed device returns a *Response* message carrying current values. The SNMPv3 PDU Class is *Read*.
- *GetNextRequest*: This (manager-to-agent) message is sent to retrieve the value of the next OID in the MIB tree in the managed device. Application of the *GetNextRequest* message can be to walk through, iteratively, the entire MIB of an SNMP agent starting at OID 0. The SNMPv3 PDU Class is *Read*.
- *SetRequest*: This (manager-to-agent) message is used by an SNMP manager to modify or assign a value to a variable or list of variables in the MIB of the managed device. The SNMP agent returns a *Response* message with current new values for the variables that have been modified/assigned. The SNMPv3 PDU Class is *Read-Write*.
- *Response*: This message is a command used to convey the value(s) or signal actions that have been executed as directed by an SNMP Manager. An SNMP agent returns an acknowledgment and a result as a response to the following SNMP messages: *GetRequest*, *SetRequest*, *GetNextRequest*, *GetBulkRequest*, and *InformRequest*. The SNMPv3 PDU Class is *Response*.
- *Trap*: This message is sent (as an asynchronous notification) by an SNMP agent to signal to an SNMP Manager the occurrence of an event. The SNMP agent sends these messages to the SNMP manager without receiving an explicit request. An SNMP agent sends traps (as an unsolicited SNMP message) to notify a management station of significant events. The SNMP *Trap* message format was changed in SNMPv2, with the PDU renamed *SNMPv2-Trap*. The SNMPv3 PDU Class is *Notification*. An SNMP *Trap* may be a *generic trap* or *enterprise-specific trap*.

SNMPv2 adds two more PDUs:

- *GetBulkRequest*: This message was introduced in SNMPv2 to minimize the number of SNMP message exchanges required to retrieve a large amount of MIB information. This (manager-to-agent) message is used to retrieve a large volume of values from an MIB that is possibly very large in itself. This message implements an optimized operation of *GetNextRequest*, in essence, implements multiple iterations of *GetNextRequest*. The use of *GetRequest* and *GetNextRequest* would otherwise result in the return of many small blocks of data. The SNMP agent returns a *Response* that contains multiple variables that have been listed in the request. The SNMPv3 PDU Class is *Read*.

- **InformRequest**: This message is similar to the SNMP *Trap* message that an SNMP agent initiates but with some differences. An SNMP manager that receives an *InformRequest* message from an agent acknowledges or confirms receipt of the message by sending an SNMP *Response* message back to the SNMP agent. Given that SNMP running over UDP does not assure message delivery, and dropped messages are not reported (meaning delivery of an SNMP *Trap* message is not guaranteed), an SNMP agent can send an *InformRequest* message as a way of demanding an acknowledgment from the receiver upon receipt of the information sent. If no response is received for an *InformRequest* message, the sender can send the message again (until a response is received). SNMP *Trap* messages are less reliable than *InformRequest* messages because the sender does not receive an acknowledgment when a Trap is sent and actually received. While SNMP traps are sent as "unconfirmed" notifications, SNMP *InformRequest* messages are acknowledged asynchronous notifications (or "confirmed" notifications). The SNMPv3 PDU Class is *Notification*.

SNMPv3 added one more PDU:

- **Report [RFC2572] [RFC3412]**: The use of this message is not as regular or common as the other seven SNMP messages and is often not listed when looking at regular SNMP messages. This PDU is mainly used for internal SNMP communication (engine-to-engine communication [RFC2572]), for example, an SNMP agent may send a *Report* PDU to communicate SNMPv3 security violation.

2.3.5 SNMP NOTIFICATIONS

As discussed above, SNMP notifications can be sent as SNMP *Trap* or *InformRequest* messages. SNMP traps are defined as either standard MIB traps or enterprise-specific MIB traps. Standard traps are defined in various IETF RFCs while enterprise-specific traps are defined by a specific equipment manufacturer.

With SNMP traps, the sender does not receive any acknowledgment when a *Trap* is received, and the sender cannot ascertain if the *Trap* was even received. So, to increase the reliability of SNMP notifications, SNMP *InformRequest* messages can be used. An SNMP manager that receives an *InformRequest* message acknowledges receipt of the message by sending an SNMP *Response* message (Figure 2.2). SNMP *InformRequest* is similar to an SNMP *Trap* except that, a copy of the former is stored and retransmitted at regular intervals until one of the following conditions occurs:

- The receiver (SNMP manager) of the message sends an acknowledgment to the sender (SNMP agent).
- The sender has attempted a specified number of unsuccessful retransmissions and then discards the *InformRequest* message.

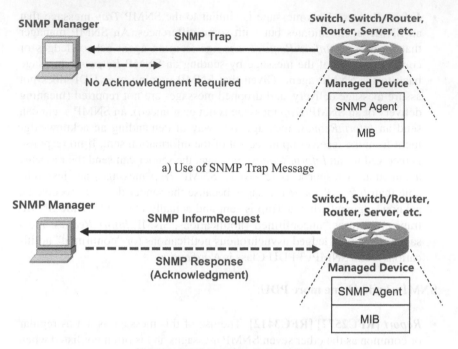

a) Use of SNMP Trap Message

b) Use of SNMP InformRequest Message

FIGURE 2.2   Use of SNMP Trap versus SNMP InformRequest message.

Because the sender can retransmit an SNMP *InformRequest* message when a response is not received, such messages are more likely to reach their intended target than SNMP *Trap* messages. SNMP *InformRequest* messages use the same communications channel as SNMP *Trap* messages (same UDP port number) but the two have different PDU types.

Even though SNMP *InformRequest* messages provide more reliable notifications than SNMP traps, they consume more network resources. Unlike an SNMP *Trap*, the sender holds a copy of an *InformRequest* message in memory until it receives a response or the specified timeout is reached. Also, a sender transmits an SNMP *Trap* only once, whereas it may retransmit an *InformRequest* message several times. It is therefore recommended to use *InformRequest* messages for notifications when it is important to ensure that the SNMP manager receives all such notifications. However, if there is the need to minimize such control and management traffic, or network device memory, then SNMP traps are more appropriate.

2.3.6   SNMPv1/v2 Security: SNMP Agent and SNMP Manager Authentication and Communication

SNMPv1 and SNMPv2 provide only very limited security by defining *communities* (traditionally referred to as *community strings*) to establish access rights between SNMP managers and agents in a network. A community string is an administrative name used to group a collection of managed devices (and the SNMP agents they are running) into a common management domain. If a given SNMP manager and agent share the same community (string), they can communicate and exchange information with each another. Because SNMP community strings and passwords/keys perform similar functions, they tend to be associated or mixed with one another.

So, in a simple sense, a community string (sent along with a *GetRequest* message) is usually seen as functioning as a type of shared user ID or password between the SNMP manager and managed device, and is used to authenticate the SNMP manager. A community field is defined as a part of each SNMPv1/v2c message and is a printable octet string (sent in cleartext). A community string functions in a way as an embedded password carried (unprotected) in the SNMP message. An SNMP agent may implement different community strings for information retrieval and modification operations.

The SNMP manager(s) and agent(s) have to be preconfigured with (using other non-SNMP means) the community strings they will use to authenticate SNMP messages and to access MIB objects. When an agent receives an SNMP message, it compares the community string in the incoming message with its local configured values and, if a match is found, it accepts the message as properly authenticated and continues to process it. If the agent does not find a match, it stops processing the message, and does not send a *Response* message.

For example, an SNMP community string allows an SNMP manager to access statistics within a managed device in the network. An SNMP manager can access data in managed devices with the correct community string. If the community string supplied is incorrect, then the managed device simply ignores and discards the *GetRequest* message. Most SNMP agents support the following three community strings, one for each of the following: *read-only*, *read-write*, and *notification* (or *trap*).

The community strings control the types of activities that can be performed within managed devices in the network. The *read-only* community string applies to *GetRequest*, *GetNextRequest*, and *GetBulkRequest* messages, and allows an SNMP manager to retrieve *read-only* information from a managed device. The *read-write* community string applies to *SetRequest* messages and allows an SNMP manager to retrieve information and change the configuration of a managed device. The *notification* or *trap* community string applies to the receipt of *Trap* or *InformRequest* messages.

Security strings or parameters are also used in SNMPv3, but the implementation here (depending on the security model used) allows more secure authentication and communication between SNMP managers and agents. Security parameters are encoded as an octet string and carried in each SNMPv3 message. The encoding and use of the security parameters depend on the type of SNMPv3 security model used.

2.3.7 SNMPv3 SECURITY

SNMPv3 defines security models and security levels that address authentication, message integrity, and encryption. A *security model* defines a security strategy set up for a user and the group to which the user belongs. A *security level* defines the level of security permitted within the selected security model. A combination of the two determines the security mechanism to be used when exchanging SNMP messages. User-based Security Model (USM) and View-based Access Control Model (VACM) are two of the security models defined by SNMPv3.

2.3.7.1   User-Based Security Model

USM is used for message security and specifies authentication and encryption functions. USM operates at the message level and provides data origin authentication, message replay protection, data integrity, and protection against message payload disclosure. With USM, both the SNMP manager and agent are configured with security parameters that define the security level (*none*, *authentication*, or *privacy*), authentication type and authentication password, and privacy type and privacy password.

USM provides better protection for SNMP messaging than communication with SNMPv1/v2 community strings, where passwords (strings) are sent in cleartext. USM provides data integrity checking and data origin authentication functions when messages are exchanged between the SNMP manager and agent. USM reduces and protects against message delays using time indicators to enforce timeout limits, and against message replays by using request IDs to check for duplicate message request IDs.

2.3.7.2   View-Based Access Control Model

VACM is used for access control and specifies access-control rules. VACM operates at the PDU level and provides access control to management information, and determines what specific type of access (*read* or *write*) is allowed. VACM defines whether a given SNMP entity is allowed access to a particular MIB object to perform specific operations.

VACM defines groups of data users, collections of MIB objects (into units referred to as "*views*"), and defines access rights that specify which views a particular group of users can perform *reads*, *writes*, or receive *notifications* (*traps*). VACM provides highly granular access control for applications and allows security policies to be applied to the name of the groups (of data users) querying an SNMP agent. The SNMP agent is then able to decide whether the group is allowed to view/read or modify specific MIB objects.

By defining MIB views, an SNMP agent can have better control over which SNMP entities can access specific branches and objects within its MIB tree. Each OID represents a subtree of an MIB object hierarchy. So, each object of an MIB view has a common OID prefix. As well as having a unique name, an MIB view consists of a collection of SNMP OIDs, which are either explicitly included (in the MIB view) or excluded. Once created, the MIB view is assigned to an SNMPv3 group (of data users) or SNMPv1/v2c community (or multiple SNMPv1/v2c communities), and automatically masking which parts of the MIB tree belonging to the SNMP agent, the group or community members can (or cannot) access. A group identifies a collection of SNMPv3 or SNMPv1/v2c data users that share the same VACM access privileges or policy.

VACM can be used to configure the access privileges granted to a group (of users as defined by USM, or community strings as defined in the SNMPv1/v2c security models). Using a predefined MIB view, VACM controls access by filtering the MIB objects available for a specific operation. MIB views can be assigned to determine the MIB objects that are visible to a particular group for *read*, *write*, and *notify* operations, using a particular context, security model (SNMPv1, SNMPv2c, or SNMPv3-USM), and security level (*none*, *authentication*, or *privacy*).

2.4   SECURE ACCESS

The security features of a network device may include SSHv2, Secure Copy, and SNMPv3 to restrict and encrypt management communications to the device. SNMPv3 provides secured SNMP management with authentication and privacy services. These tools provide secure access to the administration and management interface of the device over the network. Additionally, support for TACACS+ [CISCID13838] [DRAFTTACACS+] and RADIUS [RFC2865] [RFC2866] [RFC3576] authentication ensure secure operator access. In addition to RADIUS, TACACS, and TACACS+, user authentication with IEEE 802.1X [IEEE 802.1X] prevents unauthorized network access.

2.4.1   AUTHENTICATION, AUTHORIZATION, AND ACCOUNTING

Authentication, authorization, and *accounting* (AAA or "Triple A") refers to the family of techniques used for controlling access to network devices and network resources, enforcing access policies, auditing user activities and resource usage, and providing the information necessary for network management and service billing. AAA is important for network management and security. Specifically, the use of AAA mechanisms allows verification user identity, granting access to users, and tracking of users' actions in a network device and the network itself.

AAA can be based on either TACACS+ or RADIUS, each one with its strengths and weaknesses. Using a combination of user ID and passwords, a network device may perform local authentication/authorization of users using a local database, or remote authentication/authorization using one or more AAA servers. A pre-shared secret key may be used to provide security when the network devices and the AAA servers exchange information. An AAA security framework provides the following services:

*   **Authentication**: This is the process of verifying the identity of the entity seeking access to a network device or resource. If a user has been authenticated, then it means that the device or resource recognizes who the user is. This process provides user identification (using a unique set of criteria for system access) and includes a login and password dialog. It may also include challenge and response, encryption (depending on the security protocol used), and messaging support. The verification can be based on a combination of the user ID and password provided by the entity trying to access the network device. Typically, the network device uses a local lookup database to perform local authentication, or uses one or more TACACS+ or RADIUS servers to perform remote authentication. In the latter, a user's authentication credentials are received by an AAA server which compares them with user credentials stored in a database. The user is granted access to the device/network only if there is a matching credential in the database.

- **Authorization**: This is the process of determining the set of activities, resources, or services that an authenticated user is authorized to perform or use. Authorization in a network device can be performed using attributes that are provided by AAA servers (Figure 2.3). RADIUS and TACACS+ servers are typically remote security servers that can be used to authorize users for specific activities/resources/services in a network device or the network. This is done by associating attribute-value (AV) pairs with users which define what these specific activities comprise. Authorization can be used to limit a user to certain commands, or limit or define network services for different users. Authentication has to take place before authorization.
- **Accounting**: This is the process of tracking user activities, measuring the resources a user consumes during system access, and record-keeping in the system. Accounting involves collecting information about user activities, logging the information in a local database, or sending the information to an AAA server for activities such as auditing, reporting, profiling resource utilization, trend analysis, billing, and capacity planning. Accounting may track and maintain a log of every management session used to access a network device as well as log individual session statistics and resource usage information. This information can be used to generate reports for auditing and troubleshooting purposes. The device can be configured to store accounting information and logs locally, and/or send them to one or more remote AAA servers.

A system may be designed to support authentication, authorization, and accounting independently. Also, the system may be configured to support authentication and authorization without accounting.

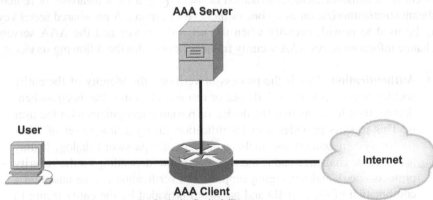

AAA Server

User

Internet

**AAA Client
(in Switch, Switch/Router,
Router or NAS)**

Client/Server Model of AAA

FIGURE 2.3 AAA architecture.

The advantages and benefits of using remote AAA services over local AAA services include the following:

- Increased control and flexibility of network device and network access configuration. This makes it easier to manage user attributes and access information for each device in the network than maintaining local databases on the devices.
- Facilitates centralized management of user information including easier management of user password lists for each device in the network. This allows a network administrator to centrally manage the accounting logs for all devices in the network. Using a single, centralized, and secure database for AAA (or its components) is much easier to administer than using information that is distributed across numerous network devices.
- Allows easy addition of features for network growth and scalability. AAA servers can be deployed widely across a network, providing easy access to AAA services.
- Allows the use of standardized authentication methods such as TACACS+ and RADIUS.
- Facilitates the use of multiple devices for backup.

AAA server groups can be configured for remote AAA services in a network. An AAA server group refers to multiple remote AAA servers that have been configured to run the same AAA protocol (TACACS+ or RADIUS). The use of a server group is to provide failover to working servers when any one remote AAA server is unreachable or fails. When any contacted remote server in the group is not responding, the next remote server in the group is contacted until one of the servers in the group responds. If all the AAA servers in the server group are unresponsive, then that server group is declared to have failed. A network can be designed to support multiple server groups for added AAA service availability.

2.4.2 TERMINAL ACCESS CONTROLLER ACCESS CONTROL SYSTEM PLUS

TACACS+ and RADIUS are two prominent security protocols that work through a centralized server to provide remote user authentication and access control into network devices or a network (that provides resources and services). Both protocols operate through an access server and support AAA features that include validation of users attempting to gain access to a network (guarding against unauthorized access), as well as secure remote access to network resources and services. TACACS+ is mainly used for network device administration, particularly, for authenticating access to switches, routers, and switch/routers, and implementing centralized authorization and auditing of user operations in such devices.

TACACS+ operates on a client/server model and uses TCP (which offers a reliable connection-oriented transport of protocol data). The TACACS+ server uses TCP port 49 for TACACS+ traffic. Both TACACS+ and RADIUS have the following key components (Figures 2.4 and 2.5):

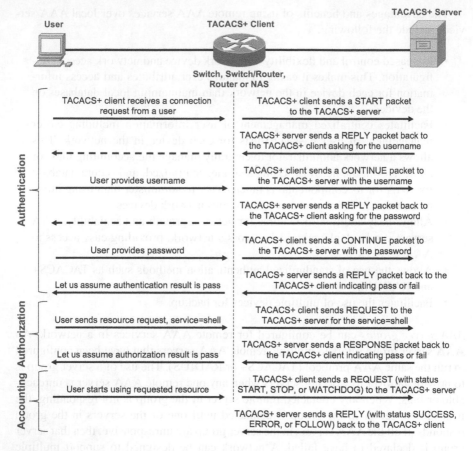

FIGURE 2.4   Using TACACS+ for secure access.

- **TACACS+ (or RADIUS) Client**: This is a device that provides a connection to a user seeking access to a network device or network services. The client can reside in a network node such as a switch, switch/router, router, or a dedicated access control server called a *network access server* (NAS). In general, a switch, switch/router, router, or any other network device running a TACACS+ or RADIUS client is considered an NAS. An NAS is simply a device (dedicated server, switch, switch/router, router, etc.) that can recognize and handle connection requests from a user.
- **TACACS+ (or RADIUS) Server**: This is a software program that runs on a security server to provide AAA services using the TACACS+ or RADIUS protocol. In TACACS+, the authentication, authorization, and accounting processes can be implemented as separate and independent modules. In RADIUS, authentication and authorization are combined and not considered as separate modules. During authentication (using either protocol), the server might respond to the client with a request for additional user information, such as a user password.

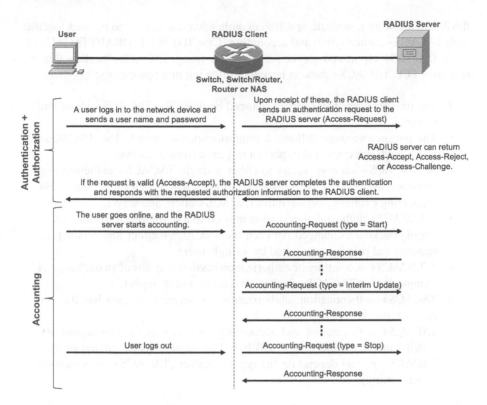

FIGURE 2.5 Using RADIUS for secure access.

The discussion now focuses on TACACS+. When the NAS receives a connection request from a user (using, for example, Point-to-Point Protocol (PPP)), it may perform an initial access negotiation with the user. The negotiation may involve establishing certain user information (such as username, password, client port number on NAS, etc.). The TACAS+ client (on the NAS) will then pass this information to the TACACS+ server, requesting user authentication. The TACACS+ server receives this information and authenticates the user request and (if successful) will authorize the services requested by the user over the user connection. The TACACS+ server performs user authentication by matching the user information received from the TACACS+ client request, with entries in a well-known, trusted database.

As noted above, TACACS+ supports an AAA architecture but implements the three AAA components as separate and modular processes. TACACS+ allows each service (authentication, authorization, and accounting) to be implemented and activated independently but still allows them to operate through a single access control server to control access to a network. An AAA implementation or configuration is not required to have all three TACACS+ AAA components. Authentication is not mandatory when using TACACS+. A network may support it for certain services and may not for other services. Also, each of the AAA services (e.g., authentication) can have its own database and still be allowed to interface and interact with the other services available on the access control server or on the network. This architecture provides flexibility, allowing a separate or different authentication method to be used together with TACACS+ authorization and accounting [CISCID13838] [DRAFTTACACS+].

A TACACS+ client and server communicate using TACACS+ packets that are sent over TCP. TACACS+ packets have the following main properties:

- The packets carry messages exchanged between the TACACS+ client and server.
- The packet exchange follows a request/response model: The TACACS+ client sends a request and expects a response from the server.
- A "TACACS+ session" refers to either a single TACACS+ authentication sequence, a single TACACS+ authorization exchange, or a single TACACS+ accounting exchange between the TACACS+ client and server.
- A TACACS+ authentication session may consist of multiple (arbitrary number of) packets exchanged between the TACACS+ client and server (i.e., requests and responses initiated by a single user).
- A TACACS+ accounting or authorization session consists of an exchange of a single pair of TACACS+ packets (a request and its reply).
- TACACS+ authentication, authorization, or accounting each has its own packet type.
- TACACS+ authorization and accounting packets may contain arguments, called "*attribute-value pairs*" (AVPs). The specific AVPs carried in each TACACS+ packet depend on the type of packet (TACACS+ authorization or accounting).

Because TACACS+ supports authentication exchanges that can be of arbitrary length and content, it allows a wide range of authentication mechanisms to be used such as PPP Password Authentication Protocol (PAP), PPP Challenge Handshake Authentication Protocol (CHAP), Extensible Authentication Protocol (EAP), Kerberos, and token cards.

The following three types of packets are used in TACACS+ authentication:

- **Authentication START**: START packets are always sent by the TACACS+ client. This message is only used to begin the authentication process and is the first message sent by the TACACS+ client in an authentication session, or as the first packet immediately after a system restart. The START packet carries information that describes the type of authentication to be carried out and may include the username and other authentication data.
- **Authentication REPLY**: REPLY packets are always sent by the TACACS+ server. The server may send a REPLY message indicating if the authentication is finished, or if it should continue. If the server's REPLY message indicates that authentication should continue, then the server will also indicate the kind of new information it is requesting.
- **Authentication CONTINUE**: CONTINUE packets are also (and always) sent by the TACACS+ client. The TACACS+ client sends a CONTINUE message to the server following the receipt of a REPLY message, and may

possibly contain information requested by the server. The TACACS+ client sends this message when responding to a request by the TACACS+ server for username and password.

Every access request received by the TACACS+ client that requires authentication is authenticated by a remote authentication TACACS+ server. The TACACS+ client sets up a TCP connection to the TACACS+ server, and then sends a START packet. The START packet contains the type of authentication to be carried out (PAP, CHAP, etc.) and may contain the username plus other authentication data. The TACACS+ server responds with a REPLY packet to the client and user, indicating either access *granted*, access *denied*, reporting an *error*, or *challenging* the user for more information.

The TACACS+ server (via the client) may challenge the user to provide a username, password, passcode, or other relevant information. Once the user has provided the requested information, the TACACS+ client sends a CONTINUE packet over the existing connection to the TACACS+ server (Figure 2.4). The TACACS+ server then sends a REPLY packet back to the client indicating the authentication result. Once the TACACS+ authentication is completed, the authentication session is closed.

TACACS+ authorization uses two types of packets, REQUEST and RESPONSE:

- **Authorization REQUEST**: The TACACS+ client sends this message to the server containing a fixed set of fields plus a variable set of arguments. The fixed set of fields describes the method the client used to obtain the user information, the type of user authentication that was performed, and the service that requested the authentication. The variable set of arguments (consisting of attribute-value pairs (AVPs)) describes the services and options for which the user is requesting authorization.
- **Authorization RESPONSE**: The TACACS+ server sends this message to the client and contains a variable set of arguments (also AVPs) that can modify or restrict the TACACS+ client's actions.

Generally, TACACS+ authorization is preceded by authentication which may not be required for some services – only authorization is performed. In such a case, the TACACS+ authorization request may indicate that the user was not authenticated and it is up to the authorization agent to decide whether the unauthenticated user is permitted to use the requested service.

TACACS+ accounting uses two types of packets, REQUEST and REPLY:

- **Accounting REQUEST**: The TACACS+ client sends this message to the server, containing the information that is used to perform accounting for the service provided to the user.
- **Accounting REPLY**: The TACACS+ server sends this message to the client to indicate that it has completed the accounting function and has securely committed the accounting record.

In TACACS+, the following three types of accounting records are supported:

- **Start records**: These indicate that a particular requested service is about to start.
- **Stop records**: These indicate that a particular requested service has just been terminated.
- **Update records**: These serve as intermediate notifications to indicate that a particular requested service is still being performed.

The TACACS+ accounting function can be used to create an audit trail of command-line interface (CLI) commands and User Exec sessions that a user has executed within these sessions. For example, TACACS+ accounting can be used to track user CLI connects and disconnects, the entry and exit times of configuration modes, and the configuration and operational commands that have been executed.

Figure 2.4 shows an example of the client/server message exchange process in TACACS+. TACACS+ provides security by encrypting all transactions between the TACACS+ client and server. Encryption is based on a shared secret key that is known to both the TACACS+ client and server. The shared secret key may consist of any alphanumeric string and both client and server sides must be configured with this shared secret key.

In TACACS+ [CISCID13838], the entire body of the packet is encrypted excluding the standard TACACS+ header. The header carries a field that indicates whether the accompanying body has been encrypted or not. During normal TACACS+ operation, it is recommended to have the entire packet body fully encrypted for more secure client/server communications. However, during times of debugging TACACS+ problems, it is useful to leave the body of the packets unencrypted.

The separation or decoupling of authentication, authorization, and accounting functions in TACACS+ is a fundamental feature of its design. For example, an implementation may choose to use Kerberos authentication together with TACACS+ authorization and accounting. In such a case, an NAS may authenticate a user on a Kerberos server, and then send a request to a TACACS+ server to obtain user authorization information without having to authenticate the user again. Upon successfully user authentication on the Kerberos server, the NAS simply informs the TACACS+ server about the outcome, and the TACACS+ server in turn provides the required user authorization information.

When a session is established and in progress, and there is the need for additional authorization checking when the user wants to use a particular command, the NAS checks with the TACACS+ server to determine if the user has permission to use that command. TACACS+ provides granular or fine-grained access control most useful and flexible for device management or for terminal services (granular command by command authorization). TACACS+ supports such features to allow it to provide greater control over the commands that can be executed on the NAS while decoupling such operations from the user authentication process. RADIUS, on the other hand, is not designed to provide fine-grain control over which commands users can or cannot execute on the switch, switch/router, or router.

Two methods are available in TACACS+ to control the authorization of commands on a per-user or per-group basis on Cisco routers [CISCID13838]. In the first method, *privilege levels* are assigned to commands, and then the router verifies with the TACACS+ server to determine if the user has permission at the specified privilege level. In the second method, the commands that are allowed, on a per-user or per-group basis, are explicitly specified on the TACACS+ server.

2.4.3 REMOTE AUTHENTICATION DIAL-IN USER SERVICE

RADIUS [RFC2865] [RFC2866] [RFC3576] follows a client/server model and sends protocol data over UDP unlike TACACS+ which uses TCP. Because UDP offers best-effort delivery of data and lacks the inbuilt reliable data transport that TCP offers, RADIUS implements additional programmable protocol parameters (such as number of data re-transmit attempts and time-out mechanisms) to compensate for the best-effort UDP transport. Note that the RADIUS (or TACACS+) server runs on a centralized platform typically in the network whose resources are being accessed, while the RADIUS (or TACACS+) clients reside in access devices or servers that can be distributed throughout the network.

RADIUS (or TACACS+) provides a method for managing users attempting to gain access to a network through multiple network access points from a single access control server (RADIUS or TACAS+ server). The RADIUS (or TACACS+) client can reside and operate in a switch, switch/router, router, or in a device implemented specifically as a network access server (NAS). As illustrated in Figure 2.5, the RADIUS client interfaces with a user and is responsible for passing the information sent by the user to designated RADIUS servers in the network.

The client is also responsible for acting on the response the RADIUS server returns and where necessary, passing any resulting response/actions to the user. The RADIUS server is responsible for receiving and processing user access requests, performing user authentication, and returning to the client, the required configuration information necessary for it to grant access and deliver service to the user.

A RADIUS server may be required to act as a proxy client to other specific AAA servers in the network. Any one of the three functions of a RADIUS server (i.e., authentication, authorization, or accounting) can be delegated to another RADIUS server. The first RADIUS server then becomes a proxy server for the second server. Proxying RADIUS functions to other servers enables a network to delegate some of the AAA functions of a RADIUS server to other servers. Thus, RADIUS allows roaming and distributed authentication, authorization, and accounting because of a RADIUS server's ability to proxy requests to other servers regardless of the location of the clients originating the requests.

The properties of RADIUS are similar to those of TACACS+ discussed above. A RADIUS client and server communicate via RADIUS packets which are formatted as described in [RFC2865] for Authentication/Authentication, and [RFC2866] for RADIUS Accounting. RADIUS supports the following six standard packet types which are all sent over UDP:

- **Access-Request**: The RADIUS client sends this message to the RADIUS server carrying information the server uses to determine whether a user is allowed access to the specific NAS and any specific services the user has requested.
- **Access-Reject**: The RADIUS server sends this message to the client if any value of the received attributes from the user is not acceptable.
- **Access-Accept**: The RADIUS server sends this message to the client containing specific configuration information necessary for the client to begin providing the requested service to the user.
- **Access-Challenge**: The RADIUS server sends this message to the client to request additional information from the user in order to determine if it can grant access to the requested service. The additional information required from the user may include a secondary user password, Personal Identification Number (PIN), token, etc. The RADIUS client, after communicating with the user, responds with another Access-Request message to the server.
- **Accounting-Request**: A RADIUS client, typically in an NAS or its proxy, sends this message to a RADIUS accounting server carrying information used by the server to provide accounting for a service provided to a user.
- **Accounting-Response**: A RADIUS accounting server sends this message to the client to acknowledge that it has received and recorded successfully the Accounting-Request.

In addition to these packets, RADIUS describes other non-standard packet types as described in [RFC2882].

Figure 2.5 shows an example of the client/server message exchange process in RADIUS. Typically, a user login triggers a query (Access-Request) from the RADIUS client to the RADIUS server that holds user authentication plus network service access information. The server processes the request and sends a corresponding response (Access-Accept, Access-Reject, or Access-Challenge) to the client. The Access-Request packet sent by the client to the server contains user information such as the username, encrypted password, NAS IP address, and NAS port.

The RADIUS server receives the Access-Request from the RADIUS client and searches a secure trusted database for an entry that matches the user information in the RADIUS client request. If the RADIUS server does not find any matching user information in the database, it either loads a default profile, or it immediately sends an Access-Reject message to the RADIUS client. The Access-Reject message may include additional information indicating the reason for the access denial.

If on the other hand, the RADIUS server finds matching user information, it will return an Access-Accept message which includes a list of attribute-value pairs (AVPs) that describe the parameters to be used for the session. Typical parameters include the service type, protocol type, static or dynamic IP address to be assigned to the user, access list to be applied, a static route to be installed in the routing table of the NAS, and so on. The configuration information supplied by the RADIUS server determines what will be installed in the NAS for the user's sessions.

In RADIUS (similar to TACACS+), transactions between the RADIUS client and server are authenticated using a shared secret that is only known by both sides but never sent over the network. Furthermore, all user passwords sent between the RADIUS client and server are encrypted. Note that only the user password carried in the Access-Request packet sent from the client to the server is encrypted by RADIUS. This prevents the snooping of user password over this network portion as could be possible when they are passed over an unsecured network. The rest of the Access-Request packet is left unencrypted (transmitted in cleartext). This means, other user information, such as username, authorized services, and accounting, in the packet can easily be read by any snooping party, and are vulnerable to different types of malicious attacks.

In RADIUS, both user authentication and authorization are combined, making it difficult to decouple these two processes. As illustrated in Figure 2.5, the Access-Accept packets the RADIUS server sends to the client contain authorization information. The RADIUS server may support a variety of methods for user authentication. Methods such as PPP PAP, CHAP, UNIX login, and others, may be specified at the time the username and original password are provided by the user. During an authentication transaction, RADIUS transmits the password information securely between the RADIUS client and the server. The password is encrypted and decrypted using the shared secret. RADIUS accounting can be used in an implementation-independent of the RADIUS authentication and authorization functions.

RADIUS does not use encryption when transmitting accounting data between a RADIUS client and server. However, each side uses the accounting shared secret to verify the "trustworthiness" of any RADIUS communications it receives from the other side. RADIUS may "sign" RADIUS accounting packets using a different shared key from the shared key used for RADIUS authentication packets. RADIUS does not specify how the shared secrets should be stored, and this has become implementation-dependent using different methods.

When RADIUS was first standardized, UDP ports 1645 and 1646 were used for RADIUS authentication and accounting packets, respectively. It was later found that these UDP ports had already been assigned to another standard. So, the RADIUS standards UDP port assignments were changed to UDP ports 1812 and 1813 (for RADIUS authentication and accounting packets, respectively). However, many organizations still use the old UDP port assignments, thereby, requiring some caution when configuring RADIUS services. This means any two RADIUS devices must use compatible UDP port numbers to be able to exchange RADIUS packets.

For example, if an NAS is being configured to exchange authentication packets with a RADIUS server, the port on which the server receives authentication packets from its clients must be precisely determined (let us assume this to be 1812). Then the NAS must be configured to send authentication packets to the RADIUS server on the same UDP port (1812). This configuration procedure also applies to RADIUS accounting.

2.4.4 IEEE 802.1X

In simple words, authentication means making sure that an entity is actually what it claims to be. IEEE 802.1X [IEEE 802.1X] is a networking protocol that enables the authentication of devices (e.g., portable wired/wireless devices) seeking to connect to a LAN or wireless LAN (WLAN). IEEE 802.1X specifies mechanisms for port-based network access control in LANs and WLANs. It is a client-server-based access control and authentication protocol that allows only authorized clients to connect to a LAN or WLAN usually through network device ports that are publicly accessible (switches, switch/routers, routers, network gateways, WLAN access points, etc.). An authentication server authenticates each client before it is allowed to connect to a network device port.

Until a client is authenticated, IEEE 802.1X allows only access control traffic (i.e., Extensible Authentication Protocol over LAN (EAPOL) traffic) to pass through the LAN or WLAN network device port to which the client is logically connected. Only after the client has been successfully authenticated as permitted to fully connect to the network will normal client traffic be allowed to pass through the LAN or WLAN network device port. Usually, clients connecting through publicly accessible LAN or WLAN ports are authenticated via IEEE 802.1X before they can logically connect and send client traffic into the network.

In the case of a wired network such as an Ethernet network, the network device ports are physical entities to which the clients plug into, while in a WLAN, the network device ports are logical entities known as associations. EAPOL data is sent over the Data Link Layer, and in Ethernet, the EtherType has the reserved value of 0x888E.

IEEE 802.1X has become increasingly indispensable for network security, because guests, contractors, and consultants of an organization typically require access to the network and its resources over the same LAN or WLAN connections used by regular company employees, who may themselves bring into the workplace unmanaged devices that need to be authenticated. IEEE 802.1X enables the authentication of clients to minimizing the possibility of unauthorized devices gaining access to controlled or confidential company information.

IEEE 802.1X provides Layer 2 (or media-level) access control which allows the network to permit or deny network connectivity based on the identity of the client device. Before authentication (when the identity of the client is unknown), all client traffic on the connected port is blocked. After authentication and if the identity of the client is known and authorized to access the network, all client traffic from that connected port is allowed.

IEEE 802.1X defines the following three main logical entities which have specific roles (Figure 2.6):

- **Supplicant**: This is a client device such as a portable network device that requests access to a LAN or WLAN and responds to requests from the access control point (see authenticator). This device must be running 802.1X-compliant client software such as that offered in the operating systems of common end-user devices.
- **Authentication Server**: This is a trusted server that performs the actual authentication of the supplicant on behalf of the authenticator. The authentication server receives information from the authenticator and validates the identity of the supplicant. The server then notifies the authenticator whether the supplicant is authorized to access the LAN or WLAN and its services, in addition to various settings that should apply to that supplicant's connection or setting. The authenticator acts as a proxy, making the authentication service transparent to the supplicant. The authentication server is a

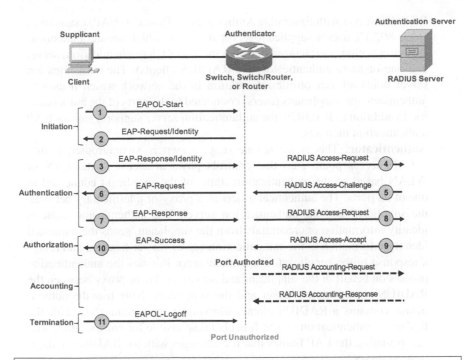

In this example, we assume the client initiates the authentication and uses MD5-Challenge EAP authentication:
1. The client physically connects to the network which in turn starts the 802.1X supplicant software. Client enters the username and password, and sends an EAPOL-Start frame to the authenticator.
2. The authenticator receives the EAPOL-Start frame and returns an EAP-Request/Identity packet, requesting the supplicant to send the entered username.
3. The supplicant responds with an EAP-Response/Identity frame containing the username to the authenticator.
4. The authenticator encapsulates the EAP-Response/Identity frame into a RADIUS Access-Request packet and sends it to the RADIUS server.
5. The RADIUS server receives the username from the authenticator and searches the username in the local database for the password corresponding to the username. The server encrypts the password with a randomly generated MD5-Challenge value, and sends the MD5-Challenge value in a RADIUS Access-Challenge packet to the authenticator.
6. The authenticator forwards the RADIUS Access-Challenge packet containing the MD5-Challenge value to the supplicant.
7. The supplicant receives the MD5-Challenge value from the authenticator and also encrypts the password with the MD5-Challenge value. The supplicant then generates an EAP-Response/MD5 Challenge packet, and sends it to the authenticator.
8. The authenticator encapsulates the EAP-Response/MD5 Challenge packet into a RADIUS Access-Request packet and sends it to the RADIUS server.
9. The RADIUS server compares the encrypted password in the received RADIUS Access-Request packet with the local password encrypted using the MD5 algorithm. If the two passwords match, the RADIUS server considers the client as authorized and responds with a RADIUS Access-Accept packet.
10. The authenticator receives the RADIUS Access-Accept packet and then sends an EAP-Success frame to the supplicant.
   - The port on the authenticator to which the supplicant is connected then enters the authorized state, allowing the client to access the network.
   - The authenticator periodically sends handshake packets to the supplicant to monitor the online status of the client.
11. To logout, the supplicant sends an EAPOL-Logoff frame to the authenticator.

FIGURE 2.6 IEEE 802.1X message exchange example.

RADIUS server with Extensible Authentication Protocol (EAP) extensions. IEEE 802.1X uses a supplicant-server model in which secure authentication information is exchanged between the RADIUS (authentication) server and one or more authenticators (or RADIUS clients). The authentication server holds all user profile information in the network which it uses to authenticate the supplicants (users) connected to the ports of the authenticator. In addition to RADIUS, the authentication server supports various EAP authentication methods.

- **Authenticator**: This is the device (e.g., a switch, switch/router, router, WLAN access point, etc.) that controls physical access to the LAN or WLAN based on the authentication status of the supplicant connected to one of its ports. The authenticator acts as a proxy or intermediary between the supplicant and the authentication server. The authenticator requests identity information or credentials from the supplicant, sends the requested identity information to the authentication server for verification, and relays a response to the supplicant. The authenticator initiates the authentication process on behalf of the supplicant and acts as a relay or proxy between the RADIUS authentication server and the supplicant. Note that the authenticator contains a RADIUS client, allowing it to communicate with the RADIUS authentication server. It is also responsible for encapsulating and decapsulating the EAP frames that it exchanges with the RADIUS authentication server.

The authenticator running in a network access device such as a switch, switch/router, or router acts/serves as a security barrier to a protected network. The authenticator only allows the supplicant access to the protected side of the network only when the supplicant's identity has been validated and access is authorized (Figure 2.6). The supplicant initially provides the required credentials to the authenticator (using methods that have been specified in advance by the network administrator). The requested credential could include a username plus password, or a permitted digital certificate. The authenticator then forwards these credentials to the authentication server which decides if access should be granted or not. If the authentication server verifies the supplicant's credentials to be valid, it sends a response to the authenticator informing it to grant access to the supplicant to the resources located on the protected side of the network.

The various protocols supported by IEEE 802.1X are summarized as follows:

- **Extensible Authentication Protocol (EAP) [RFC3748]**: This is (simply) an authentication framework (not a specific authentication mechanism by itself). EAP only defines message formats that can be used by other protocols for formatting and transporting information/parameters required by these protocols for authentication purposes. Each protocol (such as IEEE 802.1X) that uses EAP, defines its own way of encapsulating EAP messages within that protocol's messages. The message format and framework defined by IEEE 802.1X provide a way for a supplicant and the authenticator to negotiate an authentication method (referred to as the EAP method).

EAP provides some common functions that can be used by other protocols to negotiate the type of authentication methods (or EAP methods) to be used. IEEE 802.1X uses EAP (via EAP over LAN (EAPOL) frames) to facilitate communication between the supplicant and the authenticator, and between the authenticator and the authentication server. EAPOL frames are sent over Layer 2 protocol frames.

- **EAP Method:** This defines the specific authentication method used between the supplicant and the authentication server. It defines the type of credentials used and how they are exchanged between the supplicant and the authentication server (based on the EAP framework). Some of the EAP methods include EAP Transport Layer Security (EAP-TLS), EAP Tunneled Transport Layer Security (EAP-TTLS), EAP-MD5, Lightweight Extensible Authentication Protocol (LEAP), EAP Flexible Authentication via Secure Tunneling (EAP-FAST), Protected Extensible Authentication Protocol (PEAP), PEAP-MSCHAPv2, etc. IEEE 802.1X supports various authentication methods and a client seeking access to a network will have to use an authentication method supported by the authentication server. The network administrator needs to configure which methods the authentication server will use. EAP-TLS is widely used and uses a public key infrastructure (PKI) (e.g., X.509 digital certificate) to authenticate the supplicant and authentication server. That is, EAP-TLS requires digital certificates for client-side and server-side mutual authentication.
- **RADIUS**: This is currently the de facto standard in IEEE 802.1X for communication between the authenticator and the authentication server. The authenticator extracts the EAP payload from the Layer 2 EAPOL frame sent by the supplicant and encapsulates the payload inside a RADIUS packet (Application Layer) to be forwarded to the authentication server.

A key benefit of IEEE 802.1X is that the network access devices to which users connect (e.g., switches, wireless access points, etc.) do not in themselves need to fully know how to authenticate or support mechanisms for authenticating a client requesting access to the network. All they need to do is to, simply, receive user authentication information and pass that information to the authentication server. The authentication server is the entity responsible for handling the actual verification of the user's credentials. This simple feature enables IEEE 802.1X to support a wide range of authentication methods, from simple username and password to challenge and response, hardware token, and digital certificates.

Authentication data exchanged between the three IEEE 802.1X entities is done using EAP packets which in themselves can be encapsulated in other protocol packets. When the authenticator receives EAPOL frames from the supplicant, it strips off the Ethernet header, and then encapsulates the remaining EAP frame in the RADIUS format before relaying the RADIUS packet to the authentication server (i.e., if the authenticator and authentication server reside on different devices). The authenticator does not modify or examine the EAP frame during the encapsulation process, and the authentication server is required to support EAP within the native RADIUS frame format. In turn, when the authenticator receives RADIUS packets from the authentication server, it strips off the server's RADIUS packet header, leaving the EAP frame, which the authenticator then encapsulates in Ethernet frames and sends to the supplicant.

As illustrated in Figure 2.6, the operation of IEEE 802.1X NAC can be divided into the following five phases:

- **Session Initiation**: An IEEE 802.1X authentication can be initiated by either the supplicant or the authenticator.
  - o *Authenticator Initiates*: The authenticator initiates the authentication session when it detects a link-up condition on one of its ports. In this case, the authenticator initiates the authentication by transmitting an *EAP-Request/Identity* message to the supplicant and if no response is received, another request message is retransmitted at periodic intervals. The authenticator initiates authentication by periodically transmitting *EAP-Request/Identity* frames to the special Layer 2 (Ethernet) address (01:80:C2:00:00:03) on the local network segment. The supplicant also listens on this special reserved Layer 2 address for such messages and when it receives one, it responds with an *EAP-Response/Identity* frame. On the detection of a new supplicant on a port, the authenticator enables that port and sets it to the "*unauthorized*" state. In this state, the port accepts only IEEE 802.1X traffic; other non-IEEE 802.1X traffic from the supplicant is dropped.
  - o *Supplicant Initiates*: The supplicant initiates an authentication session by sending an *EAPOL-Start* frame to the authenticator which prompts it to request the supplicant's identity via an *EAP-Request/Identity* message. The sending of an *EAPOL-Start* message in this case helps to speed up the authentication process without the supplicant having to wait for the next periodic *EAP-Request/Identity* from the authenticator. *EAPOL-Start* messages are most useful in situations where a supplicant may not be ready to process an *EAP-Request* sent by the authenticator. The supplicant may not be ready because its operating system is still booting or no physical link state change has been detected on the authenticator.
- **Session Authentication**: Upon receiving an *EAP-Request/Identity* frame from the authenticator, the supplicant responds with an *EAP-Response/ Identity* frame containing an identifier for the supplicant such as the username. The authenticator then encapsulates this *EAP-Response/Identity* frame in a *RADIUS Access-Request* packet and forwards it to the authentication server. The supplicant may also restart the authentication session by transmitting an *EAPOL-Start* frame to the authenticator, which will then respond with an *EAP-Request/Identity* frame.

  During this stage, the authenticator relays EAP messages exchanged between the supplicant and the authentication server. The authenticator encapsulates each EAP message contained in the EAPOL frame sent by the supplicant in an attribute value (AV) in a RADIUS packet. The supplicant sends an *EAP-Response/Identity* message to the authenticator, which encapsulates this message in a RADIUS packet, and then forwards it to the authentication server. Note that, at the beginning of this exchange, the supplicant and the authentication server have to agree on the EAP method to be used. The EAP messages exchanged between the supplicant and the authentication server via the authenticator are used to authenticate the supplicant.

The exact message sequence in this exchange is defined by the specific EAP method agreed on. As noted above, the EAP method defines the credential type to be used by the authentication server to authenticate the supplicant, and how these credentials are submitted to the server. The EAP method may request the supplicant to submit a password, certificate, token, plus other information. The credentials may be submitted via a TLS-encrypted tunnel as a hash or using some other information protection method.

Upon receiving a *RADIUS Access-Request* message, the authentication server sends a *RADIUS Access-Challenge* packet back to the authenticator, specifying the EAP method and additional information it wishes the supplicant to perform and provide. The authenticator copies this information into an *EAP-Request* message and transmits it in an EAPOL frame to the supplicant. The supplicant can opt to start using the requested EAP method, or respond with a *Negative Acknowledgement* (NAK) containing the EAP methods it is willing to use.

- **Session Authorization**: Using the EAP method agreed upon as described above, the supplicant and the authentication server exchange *EAP-Request* and *EAP-Response* messages that are translated by the authenticator:
  - *Valid Credentials*: If the credentials submitted by the supplicant are valid, the authentication server returns an *EAP-Success* message encapsulated in *RADIUS Access-Accept* message through the authenticator to the supplicant. This message indicates to the authenticator that the supplicant should be granted access to the port it is connected to. Optionally, the authentication server may include in the *RADIUS Access-Accept* message to the authenticator, dynamic network access policy instructions (e.g., an access control list (ACL) or a dynamic VLAN). When no such dynamic policy instructions are included, the authenticator simply opens the port for the supplicant. Upon successful authentication, the authenticator sets the port to the "*authorized*" state, allowing normal traffic from the supplicant.
  - *Invalid Credentials*: If the credentials submitted by the supplicant are invalid, or the supplicant is not allowed to access the network for policy reasons, the authentication server returns an *EAP-Failure* message encapsulated in a *RADIUS Access-Reject* message through the authenticator to the supplicant. This message indicates to the authenticator that the supplicant should not be granted access to the port it is connected to. Depending on how the authenticator is configured, the supplicant or authenticator may retry the authentication, try an alternative authentication method, or deploy the authenticator's port into the *Auth-Fail VLAN*. An Auth-Fail VLAN allows users (typically, visitors to an organization)

without valid credentials on the authentication server to access a limited set of services. Upon an unsuccessful authentication, the authenticator's port remains in the "*unauthorized*" state.

The port state on the authenticator determines if a supplicant has been granted access to the network. A port starts in the *unauthorized* state where it denies all ingress and egress traffic except for IEEE 802.1X protocol packets. The port transitions to the *authorized* state when the supplicant is successfully authenticated, thus allowing all traffic from the supplicant to flow normally into the network.

If a client that does not support or have IEEE 802.1X enabled on it is connected to an IEEE 802.1X port that is in the *unauthorized* state, the authenticator will request the client's identity via an *EAP-Request/Identity*. However, if the client does not respond to the identity request, the authenticator port remains in the *unauthorized* state, and the client will not be granted access to the network.

If on the other hand, an IEEE 802.1X-enabled client connects to a port that does not support or is not IEEE 802.1X-enabled, the client initiates the authentication process by transmitting an *EAPOL-Start* frame. When the client does not receive a response via *EAP-Request/Identity* message after a fixed number of attempts to start authentication (e.g., three tries), the client will transmit traffic as if the port is in the *authorized* state.

- **Session Accounting**: If the authenticator is able to successfully apply the authorization policy as instructed by the authentication server, it can send a *RADIUS Accounting-Request* message to the server, carrying details about the authorized supplicant's session. RADIUS accounting is used to keep records about the session including details on the user and device (such as correlating a username with MAC address, IP address, authenticator port), session types, service details, usage statistics, and so on.
- **Session Termination**: A session can be terminated by any one of the following methods:
  - *Physically disconnecting or unplugging the endpoint device causing a link down on the connected port on the authenticator*: When the authenticator detects that the link state of the port on which the supplicant is connected goes down, it will completely clear the IEEE 802.1X session. If the original endpoint device or a new endpoint device plugs into that port, the authenticator will restart the authentication process from the beginning.
  - *The supplicant logs off by sending an EAPOL-Logoff message to the authenticator*: The *EAPOL-Logoff* message informs the authenticator to terminate the existing IEEE 802.1X session, and on receiving the *EAPOL-Logoff* message, the authenticator terminates the session. A *proxy EAPOL-Logoff* message can also be used where, for example, an IP phone transmits a *proxy EAPOL-Logoff* message when it detects that an IEEE 802.1X-authenticated endpoint device connected to it has unplugged from it. The IP phone can substitute the MAC address of the

endpoint device with its own MAC address, so that the *proxy EAPOL-Logoff* message will be indistinguishable from an actual *EAPOL-Logoff* message from the endpoint device itself. The authenticator upon receiving such a *proxy EAPOL-Logoff* message will immediately clear the session.

o **By using management software such as the activation of an inactivity timer**: When an *inactivity timer* is used, the authenticator monitors the activity from the authenticated endpoint. When the inactivity timer expires, the authenticator will remove the authenticated session. The authenticator uses the inactivity timer as an indirect mechanism for inferring if an endpoint device has disconnected. The inactivity timer can be manually configured on the authenticator port or it can be dynamically set using the *RADIUS Idle-Timeout* attribute. Setting the inactivity timer via the *RADIUS Idle-Timeout* attribute is advantageous because this provides control over which endpoints on the authenticator are subject to a particular inactivity timer and the length of the timer setting for each class of endpoints.

When any of the above happens, the authenticator then sets the port connected to the supplicant to the *unauthorized* state, once again blocking all non-EAP traffic from the supplicant. In IEEE 82.1X, authenticated sessions must be cleared when the authenticated endpoint device disconnects or unplugs from the authenticator/network. Session termination is important in the IEEE 802.1X authentication process to further ensure the integrity of the authenticated session. Authenticated sessions that are not properly terminated immediately can create security holes and lead to security violations.

2.4.4.1 What Is MAC Authentication Bypass?

Some devices such as network printers, cameras, wireless phones, and Ethernet-based electronics like environmental sensors may not support IEEE 802.1X authentication. To allow such devices to be used in a protected network environment, alternative mechanisms such as MAC Authentication Bypass (MAB) have been developed to authenticate them.

A port on an access device (switch, switch/router, router, etc.) that has MAB configured on it will first try to check if a connected endpoint client is IEEE 802.1X compliant, and if it receives no reaction from the connected client, it will try to authenticate the endpoint with the authentication server using the client's connected MAC address as username and password. To do this, the network administrator must configure the RADIUS server to authenticate these MAC addresses, either by adding them in the server as regular users or implementing additional processing steps to resolve the MAC addresses in a network inventory database.

The access device will use the MAC address as the client's identity (a method referred to as MAB). The authentication server must have a database of client MAC addresses that are allowed network access. After detecting a client on a port, the access device will wait for an Ethernet packet from the client. The access device will then send to the authentication server, a *RADIUS Access-Request* message with a username and password based on the client's MAC address. If authentication server successfully authenticates the client, the access device will grant the client access to the network. If authentication fails, the access device may assign the client to a *guest VLAN* if one is configured.

2.4.5 SECURE SHELL

Secure Shell (SSH) is a network protocol that supports a number of cryptographic features and is used for accessing network services securely over a network that is not secured. Some of the typical and everyday applications of SSH include remote machine login, remote command-line, and remote command execution. SSH can also be used to secure a wide range of network services.

The security features of SSH have made it a more secure replacement for TELNET and other remote shell protocols such as the Berkeley r-commands `rlogin` (remote login), `rsh` (remote shell), and `rexec` (remote execution) that are all considered to be unsecured (Figure 2.7). For example, TELNET or `rlogin` are insecure protocols and do not use any security features such as authentication and encryption when transmitting or receiving data. These legacy protocols send information, particularly, passwords, in plaintext, making them more susceptible to interception and disclosure using, for example, packet sniffers. SSH's encryption allows it to provide data confidentiality and integrity over open and unsecured networks, for example, the public Internet.

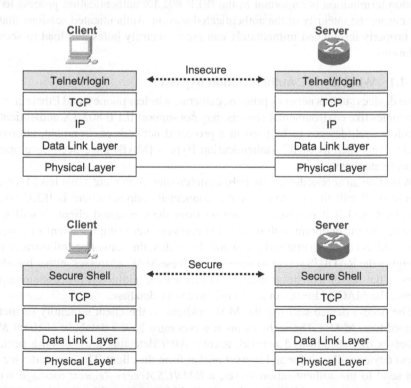

FIGURE 2.7 Connection via TELNET (or rlogin) versus SSH.

SSH also supports other features such as X11 (or X Window System) forwarding, and application port forwarding (also called port tunneling) [RFC4254]. SSH is also used in secure transfer file mechanisms such as Secure Copy (SCP), rsync, SSH File Transfer Protocol (SFTP), Files Transferred over Shell Protocol (a.k.a. FISH), Fast and Secure Protocol (FASP), etc.

SSH uses a client-server architecture, connecting an SSH client application with an SSH server, and provides a secure channel over an unsecured network between the client and server. The TCP port number assigned for contacting an SSH server is 22. This section focuses on the SSH protocol (referred to SSH-2) as defined in [RFC4251] [RFC4252] [RFC4253] [RFC4254].

2.4.5.1 SSH Sub-Layers

SSH consists of three main sub-protocols (or components) [RFC4251] as illustrated in Figure 2.8:

- **SSH Transport Layer Protocol [RFC4253]**: This component is a low-level SSH protocol that provides a confidential channel over an insecure network for the exchange of SSH data. In addition, this layer is responsible for handling SSH server authentication (not client authentication), initial key exchange, encryption (for data confidentiality), and message integrity protection (via the addition of a message authentication code (MAC) to each message to allow the receiver to perform integrity checks to verify that the received data has not been tampered with). Optionally, this layer may also support data compression functions. This protocol typically runs over a TCP/IP connection, but may also be used over any other reliable data transport mechanism. The Transport Layer Protocol also derives a unique *session identifier* (ID) that may be used by the other higher-level SSH protocols. Leaving out other less pertinent details, the SSH Transport Layer Protocol packet has a format that includes the following fields: packet length, padding length, payload, padding, and message authentication code (MAC).

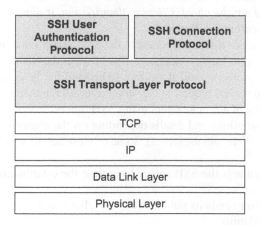

FIGURE 2.8   SSH sublayers on TCP/IP.

- **SSH User Authentication Protocol [RFC4252]**: This component runs over the SSH Transport Layer Protocol and handles client authentication (during which the SSH server uses a specified authentication method to authenticate the client). This protocol provides client-side authentication and is client-driven. For example, when a particular user is prompted for a password, this may be a prompt issued by the SSH client and not the SSH server. The SSH server simply responds to authentication requests issued by the SSH client. The user authentication protocol may support a number of authentication methods that the server may use to authenticate the user's identity. The particular authentication method specified by the authentication protocol uses the *session ID* provided by the SSH Transport Layer Protocol. The authentication method also depends on the security and integrity guarantees of the SSH Transport Layer. The User Authentication Layer has been designed to be highly extensible, allowing a wide range of authentication methods including custom authentication methods to used.
- **SSH Connection Protocol [RFC4254]**: This protocol defines the concept of channels and handles the channels over which data is transmitted between the SSH client and the server. The SSH Connection Protocol consists of a mechanism that multiplexes multiple channels (or distinct data streams) over the confidential and authenticated Transport Layer mechanism. Several logical channels can be multiplexed over the encrypted tunnel/connection between the client and the server. A single SSH connection can carry multiple logical channels simultaneously with each channel transferring data in both directions. The channels provided by the Connection Protocol can be used for a wide range of applications. This protocol provides standard methods for establishing secure interactive shell sessions, as well as, for forwarding (i.e., tunneling) TCP/IP ports and X11 connections.

These three protocols provide unique functions that allow the SSH protocol to secure connections, authenticate the other party, encrypt data, and transfer data across different channels. *As noted above, the SSH Transport Layer Protocol only authenticates the SSH server and not the client. Client authentication, if required, is performed by the SSH User Authentication Protocol.*

The SSH client sends a service request to the SSH server once the two have established a secure Transport Layer connection. This involves the client and server negotiating the parameters, protocols, and algorithms to be used for the connection. The client then sends a second service request to the server after user authentication is complete.

The login process of a remote client using SSH can be described by the following basic steps (with variations and details depending on the algorithms and parameters agreed upon by the client and the server). Some of these details will be discussed later:

1. The client contacts the SSH server to initiate the establishment of an SSH connection.
2. The SSH server sends its public key to the client so that it can authenticate the server's identity.

3. The client and the server negotiate the parameters and algorithms for the connection, and if successful, they proceed to establish a secure channel based on these parameters/algorithms.
4. The client then logs into the server host to access the resources and services supported and can now perform its tasks remotely.

The connections set up between an SSH client and an SSH server follow a client-server model. After the client and the server have successfully set up a connection, SSH uses strong symmetric encryption and message authentication code (MAC) algorithms to ensure the confidentiality and integrity of the data that is exchanged between them. The data transmitted is encrypted according to the algorithms and parameters negotiated during the setup phase.

Each SSH server host must have a *public host key* [RFC4251]. The SSH client uses the *server public host key* during the *key exchange* phase to verify that it is really communicating with the correct SSH server. To make this possible, the SSH client must have a priori knowledge of the public host key of the SSH server (as discussed below). Also, hosts may have multiple host keys each using a different public key algorithm. Furthermore, a single host key may be shared by multiple hosts. If a host has multiple host keys, the host must have at least one host key that is associated with each required public key algorithm [RFC4251].

2.4.5.2 Key Features and Functions of the SSH Transport Layer Protocol

At the SSH Transport Layer [RFC4253], the SSH client is the entity that initiates the connection. Then the client and the server negotiate the key exchange method, the public key algorithm to be used, the symmetric-key cipher to be used to encrypt data, the message authentication algorithm to be used to check for data integrity, and the compression method (if any) to be used. The public key, encryption, data integrity, and compression algorithms can be different for each direction of an SSH connection. Once the connection is initiated, both the SSH server and the SSH client exchange an identification string, which includes the SSH protocol version used (e.g., 2 for SSHv2.0).

2.4.5.2.1 Algorithm Negotiation

To set up the parameters of the SSH connection at the beginning of the key exchange process, each side sends a packet to the other listing the parameters that it is willing to accept in the connection setup, starting with the preferred algorithms. The packet sent contains a list of parameters that include the following [RFC4253] (citing here only the major parameters):

- SSH_MSG_KEXINIT
- Cookie (random bytes)
- Name-list of key exchange algorithms
- Name-list of server host key algorithms
- Name-list of encryption algorithms (client-to-server)
- Name-list of encryption algorithms (server-to-client)

- Name-list of MAC (message authentication code) algorithms (client-to-server)
- Name-list of MAC algorithms (server-to-client)
- Name-list of compression algorithms (client-to-server)
- Name-list of compression algorithms (server-to-client)
- Name-list of languages (client-to-server)
- Name-list of languages (server-to-client)

The "SSH_MSG_KEXINIT" in the message is used to signal the beginning of the key exchange process (i.e., the initial message signaling the start of the key exchange or key re-exchange process). When a party receives this message, it must respond with its own SSH_MSG_KEXINIT message, except when the received SSH_MSG_KEXINIT is a reply. The "*cookie*" represents a random value generated by the sender (SSH client or server) and serves as a mechanism for making it impossible for either side to fully determine the keys and the session ID. The *name-list of languages* contains a list of language tags in the order of preference, where a language tag indicates the language used in an information object [RFC5646]. Both parties may choose to ignore the language name-list, and if no language preferences are specified, this name-list should be empty. The rest of the parameters are described in the sections below.

Each name-list contains the parameters/algorithms the side is willing to accept in the connection, separated by commas. Each side lists the algorithms it supports (or willing to allow) in order of preference (from most to least preferred) in each name-list. Each name-list lists the preferred algorithm first, and each list must contain at least one algorithm name.

2.4.5.2.2   Key Exchange Method

This is the method to be used for SSH server authentication. The key exchange process also sets up the keys that will be used to secure the SSH connection. The key exchange method specifies the process that both parties go through to generate one-time *session keys* that are to be used for authentication and encryption, and how the server authentication is performed. SSH binds each session key to a session by including randomly generated session-specific data in the hash used to generate the session key. For each session or connection, SSH uses a *key exchange algorithm* for generating unique keys. A typical key exchange algorithm is the *Diffie-Hellman algorithm* [RFC4419] which can be used to generate a *shared secret key* that is only known to the communicating parties.

At the beginning of the key exchange process, each side sends a name-list of supported algorithms to the other (with a preferred algorithm in each list). A connection attempt failure occurs if the two parties do not have a mutually supported algorithm that satisfies the connection requirements.

SSH server authentication falls under one of the following categories:

- **Explicit server authentication**: A key exchange method is considered to use *explicit server authentication* if the key exchange messages include a *signature* or some other proof of authenticity or legitimacy of the SSH server. The key exchange method defined in [RFC4253] falls under explicit server authentication.

- **Implicit server authentication**: In a key exchange method that uses *implicit server authentication*, the SSH server is required to show that it knows a shared secret, K, by transmitting a message plus a corresponding MAC to the SSH client for it to verify the server's authenticity. Key exchange methods that use implicit server authentication can be combined with explicit server authentication.

After the two sides have exchanged SSH_MSG_KEXINIT messages, they each run the key exchange algorithm. The key exchange process may involve the exchange of several packets, as specified by the negotiated key exchange method.

2.4.5.2.2.1 Key Exchange Output

The key exchange results in the SSH client and the server establishing a *shared secret key K* and an *exchange hash* value H. These two values are used to derive the encryption and authentication keys. The *encryption keys* are computed as a hash of a known data value and the shared secret K. The exchange hash value which is derived from the initial/first key exchange is used additionally as a unique *session ID*, that is, as a unique identifier for the SSH connection. Once derived, the unique session ID for the connection is not changed, even if key re-exchange later takes place. The authentication method uses the unique session ID as a part of the data that it signs as a proof that it possesses a private key.

A *hash function* is specified for each key exchange method that is used in the key exchange process. The same hash algorithm is also used in the derivation of the keys. The hash function that is settled depends on the key exchange method (e.g., *diffie-hellman-group-exchange-sha256*, *diffie-hellman-group14-sha256*, *diffie-hellman-group16-sha512*, etc.) the SSH client and the sever have decided on during the algorithm negotiation process. After the key exchange process is completed, the SSH client and the server use the new set of keys and algorithms for all future communications.

Higher level protocols can use this unique session ID to bind data to a given SSH session, preventing the replay of data from prior SSH sessions. The SSH User Authentication Protocol uses this unique session ID to prevent replay of digital signatures from previous SSH sessions. The public key authentication exchanges are cryptographically bound to the initial key exchange and the SSH session, making them difficult to be successfully replayed in other SSH sessions. It should be noted that the unique session ID can be made public without hurting the security of the protocol.

To end the key exchange process, each side sends an SSH_MSG_NEWKEYS message to the other. The SSH_MSG_NEWKEYS message is sent containing the old keys and algorithms used. Each side that receives this SSH_MSG_NEWKEYS message must begin to use the new keys and algorithms for receiving. The SSH_MSG_NEWKEYS message also ensures that a party receiving this message is able to respond with an SSH_MSG_DISCONNECT message to the other side in the event that it has detected that something has gone wrong during the key exchange process.

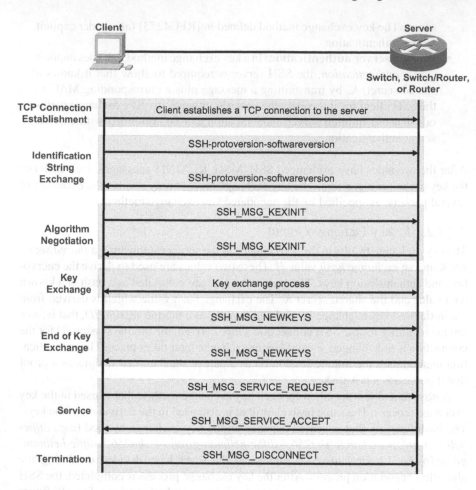

FIGURE 2.9 Summary of SSH connection initialization and key exchange operations.

Figure 2.9 highlights the main steps of the SSH connection initialization and key exchange operations.

2.4.5.2.2.2 Use of Diffie-Hellman Key Exchange in Secure Shell

The *Diffie–Hellman key exchange* is a method for two parties to securely exchange cryptographic keys over a network that is insecure [RFC4419]. Two parties can use this method to generate and establish a *shared private key* with which they can exchange information across the insecure network. This method establishes a shared secret (private key) that cannot be determined by either party alone. To provide explicit server authentication as described in [RFC4253], the key exchange messages contain a signature which is used with the server's public host key to provide server authentication.

The SSH key exchange based on the Diffie–Hellman method can be summarized by the following steps [RFC4253]:

1. All parties agree on a finite group G of order q and a generating element g in the group G. This phase is usually done well before the Diffie–Hellman method is run. The generator g is assumed to be known by all parties.

2. The client generates a random number x ($1 < x < q$) and then computes an exchange value e (which is a function of g and x), and sends e to the server in an SSH_MSG_KEXDH_INIT message.

3. The server generates a random number y ($0 < y < q$) and then computes an exchange value f (which is a function of g and y). The server receives the exchange value e and then computes the following:
   a. Shared secret K (which is a function of e and y)
   b. Exchange hash value H from a concatenation of the client's identification string, server's identification string, payload of the client's SSH_MSG_KEXINIT message, payload of the server's SSH_MSG_KEXINIT message, server's public host key, client's exchange value e, server's exchange value f, and the shared secret K. The hash algorithm that is used to compute the exchange hash value H is defined. When the connection starts, each side must send an identification string to the other (see [RFC4253] for details).
   c. Signature s by applying a signature algorithm over the exchange hash value H, and not on the original data. A second hashing operation may be used in the signing operation.
   The server then sends its public host key, exchange value f, and signature s (and certificates) in an SSH_MSG_KEXDH_REPLY message to the client.

4. The client verifies that the received public host key is indeed the public host key for the server, for example, using a local database or certificates. The client may choose to accept the key without verification; however, it must recognize the decreased security impact of doing so. The client then computes the following:
   a. Shared secret K (as a function of the exchange value f and x).
   b. Exchange hash value H also from a concatenation of the client's identification string, server's identification string, payload of the client's SSH_MSG_KEXINIT message, payload of the server's SSH_MSG_KEXINIT message, server's public host key, client's exchange value e, server's exchange value f, and the shared secret K.
   c. Signature s by applying a signature algorithm over the exchange hash H, in order to verify the signature received from the server.

Note that only the random numbers x, y, and the shared secret K are kept secret by all parties. All the other values like e, f, and signature s (of the exchange hash value of H) are sent in the clear. Once the client and the server have computed the shared secret K, they can use it to derive encryption keys (known only to them) for sending messages across the same insecure communication channel. Both client and server can use the shared secret key to encrypt and decrypt data making the process symmetric. The process allows both sides to encrypt and decrypt data while keeping their individual private keys a secret. The Encryption keys are computed using the same hash algorithm used for computing the exchange hash H as described in [RFC4253].

The exchange hash H (which should be kept a secret) is used to authenticate the key exchange. Readers can refer to [RFC4253] and other relevant RFCs for more information on the use of Diffie–Hellman key exchange in SSH.

2.4.5.2.3   Server Host Key Algorithm

Server host key algorithms are also negotiated during the key exchange process. The SSH server sends to the client a name-list of the algorithms it supports for each of its public host keys. The name-list the server sends will list the algorithms for which it has public host keys, while the client in turn will send a name-list of the algorithms that it is willing to accept. The server may support multiple public host keys, possibly with different algorithms.

Server host key algorithms are public-key algorithms used to authenticate the server as well as to securely establish the shared unique session ID. Each server has a cryptographic key pair (a private key and a public key) that identifies it. Whenever an SSH client connects to an SSH server, the server uses the cryptographic keys to identify/authenticate itself to the client. This ensures that the client and the intended server are provided encryption and integrity protection end to end.

The SSH server identifies itself to the client by sending its public host key. The client uses the server host key during the key exchange process to verify that it is indeed communicating with the correct server. This is only possible if the client has a priori knowledge of the server's public host key using one of the following trust models; the client must know the public key of the server so that it can securely authenticate the server:

- **Use of a local client database**: In this model, the client has a local database in which it associates the name of each server (as entered by the user) with a corresponding public host key. Although this method does not require a centrally administered infrastructure and third-party coordination, it has the drawback of becoming burdensome to maintain as the database of name-to-key associations grows.
- **Use of a trusted certification authority**: In this model, the client uses a trusted certification authority to certify the server host name-to-key association. In this case, the client only knows the root key of the certification authority and is able to verify the validity of all server host keys certified by accepted certification authorities. A certificate is a digital document that is used for secure authentication of communicating parties. A certificate binds the identity information about an entity to the public key of the entity for a certain validity period. A certificate is digitally signed by a certification authority who has verified that the key pair actually belongs to a particular entity. Public-key infrastructure (PKI) relies on digital certificates and simplifies the distribution of public keys used in public key authentication.

The public key algorithms are also used by SSH for encryption and/or digital signatures. The public key algorithms that are used for signing are negotiated using SSH_ MSG_KEXINIT messages. A public key type is defined by the key format, signature and/or encryption algorithms, and encoding of signatures and/or encrypted data. The SSH protocol is designed to use a wide range of public key algorithms, encoding types, and formats (e.g., *ssh-rsa-sha256*, *ssh-dss-sha256*, *x509v3-sign-dss-sha256*, etc.). An implementation may support a wide range of encoding methods, allowing SSH sessions with different data formats, padding, and byte order to be configured.

By using different key formats, keys can be encoded in different ways, as well as allowing the use of a range of certificate representations. The server may spell out to the client the host key algorithms that it has keys for, while the client may specify the specific key algorithms that it is willing to accept. Some public host key types may not support both encryption and signatures, thus, making not all public host keys valid for all key exchange methods. The selection depends on whether the key exchange method previously negotiated requires a host key that is encryption-capable or a digital signature. In some cases, policy statements (e.g., in certificates) may also restrict public host key usage. In these cases, it is recommended to define different public key types for the different policy alternatives supported.

2.4.5.2.4 Encryption Algorithms

SSH uses symmetric-key algorithms (also known as ciphers) to encrypt the data and provide confidentiality. The SSH client and the server establish the parameters and shared key used in the encryption process during the earlier phases of the SSH connection. The parties negotiate an encryption algorithm and a key during the key exchange. The parties exchange name-lists of acceptable symmetric encryption algorithms starting with the most preferred algorithm.

The first algorithm that shows up on the client's name-list that also happens to be on the server's name-list is selected to be used. If the parties do not reach an agreement, then the connection fails. Otherwise, the algorithm that is settled on is used to encrypt the SSH Transport Layer Protocol packet payload, the packet length field, the padding length field, and the padding field. SSH supports a range of different encryption algorithms (e.g., *aes128-cbc*, *aes192-cbc*, *aes256-cbc*, etc.) and keys must be a minimum of 128-bit (but with larger keys preferred).

The encrypted data in all packets transmitted in one direction (client-to-server or vice versa) is considered a single data stream. An SSH implementation must allow the ciphers that run in each direction to be independent of each other. The encryption algorithm for each direction can be independently selected if the implementation allows multiple algorithms to be used. However, in practice, it is recommended to use the same algorithm in both directions.

2.4.5.2.5 Message Authentication Code (MAC) Algorithms

SSH uses a MAC that is added to each message to allow authentication of the message, that is, to confirm that the message originates from the stated sender and has not been altered (in other words, determine its authenticity). Thus, to protect the data integrity of a message as well as its authenticity, SSH adds a MAC to each message, allowing intended receivers (who also possess the secret key) to verify and detect any changes to the content of the message. At the start of the key exchange process, each side sends to the other a name-list of acceptable MAC algorithms listed in the order of preference.

The SSH Transport Layer Protocol uses a MAC that is added to each message or packet to verify the integrity of the message data. The MAC is computed from the unencrypted packet (including the length fields, payload, and, possibly, random padding), the packet sequence number, and a shared secret (which is established in the key exchange process). The MAC is derived before packet encryption is performed. The MAC value computed from the MAC algorithm is transmitted as the last part of the packet and is not encrypted. The size of the MAC in bytes depends on the negotiated MAC algorithm. The MAC algorithm used at the sender and receiver produces a matching MAC for a message only if the same message, packet sequence number, and secret key are passed as input to the same MAC algorithm.

The receiver uses the MAC, the secret key, and the same (verifying) MAC algorithm to verify the authenticity of the received message. The receiver inputs only the message portion (excluding the received MAC) into the same MAC algorithm using the same secret key, producing a second local MAC. The receiver then compares the received MAC to the locally generated MAC. If the two MACs are identical, the receiver assumes that the message was not tampered with or altered during transmission (confirming data integrity). The receiver accepts the message when the message and the MAC are not forged or tampered with, otherwise, it rejects the message.

Although the same MAC algorithm may be used by both sides, an implementation may choose to run a different and independent MAC algorithm in each direction. A wide variety of MAC algorithms can be used with SSH; however, algorithms such as HMAC-SHA256 are used in most situations to ensure a high level of security.

2.4.5.2.6 *Compression Algorithm*

Compression is optional in SSH, and an SSH implementation must allow a connection to continue even if compression is not negotiated. At the start of the key exchange process, the parties exchange name-lists of acceptable compression algorithms starting with the most preferred algorithm. When negotiated, a compression method such as zlib (LZ77 compression) can be used. If compression is used, only the payload field of the packet is compressed using the negotiated algorithm. The MAC and the packet length field of the packet are then calculated from the compressed payload. Encryption will then take place after payload compression.

An SSH implementation can use independent compression methods for each direction, which also allows a different and independent algorithm for each direction. In practice, however, it is recommended to use the compression method in both directions.

2.4.5.3 Key Features and Functions of the SSH User Authentication Protocol

We recall from above that the SSH server is authenticated in the SSH Transport Layer Protocol during the key exchange and takes place before user authentication. This means there is no need to re-authenticate the server once that has been done. We assume that the SSH User Authentication Protocol runs over a secure Transport

Layer protocol, the SSH server host has already been authenticated, an encrypted communications channel has been established, and a unique session ID for the session has been computed.

The SSH User Authentication Protocol [RFC4252] is the protocol through which the SSH server authenticates the client (protocol for performing client user authentication). The server does this using a variety of different methods, many of which rely on the unique session identifier established in the SSH Transport Layer Protocol. Some of these authentication methods use the encryption and integrity checks provided by the SSH Transport Layer Protocol in conjunction with the unique session identifier, while others do not.

The security of the SSH User Authentication protocol depends very much on the security provisions of the underlying SSH Transport Layer Protocol. If the SSH Transport Layer Protocol cannot guarantee data confidentiality or provide data integrity checks, then this limits the security guarantees of the SSH User Authentication Protocol. For example, if the SSH Transport Layer Protocol does not provide data integrity protection, then it is not recommended to send requests such as password changes, because this could create opportunities for attackers to intercept and tamper with the data without being caught.

The SSH user authentication process begins with the server sending to the client a list of the authentication methods it supports. The client then selects from this list any method it wants to use. The user authentication process is designed to give control to the server but also introduces enough flexibility to allow the client to decide which method is the most convenient to use. We describe below the following widely used user authentication methods:

- Public Key Authentication Method [RFC4252]
- Password Authentication Method [RFC4252]
- Host-Based Authentication [RFC4252]
- Keyboard Interactive [RFC4256]
- Generic Security Service Application Program Interface (GSSAPI) [RFC4462]

2.4.5.3.1 Public Key Authentication Method

To help explain the workings of public key authentication, we define the following key terminology:

- **Public-key (asymmetric) cryptography**: Public key authentication is based on the use of public-key (asymmetric) cryptography. In this form of cryptography, the messages sent by an entity are encrypted and decrypted using different keys. Each entity has a key pair that consists of a public key and a private key.
- **Private keys**: These are secret keys that are known only to the entities that own them. They are used for proving the identity of an entity. Typically, private keys are protected by a passphrase (to provide much stronger identity checking) and are usually stored on separate cryptographic devices. In such a case, an entity must have both the correct passphrase and the private key to be able to authenticate itself to another entity.

- **Public keys**: These keys are public and are distributed to and known by all host devices with which an entity wants to securely communicate with. Before public key operations can occur, an entity must securely receive the public key of the other entity, so that no other entity can substitute the genuine public key with a wrong key. The receiving entity must then verify that the public key is correct, and the remote entity is clearly identified to be who it claims to be. Each public key has a unique fingerprint that the receiving entity can use in the verification.
  - o The private key and public key in a key pair are mathematically related, but the private key cannot be derived from the public key. Data that is encrypted with a public key can only be decrypted with the corresponding private key (and vice versa).
  - o If the entity receiving a public key is a server host, the server can be configured to associate the public key with a specific user account (i.e., a database of usernames and their public keys can be maintained). The owner entity of a key pair will be able to authenticate itself to the server using its key pair, and then gain access to the system based on the configured server policy.
  - o If the entity receiving a public key is a client host, then it can add the public key of the sending entity (server) to its local database and associate it with the IP address of the server. The client host can then trust the sending entity in the future when it authenticates itself with the same key pair.
- **Identity establishment**: This is the process of proving the initial identity of the other entity. That is the initial process of associating a public key with an entity. The initial verification of the identity of the other entity then forms the basis for all subsequent security policy decisions. If an entity cannot verify the identity of the entity that sent a public key and the validity of the received public key, then the wrong entity might end up being trusted.

Public key authentication is often used in both directions allowing a client and a server to mutually authenticate each other via their private and public keys. Typically, the key pairs of server hosts do not have their corresponding private keys protected by a passphrase, as they must be operated on non-interactively. The private keys of servers typically are not protected by a passphrase and must be kept very securely because any entity possessing the key pair of the server host would be able to pose as that server.

SSH supports a range of authentication methods, but arguably, public key authentication is one of the most important methods for automated and interactive connections. In this method [RFC4252], the client has possession of a private key that is used for authentication. The client sends a digital signature created with its private key to the server. The signature is generated by applying the client's private key over data that includes the session identifier, username, service name, public key algorithm name, and the public key.

The signature is sent in an SSH_MSG_USERAUTH_REQUEST message that also contains a number of parameters including the username, service name, public key algorithm name, and the public key. The public key is the key that the server uses to verify the digital signatures generated by the client using a corresponding private key. Public keys can also be used for encrypting messages so that only the user with the corresponding private key can decrypt the messages. As noted above, public keys and private keys are used in pairs and jointly, the two are called a key pair. This means each private key has and is used with a corresponding public key. In public key authentication, the client uses its private key for digitally signing the data it sends. In some cryptosystems, a user can also use a matching private key for decrypting data that was encrypted using a public key.

The SSH server receives a packet from the SSH client and checks if the public key is valid and is acceptable for authentication of the client, and also checks if the received signature is valid/correct. The server checks the signature sent by the client against its public key to verify the client's identity. This is to prove that the entity sending the public key also possesses the private key (i.e., identity checking and proof of possession). If the server finds that both are valid, it accepts the authentication request (by sending an SSH_MSG_USERAUTH_PK_OK message); otherwise, it rejects it (by sending an SSH_MSG_USERAUTH_FAILURE message). The server may request additional authentication of the client even after successful authentication or if the request failed. If the server successfully verifies the signature, the user is now authenticated, and the server can then proceed to authorize the user or give access to relevant parts of the system.

In a typical SSH implementation, private keys are stored at the SSH client in an encrypted form, and the client supplies a passphrase before the signature is generated. It should be noted that in a signing operation, the client sends a message containing parameters including the pubic key and public key algorithm name for querying whether authentication using the public key authentication method is acceptable. Public key authentication method uses a range of algorithms to authenticate the client through public-key cryptography such as RSA (Rivest–Shamir–Adleman), DSA (Digital Signature Algorithm), and ECDSA (Elliptic Curve Digital Signature Algorithm), and in some implementations, X.509 certificates are supported as well.

Public key authentication also allows automated logins without passwords, a feature that has become a key enabler for many secure automated processes that many organizations run within their networks. It allows users in an organization to use single sign-on across the SSH servers they intend to connect to. For example, a simpler form of public key authentication is to use a manually generated public-private key pair on SSH client hosts in a network, allowing users to log into the SSH server without each user having to specify a password.

In this scenario, any user can produce a matching public and private pair with the private keys being different. The network administrator configures a public key on all SSH client hosts that in turn allow access to the owner of the matching private key. The requirement here is the owner must keep the private key secret. In this case, user authentication is done based on the owner's private key, but the private key itself is never sent over the network during authentication. The SSH server only verifies if the same user sending the public key is also the owner of the matching private key.

This passwordless login using a public-private key pair can be summarized as follows:

1. The public-private key pair is created (typically by the user).
2. The user keeps the private key while the public key is sent to the SSH server.
3. The SSH server receives the public key and if it considers the key to be trustworthy, it stores it, and marks it as authorized. The server maintains a file of authorized keys. A public key that the server uses to grant login access to users is called an *authorized key*.
4. The SSH server will now allow access to any user who possesses the corresponding private key. The possession of the corresponding private key serves as proof of the user's identity. Private keys that the SSH server uses for user authentication are generally referred to as *identity keys*. The identity key serves as a credential that the server uses for user authentication. Jointly, *authorized keys* and *private* or *identity keys* are referred to as *user keys*.
   a. The SSH server will be able to authenticate successfully only a user in possession of a private key that corresponds to the public key the server has stored and marked as authorized. Furthermore, using a setup with public-private key pairs, any user that possesses a copy of the public key can encrypt data which can only be read by a user who possesses the corresponding private key.

For the above method to work well, it is extremely important to safeguard the private key very carefully. In most cases, the private key is encrypted with a passphrase. When the user needs the private key, the passphrase has to be supplied so that the private key can be decrypted. An SSH agent can be used to automate the handling of passphrases.

Public key authentication is used under the assumption that neither the SSH server host's private key nor the SSH client host's private key has been compromised. If the server has been compromised, an attacker can easily disclose the username and password of the client. Also, for the use of SSH to be effective, it is important for clients to verify unknown public keys, that is, associate the public keys with the correct identities, before accepting them as valid. Accepting the public key of an unauthorized user or attacker without validating it will authorize that user or attacker as a valid client.

2.4.5.3.2 Password Authentication Method

In SSH, the SSH server can also use passwords to authenticate the client [RFC4252]. In this case, the SSH client sends its password in a packet which is encrypted by the SSH Transport Layer Protocol. Client passwords are never maintained or stored at the client. The client sends an SSH_MSG_USERAUTH_REQUEST message containing a username, service name, and cleartext password, and the entire packet in turn is encrypted by the SSH Transport Layer Protocol. User passwords are sent over the network as encrypted data, making it impossible to read them by capturing network traffic.

The SSH server receives and compares the password with entries in a database to see if there is a matching value. SSH relies on the server-side operating system to provide confidentiality of all user accounts and passwords. If there is a match, the server responds (via an SSH_MSG_USERAUTH_SUCCESS message) to the client's message indicating success (i.e., client is accepted), and the SSH connection is allowed to continue. Otherwise, the server sends an SSH_MSG_USERAUTH_FAILURE if authentication fails.

To use password authentication, both the SSH server and the client must ensure that the underlying SSH Transport Layer provides confidentiality (i.e., encryption is being used). The SSH Transport Layer can use public-private key pairs to encrypt the network connection, and then clients can use password authentication to log onto the SSH server. Password authentication is simple to use and deploy; however, security is based entirely on the strength and confidentiality of user passwords. It does not provide protection against weak passwords, and also does not provide a strong user identity check (it is based only on user password).

Unlike in password authentication, identity checking in public key authentication is stronger. In password authentication, knowing only the user password is sufficient. Public key authentication requires the client (user) to know both the passphrase and the private key. The dependency of public key authentication on two separate elements (the passphrase and the private key) to ensure stronger security is referred to as *two-factor authentication*. Password authentication depends only on the password and is only a *one-factor authentication* scheme. In both methods, however, security still relies on correct identity establishment. If the wrong person gets a password, or if the wrong public key is associated with a user account, the strength of the identity checking will not prevent unauthorized users from accessing the system. If the private keys cannot be securely protected, then security with public key authentication is no better than with password authentication.

2.4.5.3.3   Host-Based Authentication

In this method [RFC4252], the SSH server performs authentication based on information maintained by an SSH client's host system (the "SSH client host") rather than using information associated with the user on the SSH client itself. This means if a host supports multiple SSH clients/users, the SSH server provides authentication for all those SSH clients/users using only the client host's information. The client host is the device hosting one or more clients or users.

An SSH client/user requests this form of authentication by sending a signature created using the private key of the client host. Note that users or clients do not have direct access to the client host's private key. The signature is generated by applying the client host's private key over data that includes the session identifier, username, service name, public key algorithm for the host key, public key (and possibly, certificates) for the client host, client host name, and the username on the client host [RFC4252]. The SSH server uses the client host key to authenticate the client host itself but not the users on that host.

The client host sends the signature in an SSH_MSG_USERAUTH_REQUEST message along with other parameters such as the username, service name, public key algorithm for the public client host key, public key (and possibly, certificates) for the client host, client host name, and the username on the client host [RFC4252]. The

SSH server receives this information and then checks it with its public key. Once the server establishes the identity of the client host, it authorizes the client host but performs no further authentication. This means, the SSH server does not directly verify the identity of the particular client or user on the client host, it only verifies the identity of the client host. The server performs authorization of the client or user itself based on the client usernames it maintains, and the client host name provided.

Thus, in host-based authentication, it is assumed that the client host has already authenticated its own clients/users and, thereby, essentially vouches for them when communicating with the SSH server. With this, the server needs to authenticate only the client host's identity, thereby simplifying the authentication of the multiple users on the client host.

The server must verify that the received public key in the SSH_MSG_USERAUTH_ REQUEST message actually belongs to the named client host, and that the particular client/user on that host is allowed to log in. The server must also verify that the provided signature is valid based on the the given public key. The server may choose to ignore the client/user's username, if it is configured to authenticate only the client host.

Although this method of authentication is not suitable when high-security is required, it is very convenient in many environments. When used, the network designer should take special care to prevent a regular user from obtaining the private key of the client host. The method of authentication places great trust on the administration of the client host and the server relies greatly on the integrity of the user account maintenance and management at the client host. The client host must not be compromised for this method of authentication to be secure. If there is a possibility of security breaches at the client host, then other authentication methods should be added. Host-based authentication can be used together with other forms of user authentication. This method of authentication is mostly used if the client host and the server host are administered by the same organization.

2.4.5.3.4 Keyboard-Interactive Authentication

Keyboard-interactive authentication [RFC4256] is a generic authentication method that can be used to implement a number of other authentication mechanisms that are based on keyboard input (or an alphanumeric input device). Keyboard-interactive authentication is a general-purpose authentication method that provides a common interface to different types of current authentication methods that use keyboard input, and strictly, is not considered an authentication method by itself. Current authentication methods such as password authentication, PAM (Pluggable Authentication Module), RSA SecurID, S/Key, and RADIUS that require only the user to input information via a keyboard can be performed using keyboard-interactive authentication.

The primary reason and advantage of using keyboard-interactive is that it makes it easier to add new authentication methods to a system without the need to update the SSH client software. This makes it easier to upgrade or transition to new and more secure authentication methods as they become available and as long as they are based on keyboard input. This is possible because the SSH client does not have to be made aware of the details or specifics of the actual (keyboard-based) authentication method being used.

- With keyboard-interactive, the SSH client does not have to know the particular authentication method being used, but only has to be aware that a keyboard-interactive authentication method is being used.
- This method allows the SSH client to interact with the SSH server without the need to know the specifics of the underlying authentication mechanisms used by the server. The method allows the SSH server to arbitrarily change or select the authentication mechanisms to be used without the need to upgrade client software.

2.4.5.3.5 GSS-API (Generic Security Service Application Program Interface) Authentication Methods

In reality, GSS-API [RFC2743] is not an authentication method by itself, but instead defines a function interface that provides security services for callers (i.e., applications) at a generic level that is independent of the particular underlying security mechanisms used (mechanism-independence). GSS-API defines services and primitives at an independent level (i.e., independent of mechanism, platform, programming language, communication protocol environment, and protocol association constructs) which allows different security mechanisms to be invoked via one single standardized API. GSS-API applies one or more security services (i.e., types of protection) to the data to be transmitted between peers.

GSS-API allows applications with security implementations to be written that do not have to be tailored to any particular platform, security mechanism, security protection type, or communication transport protocol; applications that are generic with respect to security. Using GSS-API, a program can be made to have little knowledge of the details of underlying protecting data network. A program that is written to take advantage of GSS-API is more portable when it comes to security. A typical GSS-API peer or caller calls on GSS-API in order to protect its communications with a peer that requires security services such as authentication, integrity, and/or confidentiality. Particularly, GSS-API offers data conversion (from enciphered form from the sender), error checking, user privileges delegation, display of information, and identity verification.

Kerberos [RFC4120] is the most common security mechanism often used with GSS-API. Unlike using GSS-API, existing Kerberos implementations use APIs that are incompatible. The Kerberos API has not been standardized, thus, using GSS-API allows the various Kerberos implementations to be API-compatible. GSS-API provides security services that can be implemented over a wide range of underlying security mechanisms that use secret-key and public-key cryptographic technologies. Reference [RFC4121] defines the protocols, procedures, and the conventions that peers that use Kerberos Version 5 and require GSS-API implementation can employ. Reference [RFC4462] further describes the methods used when an implementation requires the use of GSS-API with SSH key exchange and user authentication.

Key GSS-API terminology and concepts [RFC2743]:

- **Name**: This is a binary string that is used to label or tag the entity to be authenticated, that is, the security principal (which can be a service, program, process, user, etc.). The security principal is assigned a name that allows it to be referenced for identification.
- **Credentials**: This is the information that is used to prove/establish the identity of the entity that is to act as the named security principal (i.e., the identity of the other party in the communication). This information usually takes the form of secret cryptographic keys and/or passwords.
- **Security Context**: This is the state of one end of the authentication protocol. The GSS-API creates a security context in which applications pass data between themselves. A security context can be viewed somehow as a "state of trust" between two GSS-API applications. Applications that share a security context are able to identify each other, and thus, can perform or exchange data transfers as long as the security context lasts.
  - The security context may provide message protection services that can be used to create a secure communication channel.
  - Credentials provide the GSS-API peers with the information needed to establish security contexts with each other.
- **Tokens**: These are opaque messages that are exchanged between GSS-API callers (or peers), either as part of the initial authentication protocol exchange to establish and manage a security context between GSS-API peers (context-level tokens) or as part of a protected communication, that is, an established context to provide protective security services for data messages (per-message tokens). A GSS-API caller acquires tokens generated by its local GSS-API implementation and then communicates these tokens to a remote peer. The remote peer receives and passes the tokens to its local GSS-API implementation for processing.
- **Mechanism**: This is the underlying mechanism that embodies a cryptographic technology and provides the syntax and semantics of the data element exchanges between GSS-API callers. This is the mechanism used to support security services. This mechanism provides actual names, tokens, and credentials. A well-known mechanism is Kerberos.
- **Initiator and Acceptor**: The initiator is the GSS-API peer that sends the first token. The other GSS-API peer is the acceptor. Generally, the initiator is the client program while the acceptor is the server.

Key GSS-API procedure calls [RFC2743]:

- *GSS_Acquire_cred()*: This routine is used to obtain credentials (user's proof of identity, usually a secret cryptographic key) so that a security principal can initiate and/or accept security contexts.
- *GSS_Init_sec_context()*: This routine is used by the initiator of a security context and is responsible for generating a token (suitable for use by the target) to be sent to the target.

- **GSS_Accept_sec_context()**: This routine is used by a security context target to process a token provided by *GSS_Init_sec_context* and is responsible for generating a response token to be returned to the initiator.
- **GSS_GetMIC()**: This routine uses the security context, applies an integrity check to an input message, and returns a per-message token. The caller passes the message and the per-message token to the target.
- **GSS_VerifyMIC()**: This routine uses the security context, verifies that the input per-message token contains an appropriate integrity check for the input message, applies any active replay detection or sequencing features, and returns an indication of the quality of protection applied to the processed message.
- **GSS_Wrap()**: This routine returns a single output message data element that contains user data (optionally enciphered) in addition to control information. The data origin authentication and the data integrity functions of the routine *GSS_GetMIC()* are performed by *GSS_Wrap()*. Once GSS-API has established a security context, messages can be wrapped by the GSS-API for secure communication between the peers. Typically, GSS-API wrapping provides protection guarantees that include confidentiality (secrecy) and data integrity (authenticity). GSS-API can also provide locally at a peer, guarantees about the identity of the remote peer (data origin authentication).
- **GSS_Unwrap()**: This routine takes as an input message, a data element (optionally enciphered) generated by *GSS_Wrap()*, processes it, and returns plaintext data as an output message. It performs on the plaintext data, the data origin authentication checking and the data integrity functions of the routine *GSS_VerifyMIC()*.
- **GSS_Delete_sec_context()**: Either GSS-API peer in a security context can use this routine to flush context-specific information.
- **GSS_Process_context_token()**: This routine is used by a GSS-API peer to process context-tokens received from a peer once the peers have established a security context. This processing has a corresponding impact on the context-level state information.
- **GSS_Import_name()**: This call allows GSS-API callers to provide a name string, designate the namespace associated with the string, and convert the string to an internal form suitable for GSS-API routines. This call converts the string into a form that identifies the security entity.

GSS-API offers the basic security service of authentication but provides the two additional security services of data integrity and data confidentiality, if supported by the underlying mechanisms. Note that the verification of the validity of a piece of data is known as integrity because the data that is originated by the sender itself could have been corrupted or compromised.

The GSS-API provides facilities for data to be associated with a cryptographic tag (also known as an MIC (Message Integrity Code)) to prove to the receiver that the received data is the same as the data transmitted by the sender. The encryption of data is to ensure confidentiality. GSS-API allows data sent by an endpoint to be encrypted, if underlying mechanisms support this capability. Note that both the authentication and integrity functions pass any transmitted data unmodified as was sent by the end-point, making it possible for the data to be read if somehow intercepted.

2.4.5.3.5.1   Using GSS-API

As noted above, GSS-API is a framework that provides security services to callers in a generic fashion and is capable of supporting a wide range of underlying mechanisms and security technologies such as Kerberos (which is based on secret key cryptography). GSS-API defines a standard generic authentication and secure messaging interface for security mechanisms (commonly referred GSS-API mechanisms) such as Kerberos to be plugged in.

GSS-API allows program developers to add secure authentication and privacy protection to data being exchanged over a communication medium (encryption and/ or integrity checking) by passing the data to a single programming interface (GSS-API). With GSS-API, the underlying security mechanism(s) (e.g., Kerberos) is loaded at the time the end application is executed, as opposed to when it is compiled and built. GSS-API is generally preferred because it is a standardized API, whereas any one of the various Kerberos APIs is not.

GSS-API separates the operations by which peers initialize a security context between themselves (in order to achieve peer entity authentication (using the *GSS_Init_sec_context()* and *GSS_Accept_sec_context()* routines)), from the operations that provide per-message data origin authentication and data integrity protection for messages subsequently sent in that security context (using the *GSS_GetMIC()* and *GSS_VerifyMIC()* routines). When a security context is being established, GSS-API enables a context initiator to optionally permit the context acceptor to initiate further security contexts on behalf of the initiator (i.e., the initiator's credentials can be delegated).

The steps provided below describe at a high level, one example of the operations involved in the use of GSS-API by a client (initiator) and server (acceptor) in a mechanism-independent manner. GSS-API is used for establishing a security context and transferring a message with security protection. We assume here that the acquisition of credentials has already been completed. We also assume that the underlying authentication technology allows the server to authenticate the client using elements carried within a single token, and the client authenticates the server (mutual authentication) using a single returned token.

1. The client (initiator) calls *GSS_Init_sec_context()* to establish a security context to the server (acceptor), and sets a flag indicating that mutual authentication is to be performed during context establishment.
2. *GSS_Init_sec_context()* returns an output token to the client to be transferred to the server. The client then communicates the output token to the server.
3. The server receives the token and passes it as the input token parameter to *GSS_Accept_sec_context()*. *GSS_Accept_sec_context()* provides an output token to be sent to the client, and the server passes the output token to the client.

4. The client receives the token and passes it as the input token parameter to a successor call via *GSS_Init_sec_context()*. The successor call then processes the data contained in the token in order to achieve mutual authentication from the client's perspective.

5. The client then generates a data message and hands it over to *GSS_Wrap()*, which performs data origin authentication, data integrity, and confidentiality (optional) processing on the message. *GSS_Wrap()* encapsulates the result in an output message, which the client sends to the server.

6. The server receives the message and passes it to *GSS_Unwrap()* which inverts the encapsulation carried out by *GSS_Wrap()*. If the optional confidentiality feature was applied, *GSS_Wrap()* deciphers the message and performs the data origin authentication and data integrity checking functions. Upon successful validation, *GSS_Unwrap()* returns the resulting output message.

7. If we assume that this security context has no further use after the one protected message has been transferred from client to server, the server will call *GSS_Delete_sec_context()* to flush context-level information. Optionally, a token buffer may be provided by the server-side application to *GSS_Delete_sec_context()* to receive a context token that the server will transfer to the client to request for the client-side context-level information to be deleted.

8. If the server transfers a context token to the client, the client passes the context token to *GSS_Process_context_token()*, which deletes the context-level information in the client system.

2.4.5.3.5.2 *Using GSSAPI and Kerberos*

GSS-API is often used with Kerberos which is the most common GSS-API mechanism. GSS-API provides an authentication message exchange abstraction and framework that allows credentials to be exchanged between a Kerberos client and server in a Kerberos authentication exchange. Kerberos must be installed and running on the system on which the GSS-API-aware client and server programs are running. In addition to providing secure authentication, Kerberos allows privacy support (i.e., encrypted data transfer) to be offered to network applications such as `rlogin`, `ftp`, `rsh`, `telnet`, as well as other commonly used UNIX applications.

Using Kerberos authentication with GSS-API, the client and server must be preconfigured to be able to use Kerberos (i.e., both have to be able to acquire tickets). Users do not use Kerberos/GSS-API to initially log on to a network. Instead, a user must have logged on to the network, and the user must have existing logon credentials that can be accessed via GSS-API. Note that GSS-API enables the transfer of existing credentials (tokens) from the client to the server.

Kerberos/GSS-API itself does not transfer information over the network, a task which is the responsibility of the application using GSS-API. As discussed above, GSS-API provides opaque credential data that the application can send to a remote peer. In SSH, the credential data is transferred securely over the SSH Transport Layer, similar to any SSH authentication method.

At the start of the authentication method, the client sends a list of GSS-API mechanisms that it supports to the server. The server then selects from that list the first mechanism (e.g., Kerberos) that it supports. Once the client and the server have negotiated the mechanism, they begin token exchange. Token exchange may involve the client and the server exchanging several packets, depending on the negotiated mechanism.

All the token data exchanged between the client and server comes from GSS-API; SSH does not understand the token data but only relays it to the target. At the end of a successful token transfer, the server uses the token for user authentication. This includes the server using the received token to request the identity of the user from GSS-API.

2.4.5.4   Key Features and Functions of the SSH Connection Protocol

The SSH Connection Protocol [RFC4254] is responsible for combining or multiplexing multiple SSH channels of data over the secure SSH Transport Layer. It specifies the SSH channels for accessing various external services and protocols over the secure SSH Transport Layer. It also handles the parameters that the client uses to access secure subsystems on the SSH server host, in addition to proxy-forwarding and accessing shells.

The SSH Connection Protocol works with and on top of the SSH Transport Layer, and with the User Authentication Protocols to provide remote command execution, interactive login sessions, in addition to forwarded X11 connections, and forwarded TCP/IP connections. The SSH Connection Protocol multiplexes all of these channels onto a single encrypted tunnel. The Connection Protocol requires that neither the server nor the client is compromised, if not the services it provides will not be secure.

2.4.5.4.1   Global Requests

Global requests are independent of channel type and are requests that affect the state of the remote end of the SSH channel, globally [RFC4254]. For example, the SSH client may use a *global request* (indicating SSH_MSG_GLOBAL_REQUEST) to request an SSH server-side port to be forwarded, that is, request the server to start TCP/IP forwarding for a specific port. Both the SSH client and server are allowed to send global requests at any time, but the receiver is required to respond appropriately (indicating SSH_MSG_REQUEST_SUCCESS or SSH_MSG_REQUEST_FAILURE).

2.4.5.4.2   Channel Mechanisms and Requests

In addition to global requests, the SSH Connection Protocol defines the concept of *channels*, and *channel requests* for SSH services [RFC4254]. In SSH, all terminal sessions, SFTP and exec requests (including SCP transfers), forwarded TCP/IP connections, and forwarded X11 connections, and so on, are channels. The SSH Connection Protocol allows either side (i.e., the SSH client-side or server-side) to open a channel. Note that multiple channels can be hosted simultaneously by a single SSH connection, each channel transferring data in both directions.

Each side uses a unique local number to identify each channel, meaning the number assigned to a channel may be different on each side. The requests a side sends to open a channel contain the local channel number assigned by the sender. Each channel is flow-controlled, and a side may not send data to a channel until it receives a message from the other side indicating that sufficient window space is available. Each channel performs its own flow control using the receive window size. A side can send channel requests in order to communicate out-of-band channel-specific data, such as the changed window size of a terminal session.

- **Opening a Channel**: A channel serves as the basic communication path between the SH clients and the server, and also a channel can be opened by either side. When the SSH client or server side wants to open a new channel, it will assign a local number for the channel, and then send a message to the other side indicating the following: SSH_MSG_CHANNEL_OPEN, channel name (in US-ASCII only), the sender's local channel number/identifier, initial data window size, the maximum packet size that can be received, plus other channel-specific data in the message. The initial data window size specifies to the other side (receiver of this message) how many channel data bytes can be sent to the message sender without adjusting the window. After the remote side decides whether it can open the channel or not, it will respond by sending a message indicating either SSH_MSG_CHANNEL_ OPEN_CONFIRMATION (plus additional information indicating local channel parameters), or SSH_MSG_CHANNEL_OPEN_FAILURE (plus, possibly, additional information describing reasons for channel open failure). When one side receives a channel request open message, it will open the channel as long as it can accommodate the request. A side may fail to open a channel for any one of the following reasons: unknown channel type, resource shortages at the side, prohibitions imposed by the network administrator, failed connection, etc.
- **Data Transfer**: The data window size indicated by a side specifies how many bytes the remote side can send before it must wait for the data window to be adjusted. If a side wants to send more data than allowed by the current data window, then the data window has to be adjusted first; it must send a message to the other side requesting for more bytes to be added to the window. To adjust the data window, a side will send a message to the other indicating the following: SSH_MSG_CHANNEL_WINDOW_ADJUST, channel number of the side to receive this message, and the number of bytes to add.
- **Closing a Channel:** When a side decides that it has no more data to send to a channel, it will send a message to the other side indicating SSH_MSG_ CHANNEL_EOF plus the channel number of the side to receive this message. However, the channel remains open after a side sends this message, because the other side may still have more data to send in the other direction. This is to allow the other side to send data if it has not finished transmitting data. When either side decides to completely terminate the channel, it will send a message indicating SSH_MSG_CHANNEL_CLOSE plus the channel number of the side to receive this message. Upon receiving this message, the receiver is required to send back a message with SSH_MSG_ CHANNEL_CLOSE to the sender for the channel unless it has already sent this message. A side considers the channel closed when it has both sent and received SSH_MSG_CHANNEL_CLOSE messages, in which case the side

may then reuse the channel number. A side is allowed to send SSH_MSG_ CHANNEL_CLOSE message without having sent or received a message with SSH_MSG_CHANNEL_EOF.

- **Channel-Specific Requests**: Many SSH channel types have extensions that are channel type-specific. For example, a channel may request a pseudo-terminal (pty) for an interactive session. SSH channel-specific requests are sent using messages that include the following: SSH_MSG_CHANNEL_ REQUEST, the channel number of the side to receive this message, request type, and channel type-specific data. The recipient of this message may respond with either an SSH_MSG_CHANNEL_SUCCESS message indicating the receiving side's channel number, an SSH_MSG_CHANNEL_ FAILURE message indicating receiving side's channel number, or channel request-specific continuation messages.

2.4.5.4.3 Interactive Sessions

Generally, in computer and communication networking, a session refers to a temporary and interactive exchange of information between two or more communicating entities, for example, a login session between a user and computing device. In SSH, a session involves the remote execution of a program or application and maybe a shell, a system command, or an application subsystem running remotely. An SSH session may use a TTY (TeleTYpewriter) and may involve X11 forwarding. SSH allows multiple sessions to be active simultaneously.

- **Opening a Session**: In SSH, a party starts a session by sending a message indicating SSH_MSG_CHANNEL_OPEN, "session", sender channel, initial window size, and maximum packet size. Once a session channel is opened, a user may send subsequent requests to start the remote program.
- **X11 Forwarding**: The X11 or simply, X (which stands for X Window System [RFC1013] [SCHEIXWIN96] [STEINXWIN88]) is a type of graphical user interface (GUI) for bitmap display screens. SSH supports tunneling or forwarding X11 which allows graphical displays from a remote machine to be opened on the local computer. *X11 Forwarding* is a mechanism (a special case of *TCP/IP port forwarding*) that allows a user to run GUIs on a remote machine on the local machine (i.e., forward the remote display to the local machine). To achieve this, an *SSH client* and *X11 server* run on the local machine, while an *SSH server* and *X11 client* applications run on the remote machine (Figure 2.10). The X11 connections are then tunneled and automatically encrypted by the SSH server running on the remote machine to the SSH client on the local machine.
  - ○ *Requesting X11 Forwarding*: A party may request X11 forwarding for a session by sending a message indicating the following: SSH_MSG_ CHANNEL_REQUEST, the channel number of the side to receive this message, "x11-req", x11 authentication protocol/method, x11 authentication cookie, plus a few other parameters as described in [RFC4254].

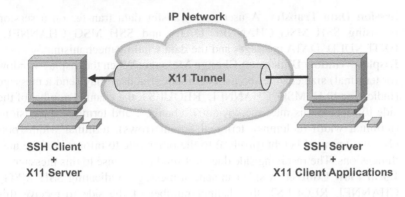

FIGURE 2.10 Illustrating X11 Forwarding.

- o **X11 Channels**: A party opens an X11 channel with a channel open request message indicating the following: SSH_MSG_CHANNEL_ OPEN, "x11", sender channel, initial window size, maximum packet size, originator IP address, and originator port. The resulting X11 channels that are created are independent of the session channel, which means closing the session channel does not result in the forwarded X11 channels being closed. The receiver of the X11 channel open request message is required to respond with a message indicating either SSH_MSG_CHANNEL_OPEN_CONFIRMATION or SSH_MSG_CHANNEL_OPEN_FAILURE.
- • **Passing Environment Variables**: A party may pass environment variables to the shell/command to be started later by sending a message indicating the following: SSH_MSG_CHANNEL_REQUEST, the channel number of the side to receive this message, "env", the variable name, and the variable value.
- • **Starting a Shell or a Command**: Once a session has been established, a user can send a request message for a program such as a shell, an application program, or a subsystem with a host-independent name to be started at the remote end. Only one of these requests is allowed to run per session channel; only one can succeed.
    - o A user sends a message indicating SSH_MSG_CHANNEL_REQUEST, the channel number of the side to receive this message, and "shell", to request that the user's default shell be started at the other end (in UNIX systems, this is typically defined in/etc/passwd).
    - o A user sends a message indicating SSH_MSG_CHANNEL_REQUEST, the channel number of the side to receive this message, "exec", and a command (which may contain a path) to request that the server start the execution of the specified command.
    - o A user sends a message indicating SSH_MSG_CHANNEL_REQUEST, the channel number of the side to receive this message, "subsystem", and a subsystem name to execute the predefined subsystem.

- **Session Data Transfer**: A user may transfer data transfer on a session by using SSH_MSG_CHANNEL_DATA and SSH_MSG_CHANNEL_EXTENDED_DATA messages and the data window mechanism.
- **Display Window Dimension Change Message**: When the display window (or terminal) size changes on the client side, the user may send a message (indicating SSH_MSG_CHANNEL_REQUEST, the channel number of the side to receive this message, "window-change", and terminal dimensions (terminal width (columns), terminal height (rows), terminal width (pixels), and terminal height (pixels))) to the other side to inform it of the new dimensions. The receiving side does not send a response to this message.
- **Local Flow Control**: A side can send a message (indicating SSH_MSG_CHANNEL_REQUEST, the channel number of the side to receive this message, "xon-xoff", plus other indicators) to inform the side originating a session if it can or cannot perform flow control (using control-S/control-Q processing). No response is required from the session originator for this message.
- **Sending Signals**: A side may send a signal to the remote process/service by sending a message indicating SSH_MSG_CHANNEL_REQUEST, the channel number of the side to receive this message, "signal", and signal name (codes as defined in [RFC4254]).
- **Returning Exit Status**: When a command running at one end terminates, that end can send a message indicating SSH_MSG_CHANNEL_REQUEST, the channel number of the side to receive this message, "exit-status", and the exit_status, to the other end to return the exit status of the command (an exit_status of zero value usually means that the command has terminated successfully). The receiver is not required to send an acknowledgment for this message. The endpoints need to close the channel by sending SSH_MSG_CHANNEL_CLOSE after the exit status message.

2.4.5.4.4   SSH Port Forwarding

SSH Port Forwarding (also called *SSH Tunneling*) [BARSILSSH01] [RFC4254] is a mechanism used in SSH for tunneling or forwarding application ports on a client host to a corresponding server host, or vice versa. Using this mechanism, one SSH side on a host, listens on a port and intercepts a service request from an application client on the same host, and sets up an SSH session carrying the encrypted request to the other side of the SSH connection. The other SSH side receives the encrypted request and decrypts it before sending the decrypted request to the application server on the remote host.

This mechanism can be used for the following:

- Going through company firewalls (i.e., to implement Virtual Private Networks (VPNs) to access a company's intranet services across firewalls)
- Adding encryption to legacy applications that do not support encryption natively

- Providing system administrators and IT professionals working remotely with a means for opening backdoors into their internal company networks from their home machines.

To improve security, IT administrators typically block certain ports on hosts in an internal company network with firewalls, preventing them from external access. SSH Port Forwarding can also be exploited by malware and hackers to gain access from the Internet to an internal company network.

2.4.5.4.4.1 Requesting SSH Port Forwarding

Using SSH, a party does not need to explicitly request for data to be forwarded from its end to the remote end. However, if a party wants connections to a port on the remote side be forwarded to its own local end, it must explicitly request this by sending a message indicating the following: SSH_MSG_GLOBAL_REQUEST, "tcpip-forward", the IP address (or domain name), and the local port number on which forwarded connections are to be accepted (i.e., the address to bind and the port number to bind, respectively) [RFC4254].

2.4.5.4.4.2 Local SSH Port Forwarding

Local SSH Port Forwarding is used to forward an application port from an SSH client host to a corresponding SSH server host, and then to the destination host port as illustrated in Figure 2.11. In this figure, the SSH client in Host A listens for connections on a configured port, and when it receives a connection from an application client, it tunnels/forwards this connection to the specified port on the SSH server on Host B. The SSH server on Host B then connects to a configured destination port serving an application server which, possibly, can be on a different machine than the SSH server on Host B. In Local SSH Port Forwarding, the SSH client usually resides on the same host as the application client, and the SSH client listens for communication from the application client.

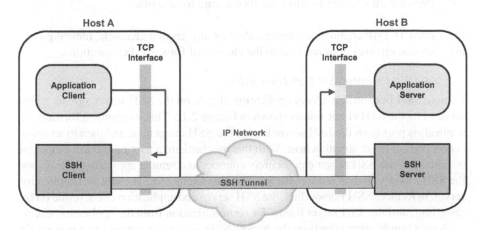

FIGURE 2.11 Illustrating Local SSH Port Forwarding.

Local SSH Port Forwarding is typically used for the following:

* Connecting a user at a remote location to a service/resource located in an internal company network
* Connecting a user to a remote file share over a service provider network and the Internet
* Tunneling sessions and file transfers through a jump server (sometimes called jump host or jump box). A jump server is a monitored and security-hardened server (possibly, with intrusion detection, logging, etc.) attached to a particular network that is used to access and manage devices located in a separate/different security zone. A jump server usually spans two dissimilar/different security zones and provides mechanisms for controlling access between the two security zones.

Local SSH Port Forwarding allows a user to connect a local client application to an external network. A user may use SSH Port Forwarding to access an IMAP (Internet Message Access Protocol) server running on a remote host in a company network from an email client running on the home machine. A user may use Port Forwarding to connect to a database that is behind a firewall or to access websites that are blocked locally.

* **Configuring Local SSH Port Forwarding:** When a connection is received on a locally forwarded port, a message is sent to the other side indicating the following: SSH_MSG_CHANNEL_OPEN, "direct-tcpip", sender's channel number, initial data window size, maximum packet size, the local host IP address and port, the IP address (or hostname) and the port of the destination host, and the remote SSH user and SSH server IP address. Note that a side may also send these messages for ports for which no forwarding has been explicitly requested [RFC4254]. The receiving side must make the decision on whether to allow the forwarding to take place.

Forwarded TCP/IP channels are independent of any session channels, meaning closing a session channel does not lead to the closing of forwarded connections.

2.4.5.4.4.3   Remote SSH Port Forwarding

Remote SSH Forwarding allows application clients on the SSH server side to access services on the SSH client side as shown in Figure 2.12. This mechanism forwards an application port from the SSH server host to the SSH client host, and then to an application port on the destination host. With this mechanism, a user on the local machine connected to the SSH server can securely connect to a remote application server, and the SSH server will then redirect the local port to a remote SSH client and application server. In Remote SSH Forwarding, the SSH server and application client reside on the same host, and the SSH server listens for communication from the application client.

A user (application client) on the remote SSH server can connect to a port on the SSH server, and the SSH server will tunnel the connection back to the SSH client host. The SSH client will then make a connection to a port on the application server.

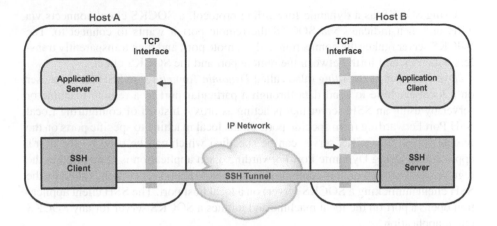

FIGURE 2.12 Illustrating Remote SSH Port Forwarding.

The SSH server listens on a given port on Host B as illustrated in Figure 2.12, and tunnels any connection to that port to the specified port on the SSH client, which then connects to a destination application port (which can be on the destination host as the SSH client or any other machine).

At the SSH client side, an IP address, `localhost`, or any other host name can be used to specify the host to be connected to [RFC4254]. A company may use Remote SSH Forwarding to provide a user outside its network, access to an internal web server, or expose an internal web service to outside users and the public Internet (e.g., an employee working from home). It should be noted that this feature may also be exploited by malicious users or attackers.

- **Configuring Remote SSH Port Forwarding**: When a connection is received on a port for which a party has requested remote forwarding, a channel is opened to be used to forward that port to the other side using a message indicating the following: SSH_MSG_CHANNEL_OPEN, "forwarded-tcpip", the sender's channel number, initial data window size, maximum packet size, the IP address and port on the SSH server, the IP address (or hostname) and port of the destination machine, and the remote SSH user and SSH server IP address.

2.4.5.4.4 Dynamic Port Forwarding Using SSH

SOCKS [RFC1928] is session-level protocol (operates at OSI Layer 5 or Session Layer) that allows a client and a server to exchange data through a proxy server. SOCKS is a versatile protocol for forwarding TCP or UDP traffic and can be used as a tool for bypassing or circumventing Internet filtering mechanisms and devices, thereby allowing users to access websites and content that would otherwise be blocked or censored (by schools, workplaces, country-specific web services, and governments). SOCKS can also provide functionality similar to a VPN, allowing connections to be forwarded to the local network of a remote server.

Using SOCKS as a dynamic forwarding protocol, a SOCKS client connects via TCP, and then indicates via SOCKS the remote port it wants to connect to. The SOCKS server makes the connection to the remote port, and then transparently transfers data back and forth between the remote port and the SOCKS client.

Dynamic Port Forwarding (also called *Dynamic Tunneling*) in SSH allows a user on a local machine to send data through a particular port to a remote machine or server by using an SSH server that is acting as proxy. Instead of configuring Local SSH Port Forwarding from specific ports on the local machine to specific ports on the remote server, a SOCKS server can be specified which can be used by the user's applications. Using Dynamic Port Forwarding, each application is configured in the regular way, except that, it is configured to use a SOCKS server (which is actually the SSH client mimicking a SOCKS server) on a local host port. The SSH client application opens a port on the local machine and mimics a SOCKS server for any SOCKS client application.

When a client application wishes to connect to a remote service, it provides the necessary information to the SOCKS server (i.e., the SSH client mimicking a SOCKS server). The SSH client (acting as a local SOCKS proxy server) uses this information to create port forwarding to the remote SSH server which relays traffic back and forth securely, as in Local SSH Port Forwarding (which is based on user-specified port forwarding). At the SSH server, the connection originating from the client application is forwarded unsecured to the destination host requested by the application.

Dynamic Port Forwarding using SSH provides more flexibility and frees a user from the limitations of connecting only to a pre-specified remote server and port. All the applications using the SOCKS proxy server (i.e., SSH client) will connect to the SSH server, and the SSH server, in turn, will forward all application traffic to their actual destinations.

Let us assume that a user enters the URL "http://mynetworkserver:1890" into the Web browser with the port 1890. The browser connects to the SSH client (mimicking the SOCKS proxy on port 1080) and requests for connection to mynetworkserver:1890. A SOCKS server accepts incoming connections from client applications on TCP port 1080 [RFC1928]. The SSH client associates the browser's connection with a new SSH session and then connects to the remote SSH server. The SSH client and the SSH server transparently pass information to connect the browser directly to the web server. Each new connection from the browser to a different website is assigned a new connection by SSH. When using Dynamic Port Forwarding, it is necessary to explicitly configure individual applications to use the SOCKS proxy server.

2.5 TRAFFIC MONITORING

Using a combination of switches, switch/routers, and routers to partition a network infrastructure into smaller and manageable network domains increases overall network performance and security but also requires additional capabilities that allow the network administrator to have a total view of network capacity, bandwidth consumption, utilization, and overall network health. Traffic monitoring mechanisms such as sFlow [RFC3176], NetFlow [RFC3954] and Remote Network MONitoring (RMON) [RFC2819] [RFC3577] can be deployed to provide real-time, network visibility, enabling network administrators to manage every traffic flow in the network. This approach also enables organizations to leverage a wide range of network traffic management, monitoring, and trending utilities.

2.5.1 sFlow and NetFlow

NetFlow is a technology developed by Cisco that is similar to sFlow and is described in [RFC3954]. Internet Protocol Flow Information Export (IPFIX), created based on NetFlow (in [RFC3954]) is a related IETF protocol described in [RFC7015]. IPFIX was created based on the need for a common, universal standard for exporting IP flow information from routers, probes, and other network devices. The flow information from sFlow and NetFlow can then be used, for example, by accounting/billing and network management systems to implement services such as measurement, accounting, and billing (Figure 2.13). Figure 2.14 describes the NetFlow lookup operations in the Policy Feature Card of the Cisco Catalyst 6500 (PFC3/PFC4) [CISC2TMANM11] [CISC2TMUL11] [CISC2TNETF11] [CISC2TQOS11].

Typically, in high-performance network devices, sFlow and NetFlow can be implemented to provide scalable, wire-speed network monitoring and accounting with no impact on device performance [FOR10ESER05]. sFlow or NetFlow can be integrated into the forwarding capabilities of the device to collect and aggregate

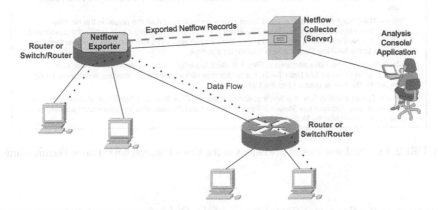

NetFlow Exporter: Processes packets flows and exports flow records to one or more NetFlow collectors (servers).
NetFlow Collector: Receives, stores and pre-processes flow data received from a NetFlow Exporter.
Analysis Console/Application: Analyzes received flow data for various applications, e.g., to assist in capacity planning, application assessment, network troubleshooting, security operations, etc..

- NetFlow is a process designed to allow a switch, switch/router or router collect information about traffic flows passing through it.
- As a control plane process, the process of exporting of NetFlow records to a NetFlow Collector can be performed by either the route processor on the network device or a processor on a line card.
- NetFlow gathers traffic statistics by monitoring packets flowing through the network device and stores the statistics in a NetFlow Table. The process of exporting data from NetFlow Table to the NetFlow Collector is often referred to as a NetFlow Data Export (NDE).
- NDE converts the NetFlow Table statistics into records which are then exported to the NetFlow Collector.
- The volume of records exported can be controlled by configuring NDE flow filters to include or exclude flows from the NDE export. When a filter is configured, NDE exports only the flows that match the filter criteria.

FIGURE 2.13 NetFlow collection and export operation overview.

Source IP = 10.1.1.10; Destination IP = 10.1.2.11;
Protocol = TCP (6);
Source Port = 33992; Destination Port = 80

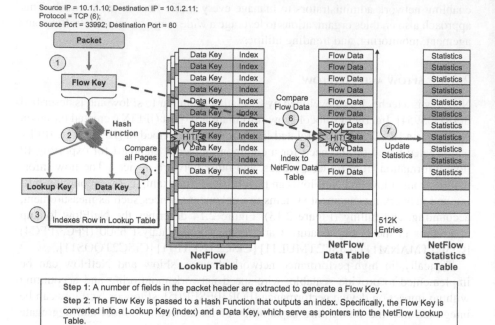

Step 1: A number of fields in the packet header are extracted to generate a Flow Key.

Step 2: The Flow Key is passed to a Hash Function that outputs an index. Specifically, the Flow Key is converted into a Lookup Key (index) and a Data Key, which serve as pointers into the NetFlow Lookup Table.

Step 3: The Lookup Key in Step 2 is used to determine the appropriate row in the Net Flow Lookup Table.

Step 4: The Data Key in Step 2 is used to perform a comparison on all the pages in the NetFlow Lookup Table to determine if this Data Key exists. If the Data Key already exists (a hit), the associated index to the NetFlow Data Table is read. If no Data Key exists in this row, the Data Key is entered, and an index to the NetFlow Data Table is created for this flow.

Steps 5 and 6: The index acquired in Step 4 is used to perform a comparison of the Flow Key information in the Net Flow Data Table. If a similar flow already exists, no learning is performed. If no flow exists, the flow is entered into the table.

Step 7: For an existing flow, the flow's statistics in the NetFlow Statistics Table are updated. For a new flow, an entry in the NetFlow Statistics Table is created and the statistics for subsequent packets in this flow then incremented by NetFlow.

FIGURE 2.14 NetFlow lookup operations in the Cisco Catalyst 6500 Policy Feature Cards.

details on traffic flows at different layers of the OSI reference model (from Layer 2 through Layer 4) and automatically deliver that information to a network management station. The network management station may then employ, for example, a Java-based network configuration and management tool to display, in graphical detail, network and application-level traffic information.

With the resulting insight, the network administrator can quickly and accurately review overall networking operations, zero in on hot spots, and quickly diagnose and troubleshoot difficulties before they develop into widespread problems. sFlow and NetFlow also automatically deliver accurate SNMP/RMON statistics (Figure 2.15), reducing the administrative burden normally associated with proactive network management, design, and capacity planning.

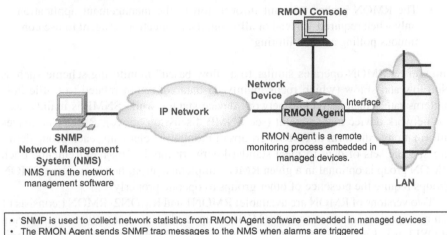

- SNMP is used to collect network statistics from RMON Agent software embedded in managed devices
- The RMON Agent sends SNMP trap messages to the NMS when alarms are triggered
- RMON MIB defines and organizes a collection of managed objects
- The NMS obtains the management information from the RMON Agent and allows the operator to control the network resources

FIGURE 2.15 Example application scenario of RMON.

2.5.2 REMOTE NETWORK MONITORING

RMON is a standard [RFC2819] [RFC3577] for monitoring the operation of network hardware and software entities through the use of remote monitoring devices known as monitors or probes (see also Chapter 5 of Volume 1 of this two-part book). RMON defines a group of functions and statistics for real-time network monitoring but is different from SNMP as explained below.

The standards-defined group of functions and statistics enable RMON-compliant network probes and console systems to exchange network-monitoring data. A network administrator can use RMON to perform extensive network-fault detection, monitoring, and protocol analysis of network entities, and to capture data that can be used for network performance tuning. Many switches, routers, and switch/routers have built-in RMON agents, even low-end devices.

Typically, RMON is implemented to operate in a client/server model. The monitoring devices (or RMON probes) are designed to contain RMON software agents that analyze packet flows and collect information at specific points and protocol layers in a network (Figure 2.15). The probes gather information and act as servers, providing information to network management applications that act as clients of the information. Although the configuration and data collection operation of RMON agents use SNMP, RMON itself operates differently when compared to other traditional SNMP-based systems:

- RMON probes dedicate their attention mainly to data collection and processing which lead to a reduction in the SNMP traffic generated and the processing load placed on the clients.

- The RMON probes transmit information to the management application only when required, instead of allowing the application or agent to use continuous polling and monitoring

In essence, RMON operates similar to a "flow-based" monitoring scheme such as NetFlow and sFlow (which focus mainly on data collection related to traffic flow patterns rather than individual network device status), while SNMP is mainly used for network device management (see SNMP discussion above). RMON organizes information in the form of RMON groups of monitoring elements, each group defining specific sets of data to meet standard network monitoring requirements. Each RMON group is optional in a given RMON implementation; however, some RMON groups require the presence of other groups to operate properly.

Two versions of RMON are available: RMON1 and RMON2. RMON1 consists of ten management information base (MIB) groups and focuses on network monitoring at OSI Layer 1 and Layer 2. RMON2 adds ten more MIB groups and extends monitoring to the OSI Layer 3 (Network) up to the Application Layer. RMON as defined in [RFC2613] (referred to as SMON) adds RMON support for switched Layer 2 networks (VLAN monitoring).

To minimize the processing load and complexity of RMON, most RMON agents implement only a minimal set of RMON MIB groups (namely, Statistics, History, Alarm, and Event) out of the total RMON1 and RMON2 MIB groups:

- **Statistics (RMON group 1)**: Collects real-time statistics on a LAN (Ethernet) interface.
- **History (RMON group 2)**: Collects history of selected statistics on an interface for a specified polling interval. The statistical samples are recorded and stored in history control tables for later retrieval.
- **Alarm (RMON group 3)**: Monitors a specific MIB object for a specified interval, triggers an alarm when statistics cross-defined thresholds. Alarms and events can be used together. An alarm can trigger an event where a log entry is generated or an SNMP trap is sent.
- **Event (RMON group 9)**: Specifies the action to be taken when an alarm triggers an event. RMON can specify an action such as to generate a log entry or an SNMP trap when an alarm occurs.

To enable high-speed implementations of RMON, network devices typically use hardware counters for data collection and processing, which also allows monitoring to be more efficient with little software processing required.

Using RMON, a managed device can monitor the values of its MIB objects or variables and compares them against defined thresholds. When the value of a variable crosses a threshold, an alarm is generated that triggers a corresponding event (e.g., make a log entry or generate an SNMP trap). A network service provider may use a fault-monitoring system or operational support system (OSS) to automatically monitor events that track a number of system variables such as faults, device environmental state, availability, and performance. The operator may be interested in when the internal temperature within a network device (router, switches, switch/routers) has crossed a pre-defined threshold, which may be as a result of a chassis cooling fan tray failure, blockage in the chassis air flow, or abnormal operation of the cooling system in the facility housing the device.

The RMON Alarm group periodically takes sample values from variables in an RMON probe and compares them against pre-defined thresholds. An event is generated if the monitored variable crosses a threshold. The Alarm group specifies the alarm type (specific MIB object being monitored), sampling method, monitoring interval (frequency of sampling), and thresholds for comparing the monitored values (start and stop thresholds).

The RMON Events group specifies how events from a device are generated and how notifications are sent (logging or SNMP traps). The Events group specifies the event type, description of event, and the time that the event occurred. Any event that is created is added as an entry in the RMON Events group table. Associated with each entry are parameters of an event that can be triggered by alarms, and what action may be taken whenever the event occurs (a log entry or SNMP trap messages).

By using thresholds, the number of notifications sent on the network is minimized. A rising and a falling threshold can be set for a monitored variable, and a notification can be sent whenever the value of the variable falls outside the defined operational range. An event is only generated when the configured alarm threshold is first crossed in any one direction instead of at each sampling interval. This means, if a rising threshold alarm occurs and the corresponding event is triggered, the system does not trigger any more threshold crossing events until a corresponding falling threshold alarm occurs.

This is done to minimize the number of events that can be generated by the system, thereby making it easier for staff overseeing the operations of the system to react when events occur. An alarm is triggered when the variable being monitored exceeds a configured rising threshold value or falls below a falling threshold value (only in one direction). An alarm is triggered when a variable exceeds its defined rising threshold value but the system does not send alarm notifications until it recovers, as defined by the falling threshold value. This means the system does not send any notifications each time a minor failure or recovery occurs.

2.6   ENHANCED SYSTEM DIAGNOSTICS AND DEBUGGING

The goal of using online and offline diagnostics and debugging tools for network devices in general (including switch/routers) is to improve system uptime and availability. This improvement can be attained by preventing the occurrence of certain types of software or hard errors (resulting in increasing system Mean Time Between Failures (MTBF)), and by improving fault isolation and resolution time (resulting in reducing Mean Time To Repair (MTTR)). The diagnostics can run in the background, triggered by events, or initiated under operator command.

Specifically, verifying the system's operational correctness using enhanced diagnostics and debugging tools has the following benefits:

- Provide early detection of conditions that could eventually lead to a hard error. If corrective action can be taken before a hard failure occurs, system MTBF measured in the field will improve significantly.
- Provide more accurate reporting of errors/exceptions with fewer false positives and better fault isolation. Pre-determination of the root cause of the Field Replacement Unit (FRU) reduces the time wasted using trial and error swap-outs to identify faulty components and results in a significant reduction in MTTR.
- Provide better detection and isolation of soft errors maximizes the availability of the network, and improves measured user response times for network applications.

Some of the functionality of the diagnostics and debugging system may include:

- Error checking that verifies the correct operation of various hardware and software subsystems, plus the capability to take automated actions when an error or exception is detected.
- Notification of network operations personnel of the detected error or exception.
- Initiation of further system checks and information logging to help isolate the condition to a specific FRU.
- Notification of network operations personnel of which FRU is causing the problem.
- Saving all relevant information in a core dump file or other crash log file in the event of an unplanned system or subsystem reset.

A switch vendor would typically provide a comprehensive suite of diagnostics and debugging tools accessible via the CLI with an appropriate show and debug commands to determine the root cause and to isolate the fault to a specific FRU. Additional mechanisms include those for system health checks that run autonomously whenever the system is in operation. These diagnostics can detect and report errors via a syslog message, and also can be configured to take action in real time to minimize the impact of an error.

MTTR is minimized by a wide range of system diagnostic and debugging features. This could include runtime monitoring of hardware and software components, advanced in-service diagnostic fault tracing and troubleshooting, and enhanced SNMP MIBs for status monitoring. The system could include proactive health monitoring functions that send real-time alerts via syslog messages or SNMP traps when out-of-range conditions are detected that can potentially lead to failures. In some cases, health monitoring functions can take automated remedial action in real time to minimize the impact of any error. When an error does occur, advanced debug commands are available to isolate the cause of the fault condition.

2.7 MANAGEABILITY, SERVICEABILITY, AND EASE OF DEVICE INTEGRATION

Managing the edge of the network has become even more challenging as more of the traffic becomes real-time versus best effort, and as the diversity of attached devices increases significantly. In order to cope with these changes, the network management systems for network devices will need two basic attributes: ease of integration and extensibility.

By leveraging industry standards, management of network devices can be readily integrated with other management frameworks, applications, and node management systems. This provides seamless end-to-end management of the network to ensure high-availability, predictable QoS, and effective security measures. Another aspect of integration involves the interaction between the fault management system and the serviceability (diagnostics and debugging) features of a network device. Fault management isolates the fault to a root cause network element, while the device diagnostics perform the complementary task of quickly isolating the root cause to a particular FRU within the device.

On the issue of extensibility, support for industry standards, such as IEEE 802.1AB LLDP (Link Layer Discovery Protocol) [IEEE802.1AB] and ANSI TIA 1057 LLDP-MED (Link Layer Discovery Protocol-Media Endpoint Discovery) [ANSI/TIA1057], allow the management system to continue to evolve to support new edge devices and new functionality that are required for efficiently provisioning, configuring and troubleshooting. The role of LLDP and LLDP-MED in system configuration and management is discussed in greater detail below.

2.8 VALUE-ADDED NETWORK SERVICES

A switch/router can support additional services such as Domain Name System (DNS) and Dynamic Host Configuration Protocol (DHCP) beyond the basic forwarding services offered. These services are generally offered in a standalone device (servers) in a network but can be included as value-added features in switch/routers. Given that switch/routers are typically used in the access and aggregation/distribution portion of enterprise and service provider networks, they can be appropriate points at which services such as DNS and DHCP can be located (Figure 2.16).

For example, if a switch/router is deployed at the access portion of an enterprise or service provider network, it may contain a DHCP server, DNS resolver, and/or DNS server(s) to readily serve the end-users. A switch/router may provide DNS service to end-users by acting as both a caching DNS server and as an authoritative name server with its own local DNS records database. Most of the switch/routers in the market today have such extended features including the other features described in this chapter and others. We describe briefly the inner working of DHCP and DNS (for IPv4) in this section.

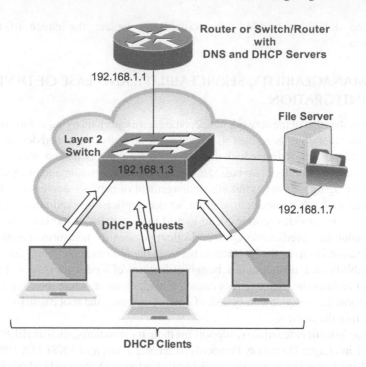

FIGURE 2.16 DNS and DHCP servers in a router or switch/router.

2.8.1 DYNAMIC HOST CONFIGURATION PROTOCOL

The limited availability of unused IPv4 addresses led to many organizations using (now very common practice) dynamic assignment of IP addresses to client devices out of the fixed pool of addresses they have been allocated. Today, most devices connected to the Internet from residential premises are assigned temporary IP addresses by the service providers. For example, when a residential user connects to the Internet, the ISP assigns the user connection a temporary IP address taken from a shared pool of IP addresses owned by the ISP. Each address taken from the pool for temporary use is known as a *dynamic IP address*. Dynamic IP address assignment is more cost-effective and easy to manage for organizations than assigning each user a *static* or *permanent IP address*. For many Internet applications, dynamic IP addresses are often sufficient for enterprise and ISP users.

2.8.1.1 Benefits of DHCP

DHCP [RFC2131] is a client/server protocol that enables devices in an IP network to dynamically obtain IP addresses and other required network configuration information (e.g., default gateway, subnet mask, domain name, domain name servers, time servers, etc.) from a DHCP server (Figure 2.17). Using DHCP, a DHCP server in a network (e.g., residential network, enterprise network, Internet service provider network, etc.) can dynamically assign IP addresses plus other network configuration information to the devices on the network so that they can properly address packets

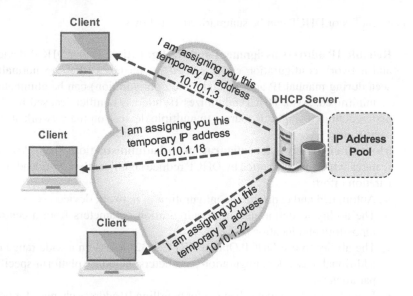

FIGURE 2.17   Role of DHCP in Internet communications.

they generate and communicate with other devices in the local and external networks. This is because all devices in an IP network must have a unique IP address to be able to communicate with other devices.

All modern operating systems include a DHCP client as part of their TCP/IP stack, and that DHCP client is generally enabled by default. With this, each DHCP client in a network during network initialization is able to request an IP address from the DHCP server. Without DHCP, devices that are moved from one subnet/VLAN to another must have their IP addresses configured manually. Also, devices that are removed from the network must have their IP addresses manually reclaimed. Note that, if a device moves to a different subnet/VLAN, it cannot use its previously assigned IP address but must use a different address belonging to the new subnet/VLAN.

However, with DHCP, the process of assigning and reclaiming IP addresses is managed centrally and automated. DHCP allows a device that has moved to be automatically assigned an IP address that is correct for the new subnet/VLAN. The network operator may deploy one or more DHCP servers in the network that maintain pools of IP addresses to be leased to DHCP-enabled clients when they start up on the network. The DHCP servers lease IP addresses dynamically rather than statically (i.e., addresses are not manually assigned). The servers reclaim IP addresses that are no longer in use automatically and return them to the address pool for future reallocation. Basically, DHCP allows a DHCP server to allocate IP addresses (from a shared pool of the limited number of IP addresses) to a group of devices that do not need permanent IP addresses.

The home router or gateway normally used in residential networks is a good example of a device that supports a DHCP server. The home gateway receives a globally unique IP address from the ISP network to enable it communicate with other devices on the ISP and the Internet. Within the local residential network, the DHCP server assigns local IP addresses to the device within that network.

The benefits of DHCP can be summarized as follows:

- **Reliable IP address assignment and configuration**: Using DHCP, device and network configuration errors caused by mistakes (that are normally seen during manual IP address entry and configuration) can be eliminated or minimized. Also, DHCP minimizes IP address conflicts caused by the assignment of a single IP address to multiple devices on the network at the same time.
- **More efficient network configuration and administration**: The following features which are supported by DHCP reduce significantly network administration effort:
  - ○ Automated and centralized configuration of network devices.
  - ○ The ability to define network configuration parameters from a central repository and location.
  - ○ The ability to use DHCP Options [RFC2132] to assign a wide range of additional network configuration parameters including platform-specific parameters.
  - ○ Provides an efficient mechanism for handling IP address changes for network clients that require frequent configuration updates, such as portable/mobile devices that move to different locations in a wireless network.
  - ○ The ability to forward DHCP messages to another subnet/VLAN using a DHCP relay agent, thereby eliminating the need for establishing a DHCP server on every subnet/VLAN.

2.8.1.2 Protocol Specifics

DHCP uses the following messages for communication (see Figure 2.18): DHCPDISCOVER, DHCPOFFER, DHCPREQUEST, DHCPACK, DHCPNAK, DHCPRELEASE, DHCPDECLINE, and DHCPINFORM. DHCP operates over UDP, and a DHCP client sends messages to a DHCP server using UDP destination port number 67 (the server listens to UDP port 67 for messages), while a DHCP server sends messages to a DHCP client using UDP destination port number 68 (the client listens for messages on UDP port 68). A DHCP broadcast message sent to UDP port 68 is read by all DHCP clients, even by clients that do not yet know their specific IP addresses.

DHCP version 6 (DHCPv6) [RFC8415] is the IPv6 equivalent of DHCP for IPv4 (discussed in this section). DHCPv6 is used for configuring IPv6 devices with IPv6 addresses, IP prefixes, and other network configuration information required for them to operate in an IPv6 network. In DHCPv6, clients send messages to servers using UDP destination port number 546 while servers send messages to clients using UDP destination port number 547. Although DHCP for IPv4 and DHCPv6 serve the same purpose, the details of each protocol are vastly different such that they may be considered entirely separate protocols.

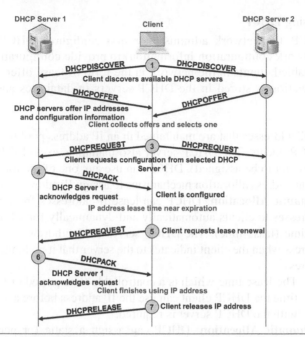

1. The client broadcasts a DHCPDISCOVER message to the limited broadcast address (255.255.255.255) on the local subnet to discovers a DHCP server.
2. Each DHCP server, after determining the client's network, selects an appropriate IP address and verifies that the address has not already been assigned to or in use by another client. The DHCP server then responds to the client by broadcasting a DHCPOFFER message that includes the selected IP address and information about services that can be configured for the client. The server temporarily reserves the offered IP address until it can determine if the client will accept it.
3. The client receives offers from several DHCP servers and selects the best offer (based on the number and type of services offered). The client then broadcasts a DHCPREQUEST message which specifies the IP address of the DHCP server that made the best offer. The message is sent by broadcast to ensure that all the responding DHCP servers know the particular server the client has chosen. The DHCP servers that have not been chosen will cancel the reservations for the IP addresses they had offered.
4. The DHCP server that is selected assigns the IP address to the client, stores the information in its DHCP database, and sends a DHCPACK (acknowledgement) message to the client. The DHCPACK message contains the network configuration parameters for the client.
5. The client starts using the assigned IP address while monitoring the lease time, and when a set period of time has elapsed, the client sends a new DHCPREQUEST message to the selected DHCP server to increase its lease time.
6. The DHCP server receives the DHCPREQUEST message and replies with a DHCPACK message to extend the lease time if it determines that the client still adheres to the local lease policy set by the network administrator. If the DHCP server does not respond within a set period of time (20 seconds), the client may broadcast a DHCPREQUEST message to see if one of the other DHCP servers can extend the lease.
7. When the client has finished using the IP address, it sends a DHCPRELEASE message to notify the owner DHCP server that it is releasing the assigned IP address.

Other DHCP messages:
- **DHCPNAK:** This message issent by a DHCP server to the client in response to a DHCPREQUEST. The DHCPNAK indicates that the server does not acknowledge the request, and does not agree to lease the specified IP address.
- **DHCPDECLINE**: This message is sent by the client to the DHCP server in response to a DHCPACK. If the client receives a DHCPACK, but, for some reason, is not satisfied with the lease time and/or network parameters in the message, it can send the server a DHCPDECLINE indicating that it will not lease the IP address.
- **DHCPINFORM**: This message is sent by the client to DHCP server, requesting only local configuration parameters such as domain name and DNS servers; the client has already obtained an IP address through other means (other than DHCP) or through manual configuration.

FIGURE 2.18 DHCP operation.

2.8.1.3   Using DHCP

To use DHCP, the network administrator first configures DHCP servers (that maintain network configuration information) to provide configuration parameters to DHCP-enabled clients in the network in the form of lease offers. The configuration information is stored in the DHCP servers in databases and includes the following:

- Valid IP addresses that are maintained in an IP address pool for assignment to DHCP clients. The DHCP server also maintains excluded IP addresses (that are not to be assigned). DHCP can use any one of the following three different address allocation mechanisms to assign IP addresses [RFC2131]:
  - o **Dynamic Allocation**: DHCP can lease IP addresses from a pool of IP addresses to clients automatically and dynamically for a limited period of time (lease time). The DHCP server can withdraw the assigned IP address when the client indicates to the server that it no longer needs that address.
    - ▪ The lease time which is a controllable time period or the length of time the DHCP client can use the IP address before a lease renewal with the DHCP server is required.
  - o **Automatic Allocation**: DHCP can assign a static (or permanent) IP address to a client automatically (e.g., email and web server, IP voice gateways, etc.). This address is selected from a pool of available IP addresses and there is no lease time associated with it, that is, the address is assigned to the device permanently.
  - o **Manual Allocation**: The network administrator can pre-allocate an IP address to a client manually and the DHCP server, in this case, only communicates that IP address to the client.
- Valid network configuration parameters for the DHCP clients on the network (e.g., default gateway, subnet mask, domain name, DNS servers, time servers, etc.).
- Reserved IP addresses from an address range defined by the network administrator to be assigned to particular DHCP clients. The DHCP server maintains a list of previous IP address assignments, so that clients can be preferentially assigned the same IP addresses that they were given previously. This feature allows the DHCP server to consistently assign a single IP address to a particular DHCP client.

Upon accepting a lease offer from a DHCP server, the DHCP client receives the following:

- A valid IP address plus network mask for the network to which the client is connecting.
- Requested network configuration information (i.e., DHCP Options [RFC2132]) which are additional parameters (e.g., default gateway, subnet mask, domain name, DNS servers, time servers, etc.) that the DHCP server has been configured to assign to DHCP clients in the network.

A DHCP client typically sends requests for network configuration information from a DHCP server immediately after it boots up, and periodically, thereafter before the expiration of the lease of the configuration information. Figure 2.18 describes the process through which hosts in a network use DHCP to obtain IP addresses and other required network configuration information from a DHCP server. The request-and-grant process described in Figure 2.18 (which is based on a lease concept with a defined lease time period) allows the DHCP server to reclaim and then reassign IP addresses that have not been renewed.

The operation of DHCP can be divided into four phases or stages:

- DHCP server discovery (by client)
- IP address lease offer (by server)
- IP address lease request (by client)
- IP address lease acknowledgment (by server)

These four phases are sometimes abbreviated as DORA, which stands for Discovery, Offer, Request, and Acknowledgment.

When a DHCP client broadcasts a DHCPDISCOVER message (intending to discover all available DHCP servers on the same subnet), it also includes its unique ID, which in most implementations, is derived from the client's MAC address (in the case of an Ethernet interface). The server sets the destination MAC address in the Ethernet frame header to the broadcast address (i.e., FF:FF:FF:FF:FF:FF), enabling it to reach all DHCP servers on the same subnet or possibly any attached DHCP relay agent (see discussion below). DHCP servers that receive the DHCPDISCOVER message will determine the client's subnet by examining the following information:

- **Determine the network interface on which the DHCP server received the DHCP request**: This indicates to the DHCP server that the client is either attached to the same network to which its interface is connected or that the client is communicating via a DHCP relay agent also connected to that network.
- **Check if the DHCPDISCOVER message includes the IP address of a DHCP relay agent**: When a DHCPDISCOVER message is relayed through a DHCP relay agent, the relay agent will insert its own IP address in the message header. When the DHCP server checks the DHCP relay agent's IP address, it will determine that the network portion of that IP address indicates the network address of the DHCP client because the DHCP relay agent must be connected to that client's network.
- **Check if the network to which the DHCP client is attached is subnetted**: The DHCP server consults its local network masks table to determine the IP subnet mask used on the network indicated by the DHCP relay agent's IP address, or the IP address of the network interface on which the DHCP request message was received. Once the DHCP server has determined the IP subnet mask used, it determines the host portion of that network address, and then selects an IP address from that host portion for the DHCP client.

The DHCP server writes the requested IP address into the YIADDR field of the DHCPOFFER message. The server also specifies the selected subnet mask and default gateway in the DHCP Options subfields, subnet mask and router options, respectively. The server also specifies other common options such as IP address lease time, DHCP server address, DNS servers, and so on. The DHCP server sends the DHCPOFFER message using the broadcast address, but will include the DHCP client's hardware address in the CHADDR field of the DHCPOFFER message, to indicate to the client that it is the intended recipient of the message.

When a DHCP client renews an assignment of configuration parameters, it typically requests the same parameters previously assigned, but the DHCP server may choose to lease out an entirely new IP address (plus other applicable parameters) based on the assignment policies set by the network administrator. A DHCP client requesting the renewal of a lease may communicate directly with the DHCP server (which offered the lease) using UDP unicast. This is because the DHCP client at that point (before the expiration of the lease) already has an established IP address. Also, DHCP messages carry a BROADCAST flag bit which the DHCP client can use (in DHCPDISCOVER or DHCPREQUEST messages) to indicate the mode of transmission (broadcast or unicast) through which it can receive the DHCPOFFER: Flag bit set to 1 (0x8000) means broadcast, and Flag bit set to 0 (0x0000) means unicast.

The BROADCAST flag bit is bit 1 in the 2-byte Flags field [RFC2131] and all other bits (to the right of bit 1) are reserved and are all set to 0. Usually, the DHCP server sends the DHCPOFFER via unicast. For devices that cannot accept unicast packets before their IP addresses are configured, the BROADCAST flag can be set to allow them to receive the DHCPOFFER via broadcast.

2.8.1.4   Use of DHCP Relay Agents

In a small network with only one IP subnet and one or more DHCP servers, the DHCP clients and DHCP servers can communicate directly with each other. However, by deploying a DHCP relay agent (also called a DHCP helper) on the local subnet, DHCP servers can be located on different subnets to provide IP addressing and other configuration information to DHCP clients on the local subnet. This is because a DHCP client in one subnet that has not yet been assigned an IP address cannot communicate directly with a DHCP server located in another subnet via IP routing.

Before IP address assignment, the DHCP client does not have a routable IP address, nor does it know the link layer address of a router (or gateway) as well as the IP address of a DHCP server. Furthermore, routers do not forward broadcasts from one subnet to other subnets, and as a result, will not propagate DHCP messages such as DHCPDISCOVER and DHCPREQUEST messages to other subnets. With no IP address initially configured, a DHCP client has to send a broadcast request to a DHCP server (on the same subnet) to obtain an IP address. So, in order to allow DHCP clients in subnets that have no direct access to DHCP servers to communicate with DHCP servers in other subnets (i.e., in another broadcast domain), DHCP relay agents can be configured on the routers that interconnect the subnets. Without the use of DHCP relay agents, a DHCP server would have to be installed on each subnet/VLAN.

On a large network consisting of multiple subnets, the network designer may deploy one or more DHCP servers (located in one subnet or even different subnets) to serve the entire network and assisted by a number of DHCP relay agents (or DHCP helpers) that are configured in the interconnecting routers. These DHCP relay agents forward DHCP messages exchanged between the DHCP clients and DHCP servers located on the different subnets.

A DHCP client in one subnet will broadcast a DHCPDISCOVER message on that subnet, and a DHCP relay agent (in an interconnecting router) receives that broadcast and transmits it to one or more DHCP servers in other subnets via unicast. The DHCP relay agent writes its own IP address (i.e., the IP address of the interface on which it received the DHCPDISCOVER message from the client) in the 4-byte GIADDR field of the DHCPDISCOVER message. A DHCP server that receives the relayed DHCP message uses the IP address carried in the GIADDR field to determine the subnet on which the DHCP relay agent received the original broadcast from the DHCP client, and assigns the correct IP address to the client on that subnet.

When the DHCP server sends a reply (DHCPOFFER) to the DHCP client, it first sends that reply to the IP address carried in the GIADDR field, again using unicast. The DHCP server sends the DHCPOFFER message via unicast on UDP port 67 back to the DHCP Relay Agent that sent the DHCPDISCOVER message. The DHCP relay agent receives the DHCP server's DHCPOFFER and then retransmits it to the DHCP client on its local subnet using either broadcast or unicast (on UDP port 68), depending on the BROADCAST flag bit set by the DHCP client. Typically, the communication between the DHCP relay agent and the DHCP server uses UDP port 67 as both the source and destination UDP port number.

2.8.1.5 DHCP IP Address Conflict Detection and Resolution

No two devices on the same network can have the same unicast IP address. So, DHCP servers or clients can use IP address conflict detection to determine whether a particular IP address is already in use in a network before leasing or using that address. The following describes the conflict detection and resolution mechanisms typically used in DHCP servers and clients.

2.8.1.5.1 DHCP Server-Side Conflict Detection

To perform conflict detection, a DHCP server can use the Ping (Packet Internet Groper) process [RFC2131] [RFC2151] to check the availability of IP addresses before leasing these addresses via DHCP offers to clients. A successful Ping by the DHCP server indicates that the IP address being checked is already in use in the network. Therefore, the DHCP server will not offer this address to a client unless it is reclaimed. The DHCP server will assign a different IP address if a Ping response is received from that IP address.

If, on the other hand, the Ping request fails and times out, then the IP address is assumed to be not in use in the network and can be offered to a DHCP client. Ping sends a series of Internet Control Message Protocol (ICMP) [RFC792] Echo messages to a target IP address to determine if the device using that address is active or inactive, and to determine the round-trip delay it takes to communicate with it.

2.8.1.5.2 DHCP Client-Side Conflict Detection

A DHCP client can use the Ping process [RFC2151] or ARP (Address Resolution Protocol) [RFC826] [RFC5227] to perform IP address conflicts:

- **Using ping**: When a DHCP client receives an IP address via a DHCPOFFER, it can use the **ping** process to make sure no other device in the network is using that address. If the IP address is available for use, then the device continues its boot-up process to join the network.
- **Using ARP Requests**: ARP is a protocol used to map a specific (known) IP address to a corresponding hardware address. A DHCP client can send an ARP request to detect if an IP address conflict exists before completing its configuration using the IP address offered by a DHCP server. To determine if an address is already in use, the DHCP client will broadcast an ARP request (also called an ARP probe) for the desired (offered) IP address by filling in the following information in the request:
  - The "sender hardware address" field of the ARP Request is filled with the hardware address of the interface over which the ARP request is to be transmitted.
  - The "sender IP address" field of the ARP Request MUST be set to all zeroes (to avoid polluting the ARP caches in other devices on the same subnet in the event the IP address is already in use by another device). An ARP Request carrying an all-zero "sender IP address" is referred to as an "ARP Probe".
  - The "target hardware address" field of the ARP Request is ignored and SHOULD be set to all zeroes.
  - The "target IP address" field of the ARP Request MUST be set to the IP address offered by the DHCP server (this is the IP address being probed by the ARP request).

A DHCP client that is offered an IP address by a DHCP server can perform an ARP test to verify that the IP address is available and no address conflicts exist. If the offered IP address is already in use by another device on the subnet, that device will send an ARP reply indicating its hardware address along with that IP address.

If the DHCP client detects a conflict in an IP address assignment, it can send a DHCPDECLINE (DHCP decline) message back to the DHCP server that offered the IP address; it can also request another IP address from the same or another DHCP server. The DHCP server typically maintains a log of all DHCP client-detected conflicts and removes the affected addresses from its DHCP address pool.

2.8.2 DOMAIN NAME SYSTEM (DNS)

The Domain Name System (DNS) [RFC1034] [RFC1035] is a hierarchical and globally distributed naming system for mapping domain names (which are generally more readily memorized) to IP addresses. A *domain name* is a text-based *label* or string that identifies Internet resources, such as computers, services, and networks.

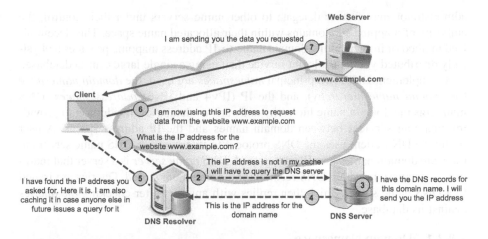

FIGURE 2.19   Role of DNS in Internet communications.

A text-based labels for Internet resources are much easy to recognize and memorize than the numerical IPv4 and IPv6 addresses used for identifying such resources. For example, *www.example.com* which we assume translates or maps to the IPv4 address 192.0.2.3, is easier to memorize than 192.0.2.3.

When a user enters a domain name like *nortel.com* into a web browser, this triggers a DNS query which ultimately leads to a DNS server that converts that domain name into an IP address (Figure 2.19). DNS is the system that orchestrates all this. Imagine trying to memorize the usually long and complex IPv6 addresses such as 2001:0db8:85a3:0000:0000:8a2e:0370:7334 when looking up information from a website using that address. The process of DNS resolution involves translating a domain name (such as www.example.com) into an IP address. Each device or interface on the Internet requires an IP address so that it can communicate with other devices on the Internet.

A domain name may represent individual instances of a resource or an entire collection of such resources. Domain names can be used for application-specific naming (e.g., website or a server hosting a website) and for various networking contexts (e.g., to identify a network domain). A *network domain* is a grouping of multiple hosts and network nodes under the authority or control of a single administrative entity. An individual instance of Internet resources can be a host computer using a specific domain name as a host identifier (also called a *hostname*).

DNS (which is one of the protocols that comprise the TCP/IP suite (see Chapter 3 of Volume 1)) together with a number of DNS resolvers and servers provide a worldwide distributed directory service (or name resolution service) for the Internet – maps domain names to IP addresses for computers and users. DNS defines in a large number of IETF RFCs, the DNS protocol (part of the Internet Protocol Suite), detailed specification of the domain name database service, data structures, and communication exchanges used in managing and mapping domain names to Internet addresses.

For each domain, DNS designates specific authoritative name servers with the responsibility of mapping domain names to Internet addresses. A network administrator may further delegate to other name servers under their control, the authority of mapping subdomains within their allocated name space. This decentralized method of implementing domain name-to-IP address mapping provides a physically distributed yet fault-tolerant service than using a single large central database.

To implement DNS, two principal *namespaces* are used, the *domain namespace* (or *domain name hierarchy*), and the IP (IPv4 and IPv6) *address spaces*. DNS maintains the domain name hierarchy and various databases, in addition to providing mapping services between domain names and the IP address spaces. A user running a DNS client uses the DNS protocol to interact with DNS name servers to translate domain names into IP addresses. A *DNS name server* is a server that maintains a database in which DNS records for a domain are stored. Queries are issued to a DNS name server which then replies with answers after checking the queries against its database.

2.8.2.1   Domain Namespace

The domain namespace is organized as a tree data structure, that is, a tree made up of domain names. Each node in the tree (i.e., also referred to as the domain name hierarchy) is assigned one or more *labels* plus zero or multiple *resource records* (RRs). The RRs are units of information added to a label and hold information associated with a domain name in the tree. RRs are the basic information blocks of domain names and Internet address information used to resolve user DNS queries.

The domain namespace is divided into distinctly managed areas called *DNS zones*. DNS zones are simply, points of delegation in the domain name hierarchical tree. A DNS zone is managed by a specific administrator or organization. The concept of DNS zones is to develop administrative spaces (i.e., portions of the DNS namespace) which allows for more granular control of the components that orchestrate the domain name system, such as authoritative name servers. DNS zones are used for delegating control in the domain namespace hierarchy. The DNS *root zone* or *domain* is at the top of the domain name hierarchy.

Starting at the DNS root zone, the tree may be further sub-divided into DNS zones. A DNS zone may be carved out of the tree to comprise only one domain, or multiple domains and subdomains (all grouped together), depending on the administrative goals of the DNS zone manager. A DNS zone contains all domains from a particular point in the DNS hierarchical tree downward, except those portions of the tree under the authority of other DNS zones – that is, just up to the point at which other zones are authoritative. Basically, a DNS zone starts at a specific domain within the DNS tree and may extend downwards to contain multiple subdomains to allow multiple subdomains to be managed by one administrative entity.

The ICANN (Internet Corporation for Assigned Names and Numbers) is the entity responsible for managing the architecture and development of the top-level Internet domain namespace. The ICANN delegates to authorized domain name registrars, the responsibility of registering and reassigning domain names for end-users.

A domain name in the tree can have one or more parts referred to as *labels*. A *label* may have the name of its parent node in the tree concatenated to it on the right, and with the two separated or delimited by a dot (.), for example, *nortel.com*. The rightmost label in a domain name designates the *top-level domain* in the naming hierarchy. For example, the top-level domain of the domain name *www.nortel.com* is *com*. From the top-level domain, the hierarchy moves down the domain from the right label to the left label in the domain name. Each label in the domain name to the left specifies a subdomain of the domain (label) to the right.

The top-level domains of the hierarchical domain name tree form the DNS root zone. So, all domain names end with a top-level domain label (the rightmost label). There are now many generic and country-code top-level domains in the Internet but some examples of the well-known ones among the generic highest-level domains are, com, net, edu, info, and org. Below the top-level domain in the hierarchy, there may be a variable number of lower levels, as required (which can be second-, third-, fourth-level domain names and so on). As noted above, each label in the domain name is delimited by a dot. These labels are placed sequentially directly to the left of the top-level domain. For example, in the domain name *bbc.co.uk*, *co* is the second-level domain.

2.8.2.2 What Is a DNS Record?

DNS records (also called *DNS zone files*) are instructions that are maintained in DNS authoritative name servers and contain information about a specific domain including the IP address associated with that domain and how DNS queries/requests must be handled for that domain. A DNS record is written using DNS syntax and consists of a series of text files. DNS syntax is made up of a string of characters that serve as commands instructing a DNS name server what actions to carry out when a request is received.

A DNS record obtained from a DNS request and maintained in a cache also has a time-to-live (TTL) assigned, indicating the length of time (the expiration time) the record will stay valid, or how often that record needs to be refreshed by the DNS name server. Caching DNS results locally in DNS servers, or in intermediate resolvers and hosts is standard practice in DNS name resolution, and is done in order to reduce the load on the DNS servers. Expiration of the TTL of a DNS record in a cache results in the record being discarded or refreshed.

Each domain maintains at a minimum, a number of DNS records that are deemed essential for users to be able to access resources in that domain (using a domain name), plus other optional DNS records that serve additional purposes. Examples of some of the essential records are the following:

- **A Record**: This (stands for "Address") record and holds the 32-bit IPv4 address of a specific domain. It is the most fundamental of all DNS record types.
- **AAAA Record**: This (stands for "quad-A") record and holds the 128-bit IPv6 address of a domain.
- **CNAME Record**: This (stands for "Canonical Name") record and holds an alias (or another name) for the name being queried and does NOT contain an IP address. The alias allows the DNS resolver to continue the DNS lookup by retrying the name resolution using the new name.
- **MX Record**: This (stands for "Mail Exchange") record and holds information that maps a domain name to a list of mail servers – directs or routes

mail to a mail server (or list of message transfer agents) for that domain (in accordance with SMTP (Simple Mail Transfer Protocol) which is the standard protocol for email). An MX record, like a CNAME record, always points to another domain name.

- **NS Record**: This (stands for "Name Server") record and stores the DNS server that is authoritative for a DNS entry or particular domain. This is the name server that contains the actual DNS records for the domain. A domain will often store in multiple NS records, the name servers that are designated primary and backup for that domain.
- **PTR Record**: This (stands for "Pointer") record and provides a domain name for a given IP address in a reverse DNS lookup. The PTR record provides the exact opposite of the A record – the PTR address points to the domain name associated with a given IP address.

2.8.2.3   Name Servers

Internet domain names are maintained in distributed databases called *name servers*. Information exchange with the name server follows a client-server model. To resolve a DNS query, DNS servers usually cooperatively interact with other servers that store DNS records. DNS name servers fall in one of following categories: root name servers, top-level domain name servers, authoritative name servers, and DNS resolvers. Figure 2.20 shows the different components a DNS query passes through when it is issued by the end-user. These components are described below.

2.8.2.3.1   DNS Resolver

A *DNS resolver* (i.e., the *"stub" resolver* in the local host) is the first point of initiating a domain name resolution and where an end-user (via a local application) sends a DNS query. The DNS resolver acts as a DNS pre-processing point between a local application and a DNS name server that may be remotely located in the Internet. The DNS resolver is a server that is usually built into the client's device to receive DNS queries from applications such as web browsers. Typically, the DNS resolver is responsible for making additional requests and interacting with external DNS name severs in order to satisfy a DNS query issued by the client.

When the DNS resolver first receives a DNS query from the client, it will either respond with the correct information cached locally, or send a request to a root name server. This request may be followed by another request to a top-level domain name server (if the root server is unable to return the correct information), and possibly to an authoritative name server. After receiving a response containing the requested IP

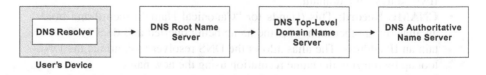

FIGURE 2.20   DNS server types and DNS record request sequence.

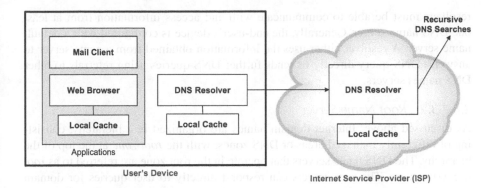

FIGURE 2.21 DNS resolution sequence involving an ISP DNS resolver.

address from the authoritative name server, the DNS resolver then sends this information to the client.

After each successful DNS lookup process, the DNS resolver will cache locally the domain name mapping information received from the authoritative name servers. The next time the local application sends a DNS query for a recently resolved domain name, the DNS resolver cuts short the DNS lookup process (by not communicating with any name servers), and just delivers the mapped IP address from its local cache directly to the client.

As illustrated in Figure 2.21, application programs (e.g., web browsers, email clients) use the name resolution services provided by DNS resolvers built into the operating system of the end-user's device (computers, tablets, smartphones, home router, etc.). Generally, the applications and the DNS resolver each support their own local caches, just like each of the other DNS name servers.

Each of these DNS entities maintains a database called a cache (used to store domain name-to-IP address mappings) which speeds up the process of translating domain names to IP addresses. The cache stores the results from previous successful DNS lookups. Upon receiving a DNS query, the DNS server will first check its local cache to see if the answer is available.

Typically, enterprise and service provider networks also support DNS resolvers capable of recursive DNS searches which the end-users use for DNS query resolution. For example, the stub DNS resolver in the end-user device may make a recursive DNS query to the DNS name server located in the user's service provider network. Likewise, DNS resolver in the ISP network could issue recursive DNS queries to other name servers in the DNS hierarchy.

Note: DNS resolvers are programs built into end-user devices (i.e., a stub resolver), or into an ISP's equipment, allowing them to extract information from sent by DNS name servers when DNS queries are received from end-user applications. A (stub) DNS resolver is typically a system routine running in the host device that is directly accessible to user applications/programs. As a result, no particular standard procedure or protocol is necessary between the user program and the resolver. DNS resolvers must be able to communicate with and access information from at least one DNS name server. Generally, the end-user's device is configured with a default name server. A resolver either uses the information obtained from a name server to answer a DNS query directly or sends further DNS queries using referrals to other DNS names servers.

2.8.2.3.2 Root Name Server

As discussed above, Internet domain names are organized in a hierarchy, consisting of differently managed areas or DNS zones, with the *root zone* at the top of the hierarchy. The DNS name servers that operate in the root zone are referred to as *root (name) servers*. The root servers can respond directly to user queries for domain name resolutions by checking resource records stored or cached locally within the root zone. The servers can also refer user queries and requests to the next most appropriate top-level domain name server (i.e., subordinate servers).

Generally, the root zone of the domain name hierarchy is served by more than one (multiple) root name servers. It should be noted that these are the name servers to which DNS queries are first sent when end-users are resolving or looking up a domain name – they are the first point of contact when a DNS resolver at the client end sends requests for DNS records. When the root server is unable to resolve a DNS request, it may direct the DNS resolver to a subordinate top-level domain name server. Usually, the root server serves as a reference to other more specific DNS name severs that can satisfy the client's DNS query. The root name servers are managed by the ICANN (Internet Corporation for Assigned Names and Numbers).

To provide resiliency in the global DNS operation in the event of network and server failures, multiple DNS servers are usually used for each domain. The root level of the global DNS hierarchy has 13 groups of root name servers distributed worldwide, with additional identical "copies" of these root servers also distributed worldwide using anycast IP addressing. Typically, every DNS resolver maintains a list of IP addresses of these 13 root servers (built into its software), and whenever the resolver initiates a DNS lookup, it will first communicate with one of those 13 IP addresses.

An anycast address [RFC1546] [RFC2526] [RFC4291] is an IP address that is assigned to a particular group of interfaces, where each interface typically belongs to a different device (e.g., server, gateway, router, etc.) in the network. Any IP packet sent to this address is forwarded to the nearest/closest interface in the group, that is, the interface that lies on the lowest-cost route as defined by the routing metric of the routing protocol in use.

Anycast addressing is generally used to distribute servers and services across a network in order to achieve robustness, redundancy, and resiliency [RFC4786]. When building a content delivery network (CDN), anycast addresses can be used to distribute the content server across the network in order to bring the content closer to the end-user. Anycast addresses can be used in the context of DNS server load balancing where DNS requests are distributed to geographically dispersed DNS servers in order to achieve faster DNS query processing, efficiently utilize server resources, and reduce the impact of denial-of-service (DoS) attacks.

2.8.2.3.3   Top-Level Domain Name Server

One step below the root servers in the DNS hierarchy are the *top-level domain name servers*. These are a group of DNS name servers that are an integral part of the DNS queries resolution process. The top-level domain name servers are the next top servers (just below the root servers) in the DNS resolution process for a specific IP address, and store information about the last portion (label) of a domain name. A top-level domain name server maintains information for all the domain names that share the same top-level domain label (e.g., *.com*, *.net*, *.edu*, *.org*).

For example, a *.com* top-level domain name server maintains information for every domain name that ends in ".com". If a user's local application triggers the sending of a DNS query for *nortel.com*, the client's DNS resolver may receive a response from a root name server that will instruct it to send a query to a *.com* top-level domain name server, which in turn may respond by referring it to the authoritative name server for that domain name. The Internet Assigned Numbers Authority (IANA), which is a branch of ICANN handles the management of top-level domain name servers.

The two main groups of top-level domain name servers are:

- **Generic top-level domains (gTLDs)**: These are domains that represent categories of business types, agencies, organizations, and multi-organizations, and are not country-specific (*.com*, *.org*, *.edu*, *.net*, *.gov*, etc.).
- **Country code top-level domains (ccTLDs)**: These are domains that are specific to a country, sovereign state, or dependent territory (e.g., *.ca*, *.us*, *.gh*, *.ph*, etc.).

2.8.2.3.4   Authoritative Name Server

Each domain in the DNS hierarchy has one or more *authoritative name servers* that own and maintain information about that domain, in addition to the name servers of any domains in the hierarchy below it. A name server is said to be *authoritative* for a portion of the domain name tree when it has complete information for that portion. A DNS zone typically has at least one authoritative name server. An authoritative name server is configured with DNS records information, or acquires that information through a *DNS zone transfer*. A zone transfer takes place when a secondary DNS server starts up and is updated with information from its primary DNS server (see discussion below).

An authoritative name server has its own local DNS records database and listens for DNS queries. It answers DNS queries using the cached and permanent entries in its local DNS records database. Also, an authoritative name server may issue zone transfer requests to other authoritative name servers in the same DNS zone for DNS records. It may also respond to zone transfer requests from other authoritative name servers in the same zone.

A DNS resolver may receive a response from a top-level domain name server that directs it to an authoritative name server. The authoritative name server is usually the DNS resolver's last point of contact in the domain name resolution process (when a

DNS query is sent for an IP address associated with a domain name that is under its zone of authority). The authoritative name server maintains and serves information specific to a domain name (e.g., *nortel.com*). It provides a DNS resolver with the IP address associated with the domain name discovered in the DNS A record, or if there is a CNAME record (alias) for that domain, it will provide the DNS resolver with an alias domain. Using the alias, the DNS resolver sends an entirely new DNS query, looking for a DNS A record from another authoritative name server (containing the required IP address).

If the authoritative name server has the requested DNS A record (after searching the permanent and cached records in its local database), it will return the IP address for the requested domain name back to the DNS resolver that issued the initial DNS query. If the DNS query is for a domain name in its zone of authority but for which it does not have the correct DNS records, it will not forward the DNS query for further lookup; it will simply reply indicating that no such records exist. The client's DNS resolver is the component involved in initiating a DNS query, while the authoritative name server is a server at the end of the DNS query resolution process.

The authoritative name server must be able to respond to DNS queries from its own stored records without necessarily having to query another server, as it is the final contact point for certain DNS records. However, in situations where a DNS query is for a subdomain such as *blog.example.com*, the authoritative name server (which is responsible for storing the CNAME record for the subdomain) may add an additional/referral DNS name server to the domain name resolution sequence (Figure 2.22).

2.8.2.4   Primary versus Secondary DNS Servers

When a DNS authoritative name server is set up, it can be designated the primary (master) or the secondary (slave) server. A *primary DNS server* stores the controlling DNS records (or zone files) for a domain. The primary server contains all the authoritative DNS records for a domain and is the trusted source for important DNS information needed to resolve DNS queries such as the IP address of the domain, plus the entity responsible for the administration of that domain. Primary DNS servers obtain all this DNS information directly from locally stored files. Changes to the DNS records of a zone can only be made on the primary DNS server, which is then responsible for updating any secondary DNS servers.

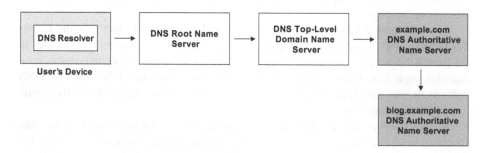

FIGURE 2.22   DNS record request sequence with CNAME record request.

Secondary DNS servers maintain read-only versions of the DNS zone file hosted in the primary DNS server. The secondary servers obtain this information from a primary server in a communication exchange process referred to as a *DNS zone transfer*. The primary server maintains the master or original copies of all DNS zone records, and the secondary server employs the special automatic DNS records updating mechanism (called *zone transfers*) to communicate with its primary server to maintain an identical copy of the primary's DNS records. Only one primary DNS server can be supported by each DNS zone, but multiple secondary DNS servers are allowed. No changes can be made to the DNS records of a zone on a secondary DNS server. In some designs, a secondary DNS server can pass any change requests issued to its primary server.

Although primary DNS servers maintain all the relevant DNS records needed to resolve DNS queries for a domain, it is standard practice (and required by many Domain name registrars) to support one or more secondary DNS servers. This provides redundancy in the event the primary DNS server fails or becomes unavailable. This also allows DNS requests to the domain to be distributed to the primary and multiple secondary servers, thus, avoiding total overload of the primary DNS server (a situation which could have been exploited for creating denial-of-service (DoS) attacks). With multiple servers, a load-balancing technique such as round-robin DNS can be used to send DNS requests to each server in the cluster, allowing each server to carry roughly equal amounts of DNS traffic.

2.8.2.5 DNS Operation

Whenever a user initiates communication with an Internet entity using its domain name and the DNS query (issued by the local application) cannot be resolved using its local cache, this triggers a DNS lookup in an external name server. All DNS lookups start at the root server at the root zone of the DNS hierarchy. Once the DNS lookup arrives at the root server, the lookup can travel down the DNS hierarchy, first receiving attention at the top-level domain name servers, then any subordinate servers handling specific domains (and subdomains), until it is received by the authoritative name server for the correct domain. This final name server determines the IP address of the particular domain name being queried and returns this to the originating application.

2.8.2.5.1 Types of DNS Queries

As discussed above, the DNS component at the end-user side is referred to as the *DNS client* or *stub DNS resolver*. The DNS resolver is responsible for issuing and sequencing the DNS queries sent to DNS name servers for domain name to IP address resolution. A DNS query can be classified as *non-recursive, recursive*, or *iterative*. A DNS resolver can also be classified according to the type of query it can process (non-recursive, recursive, or iterative). A domain name resolution process may involve the use of a combination of these different query methods.

Using a combination of these query types, the process of DNS resolution can be optimized to provide a significant reduction in DNS query processing time. The typical DNS lookup in the case of uncached DNS information involves both recursive and iterative queries. In the ideal situation, the requested DNS information is cached somewhere near the client, allowing the DNS lookup process to use the simpler non-recursive queries.

It is worthwhile noting here the difference between, for example, a *recursive DNS query* and a *recursive DNS resolver*. The *DNS query type* refers to the type of DNS request (regarding recursion) made to a DNS resolver for the resolution of the query. A *recursive DNS resolver* refers to the capability of the resolver of accepting a recursive query and processing that request by sending the necessary queries to other DNS servers. We discuss below the different DNS query types which also allows us to infer the capabilities of the DNS resolver types.

2.8.2.5.1.1 Non-Recursive Query

In a *non-recursive query*, the client's DNS resolver sends a query to a DNS server for a DNS record for which the server is authoritative (i.e., provides a full or partial result without querying other servers), or the requested record already exists in the server's local cache. Typically, DNS servers cache DNS records to prevent excessive bandwidth consumption and load on the network and servers.

When a *caching DNS resolver* receives a non-recursive DNS query, it may be able to deliver the result directly from its local DNS cache (which reduces the load on upstream DNS servers). The DNS resolver in this case caches DNS records for a period of time after receiving responses for other previous DNS queries from upstream DNS servers.

2.8.2.5.1.2 Recursive Query

The DNS resolver responds to a DNS request from the client and then takes control of the process by tracking the DNS record requested. In a *recursive query*, a DNS resolver sends a DNS query to a single DNS server which may in turn send another DNS query to other DNS servers on behalf of the DNS resolver. In this case, to resolve the initial query completely, the first DNS server will query other name servers as needed.

To handle the recursive query, each DNS server sends a DNS request to the next server until the request reaches the authoritative name server that holds the requested record (or until there is a timeout, or an error is returned if no record is found). Generally, DNS resolvers do not always need to rely on multiple recursive requests in order to discover the DNS records matching the DNS request. Caching at an earlier point in the process helps to reduce the number of requests needed. Caching allows the requested DNS record to be served early in the DNS lookup process without requiring further queries.

In the typical scenario, a DNS resolver issues a DNS query to a *caching recursive DNS server*, which may subsequently issue non-recursive queries to other DNS servers to discover the DNS record, and then send that back to the resolver. By setting appropriate bits in the DNS query message headers (RD (Recursion Desired) and RA (Recursion Available) bits), the DNS resolver, or any other DNS server performing recursion on behalf of the DNS resolver, is able to negotiate the use of recursive searches. DNS servers may support recursion but are not required to do so. ISPs typically support recursive and caching DNS servers for end-users. Even many home gateways/routers implement recursive capabilities and DNS caches to improve the efficiency of the local network.

2.8.2.5.1.3 Iterative Query

In an *iterative query*, a DNS resolver queries one or more DNS servers one after the other for a specific DNS record after an unsuccessful previous query. Each DNS server that is queried refers the DNS resolver to the next appropriate DNS server in the chain, until the resolver gets a final answer to the request. For example, a DNS query from the resolver requesting the resolution of *www.example.com* would require a query to a global root server, then possibly another query from the resolver to a "com" server, and finally a query from the resolver to a "example.com" server.

In such a situation, the DNS resolver will wait for the queried DNS server to provide the best answer it has. If the queried DNS server does not have the requested DNS records, it may return (to the resolver) a referral to a DNS server that is authoritative for the lower level in the domain name. Using this, the DNS resolver will then send a query to the referred DNS server. This process continues with the resolver sending queries to additional DNS servers down the domain name hierarchy until either a timeout or an error occurs.

2.8.2.5.2 DNS Caching

The main purpose of caching is to temporarily store recently looked-up DNS information at locations in the network to provide improvements in information availability and performance for DNS lookups. DNS caching involves storing DNS information closer to the host application so that DNS lookups can be performed early and avoiding the use of additional DNS queries further down the DNS resolution chain. This helps to improve DNS lookup times and reduce network bandwidth and name server processing resources.

Very often the information needed to resolve a DNS query will be cached either locally within the client's device sending the query, or remotely in DNS servers in the network infrastructure. When the DNS information is cached, then the processing steps required to resolve a DNS query get significantly shortened. DNS information can be cached in a number of locations in the lookup chain (Figure 2.21). As indicated earlier above, each caching point will store DNS records with an expiry time determined by a TTL (time-to-live).

2.8.2.5.2.1 DNS Caching at the Web Browser Level

All modern web browsers support caching of recently used DNS records for a set period of time. The DNS caching at the web browser shortens the DNS processing steps/times because the browser can readily check its local cache for DNS requests before attempting to query other DNS resolvers. Each time the browser makes a DNS request for a DNS record, it checks its cache first for the requested record before trying another resolver. All browsers provide very simple ways of checking the status of their DNS cache or even clearing the contents of the cache.

2.8.2.5.2.2 DNS Caching at the Operating System (OS) Level

The DNS resolver normally built into the OS of the end-user's device is the second and possibly the last stop for processing a DNS query before a decision is made to contact a DNS name sever outside the device. A process inside the OS, commonly called a *"stub DNS resolver"* or *DNS client*, is designed to handle the DNS query.

When the stub DNS resolver receives a DNS request from a local application, it will first check its own local cache to see if the requested DNS record exists. If the cache does not contain the record, the stub resolver will then send a DNS query (with the "Recursion Desired (RD)" flag bit set in the message) to a *recursive DNS resolver* in the ISP.

When the ISP's recursive DNS resolver receives a DNS query, it will also check to see if the requested DNS records already exist in its local cache (Figure 2.21). The recursive DNS resolver may also be capable of performing additional functionality depending on the types of DNS records stored in its cache:

1. If the local cache in the recursive DNS resolver does not contain the requested DNS A records but instead has the DNS NS records that point to the authoritative name servers for the queried domain name, the resolver will query those name servers directly. It will bypass several steps in the DNS lookup process, preventing lookups by the root and top-level domain name servers, thereby, helping to speed up the resolution of the DNS query.
2. If the cache in the resolver does not contain the DNS NS records, the resolver will send a query to the top-level domain name servers, bypassing the root server.
3. In the event that the cache in the resolver does not contain DNS NS records pointing to top-level domain name servers, the resolver will then query the root servers. The resolver typically takes this action after it has purged the local DNS cache.

2.8.2.5.2.3   What Is a Caching Name Server?

A caching name server caches DNS information that it has learned from other name servers in the process of answering client DNS queries. This allows the server to answer DNS queries quickly without having to query other name servers for each new DNS request it receives. A caching name server will relay DNS requests to other name servers for DNS lookup only when the local cache does not contain the required answer.

2.8.2.5.3   DNS Lookup and Resolution Sequence

In a typical DNS lookup (when the client has not cached the queried DNS mapping), the four DNS name servers described above work together to complete the task of mapping a specified domain name to an IP address for the DNS client. The DNS client is typically a simple (stub) resolver built into the operating system in the user's device.

2.8.2.5.3.1   DNS Lookup with DNS Information Cached Locally

User applications generally do not communicate directly with an external DNS name server to resolve domain names. Instead, applications such as web browsers, e-mail clients, and other applications have built-in capabilities that allow them to interact with external name servers transparently, to resolve DNS queries. When an application sends a DNS query for a domain name lookup, that request is sent to the DNS resolver in the device's operating system (Figure 2.21), which in turn handles all communications with any external name servers.

The local DNS resolver will first make a lookup in its local cache (see Figure 2.21) to see if it contains an answer from any recent lookups related to the queried domain name. If the cache contains the answer to the query, the DNS resolver will return that value to the application that sent the query. If the DNS resolver cannot find the answer in its local cache, it will propagate the DNS query to one or more designated DNS name servers in the network.

Typically, for enterprise network or ISP users, the service provider will provide the designated DNS name server to which DNS queries can be sent. The user's device is either configured manually or automatically (via DHCP), the IP address of the designated name server. In some cases, the service provider can configure the DNS resolvers in the connected client devices to point to DNS name servers that are separately maintained by the organization.

2.8.2.5.3.2 DNS Lookup with DNS Information Not Cached Locally

The designated DNS name server (or resolver for the service provider) will receive DNS queries and will follow the DNS lookup process outlined below until the query is resolved successfully or unsuccessfully. Upon a successful lookup, the correct answer is returned to the initiating DNS resolver, which in turn will pass that result back to the application that initiated the request. The resolver will also cache that result for future local DNS lookups. Figure 2.23 describes the main steps involved in resolving a DNS query when the required records are not cached.

2.8.2.6 DNS Protocol and Extensions

DNS uses two message types, *Query* and *Response* messages. Both message types have the same format, consisting of a header and four sections [RFC1034]. The message header is 12 bytes long. The four sections consist of a *Question*, *Answer*, *Authority*, and an *Additional* section. Some of these fields are set to zero in Query messages. Essentially, the typical Query message consists of a header and a Question section only. Response messages consist of the common header, plus the four sections. Flag bits in the message header are set to control the content of the four message sections.

DNS uses UDP for the majority of DNS messages because DNS Query and Response messages are often short and the application can tolerate message losses. When a sent message is not received, a new DNS Query message is simply issued. With UDP, the source and destination UDP port numbers are set to the "well-known" port number 53. The maximum size of a DNS Response message using UDP is 512 bytes. Out of this, the IP header takes a minimum of 20 bytes, and the UDP header takes another 8 bytes. This means the maximum DNS message size over UDP can only be 484 bytes. With this, the typical DNS Query consists of a single UDP message from the client which is followed by a single UDP Response message from the DNS server.

However, when the DNS message size exceeds effectively 484 bytes, and both the DNS client and server support the Extension mechanisms for DNS (EDNS), then a larger UDP message can be used. If this is not possible, then the DNS Query message

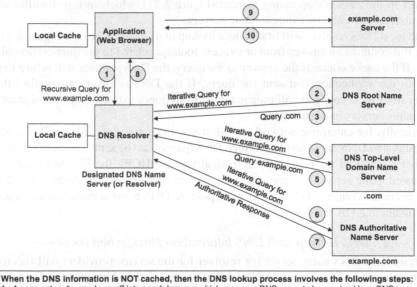

When the DNS information is NOT cached, then the DNS lookup process involves the followings steps:
1. A user enters "example.com'" into a web browser which causes a DNS query to be received by a DNS recursive resolver.
2. The DNS resolver in turn sends a query to a DNS root names erver (.).
3. The DNS root server responds supplying the resolver with the IP address of a Top Level Domain (TLD) name server (.com) which stores the DNS information for that domain (a .com TLD name server).
4. The DNS resolver then sends a DNS request to the .com TLD name server.
5. The TLD name server responds with the IP address of the authoritative name server for the domain name example.com.
6. The DNS recursive resolver then sends aquery to the specified authoritative name server.
7. The authoritative name server then returns the IP address for example.com to the DNS resolver.
8. The DNS resolver then sends a responds to the web browser carrying the IP address of the requested domain name example.com.
Once the above 8 steps of the DNS lookup have returned the IP address for example.com, the web browser is able to communicate with the example.com server:
9. The web browser makes a HTTP request to the example.com server using its resolved IP address.
10. The example.com server receives the HTTP request and returns the request webpage to the web browser.

FIGURE 2.23 DNS operation.

can be sent using TCP. DNS also uses TCP for tasks such as DNS zone transfers because these usually involve the exchange of large volumes of data that also require reliable transfer. A DNS resolver implementation may choose to use TCP for all DNS Query messages.

2.8.2.6.1 DNS Security

Standard DNS queries, although making up a majority of Internet DNS queries, lack security measures to provide DNS data integrity and user authentication. The lack of security features creates opportunities for malicious DNS attacks, for example, using attacks such as man-in-the-middle attacks and DNS hijacking. These attacks, among others, can make it easy for DNS lookups to be hijacked for malicious purposes like redirecting traffic inbound to a website to a fraudulent website, where sensitive user information can be collected and malware distributed, and exposing businesses to major liabilities.

The *DNS Security Extensions* (*DNSSEC*), defined in a number of IETF specifications [RFC4033] [RFC4034] [RFC4035], is one of the best-known methods for protecting against DNS threats and attacks. DNSSEC provides security protection against such malicious attacks by allowing DNS data to be digitally signed, helping to ensure its validity (i.e., to ensure that the DNS data has not been tampered with). To help ensure secure DNS lookups, the DNS data must be signed at every level in the DNS lookup process. DNSSEC provides a set of extensions to DNS which provide support for cryptographically signed DNS responses to DNS resolvers and severs. DNSSEC provides authentication of DNS data, and data integrity, but not confidentiality of data (no data encryption provided).

2.8.2.6.2 Dynamic DNS

The rapid growth of the Internet has created a shortage of available IPv4 addresses. With the scarcity of such addresses and given that IPv6 addresses are not yet extensively used, enterprise and service provider network have adopted the strategy of assigning dynamic IPv4 addresses to a majority of users and static (or permanent) addresses to very critical and/or important users.

Many users in enterprise and service provider network are intermittent users of network service and are therefore assigned IP addresses that change frequently. This conserves the limited pool of unused IP(v4) addresses but also creates a problem if the users of those changeable addresses also want to host web services with a specific domain name which must also have an associated IP address in DNS records. When the IP addresses are static or rarely changed, then the management of domains becomes a lot simpler.

Dynamic DNS (DDNS) is a service that keeps the DNS records of a domain name updated with the current and correct IP address, even if that IP address is subject to frequent changes. A dynamic DNS service automatically updates the DNS records of the domain name anytime its IP address changes. Dynamic DNS updates a DNS server with the current IP address of a client on-the-fly, for example, when the client moves between mobile hot spots or ISPs, or when the IP address is changed administratively. This allows other users wanting to communicate with that domain name resource to use the current correct IP address.

Large organizations have traditionally been able to acquire (or financially afford) unchanging or "static" IP addresses from their ISPs, a practice which has allowed them to operate using standard DNS practices. In contrast, residential users and small businesses (running personal websites, security cameras, digital video recorders (DVRs), small business websites, gaming servers, and so on) tend to obtain IP addresses that are changed quite frequently by their ISPs. So, the latter types of users require solutions based on dynamic DNS to keep their DNS records up to date as IP addresses are changed. Any change in IP address, when not updated in the DNS records of a domain name, can cause DNS queries to that domain name to fail, effectively making that resource unreachable.

For most users, using a dynamic IP address is less expensive and sufficient for their applications. There exist a number of dynamic DNS solutions in the market with varying degrees of features and capabilities. We discuss two of the most commonly used dynamic DNS methods below.

2.8.2.6.2.1 Dynamic DNS Coupled with DHCP Services

To help manage the assignment of the limited unused IP address pool, many organizations use DHCP which allows them to assign IP addresses to their users dynamically. As discussed earlier above, an organization will typically maintain a common pool of IP addresses and assign or "lease" these to active users as needed, for the duration of time they stay active, or until a specified maximum amount of usage time has been reached. Although IPv6 has alleviated the IP address shortage problem significantly, DHCP is still favored by organizations because it is more cost-efficient, and easy to manage than transitioning to IPv6, and using static IPv6 addresses.

Figure 2.24 illustrates one form of dynamic DNS where a DNS server is used as an extension to a DHCP system. In this system, a DNS server holds the DNS records and only authorized DHCP servers update clients DNS records in the DNS server as soon as the IP addresses of the clients change. The system in Figure 2.24 can be used to implement the standards-based dynamic DNS update mechanism described in [RFC2136]. The DHCP server automatically adds or updates the DNS records in the dynamic DNS server as it allocates IP addresses, relieving the network administrator of the task of specifically implementing these updates in the DNS server. This dynamic DNS mechanism is commonly implemented as part of many current DNS client and server software, as well as, in many common operating systems and directory services (including LDAP (Lightweight Directory Access Protocol)).

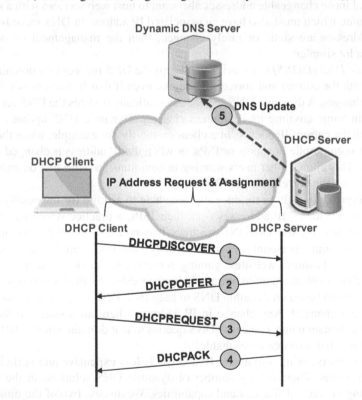

FIGURE 2.24 Dynamic DNS coupled with DHCP services

Dynamic DNS updates, as described in [RFC2136], use the DNS OpCode 5 (i.e., UPDATE) to dynamically remove or add DNS resource records from a DNS zone database maintained in a DNS authoritative name server. Such a mechanism (Figure 2.24) is useful for an authorized DHCP server to register the IP addresses of network clients in the DNS server when they boot up or become available on the network. Given that a DHCP server may assign a different IP address to a client each time it boots, implementing static DNS records for such clients is not possible or easy. Dynamic DNS updates ([RFC2136]) provide a more effective solution.

Dynamic DNS update is mostly used with managed DNS servers along with a security mechanism such as TSIG (Transaction SIGnature) [RFC2845] which is used to provide cryptographic authentication of updates to the DNS database (in the dynamic DNS server). Apart from its use to provide secure dynamic DNS updates, TSIG is also used to provide authentication between trusted DNS peers, for example, to authenticate authorize DNS zone transfer or transfers between a primary DNS server and a secondary/slave DNS server.

2.8.2.6.2.2   Dynamic DNS with Service Provider Registration and Client Software

One very common method of implementing a dynamic DNS service for users in an ISP is to provide the users with specialized DNS client software which runs on their end-systems (home router or gateway, computers, etc.). The specialized client software communicates with the dynamic DNS service provider anytime the IP addresses assigned by the ISP are changed. The dynamic DNS provider in turn almost instantly updates the dynamic DNS records with the new IP address changes.

Unlike the system in Figure 2.24 which provides a standardized method of dynamically updating DNS records, the methods described here are more proprietary, and are provided by a plethora of client and server software in the DNS software market. Figure 2.25 describes two methods for implementing this second category of dynamic DNS. Other than standard computers, the devices in the end-user's private network may consist of IP-based security appliances like IP cameras and DVRs that may require dynamic updates of DNS records in an external DNS server.

Typically, the end-user runs a software client program supplied by a dynamic DNS provider that automates the discovery and registration of the public IP addresses of the end-user's system. The client software program is run on a device or computer (home gateway, home server, etc.) in the end-user's private network. The client software may connect to the dynamic DNS provider's network with a unique/distinct login name (or username). The dynamic DNS provider in turn uses the login name of the end-user to link or associate the discovered public IP address of the private network with a hostname maintained in the domain name system.

Depending on the dynamic DNS provider and the methods used, the hostname (associated with the login name) is registered within a domain owned or belonging to the provider, or within a domain name owned by the customer. The dynamic updates in the DNS server can be performed using a number of mechanisms. Commonly, the DNS provider uses a simple HTTP-based API update service request from the end-user device (the reason being that, even environments that are usually restrictive would allow HTTP service) to send updates to the dynamic DNS server whenever the

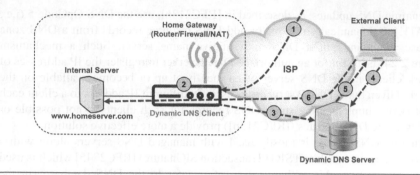

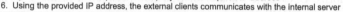

1. ISP assigns a new IP address to home gateway
2. Home gateway updates its external ISP facing interface to use the new IP address
3. Dynamic DNS client in the home gateway sends an update to the Dynamic DNS sever to update its DNS records
4. External client queries the Dynamic DNS server for the IP address of the internal server
5. Dynamic DNS server responds to the query with the required IP address
6. Using the provided IP address, the external clients communicates with the internal server

a) Option 1: Dynamic DNS Client in Home Gateway

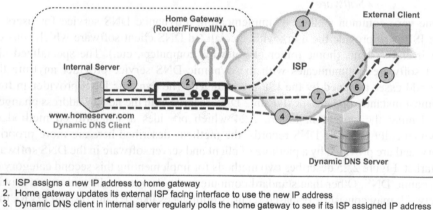

1. ISP assigns a new IP address to home gateway
2. Home gateway updates its external ISP facing interface to use the new IP address
3. Dynamic DNS client in internal server regularly polls the home gateway to see if its ISP assigned IP address has changed
4. If an address changed is detected, Dynamic DNS client in the internal server sends an update to the Dynamic DNS sever to update its DNS records
5. External client queries the Dynamic DNS server for the IP address of the internal server
6. Dynamic DNS server responds to the query with the required IP address
7. Using the provided IP address, the external clients communicates with the internal server

b) Option 2: Dynamic DNS Client in Internal Server

FIGURE 2.25 Dynamic DNS with service provider registration and client software.

client's IP address changes. In these cases, the end-user's devices (home gateway, modem/router, etc.) include client software applications in their firmware that is compatible with the type of dynamic DNS provider used. The DNS provider may also choose to use the dynamic DNS update method in [RFC2136] to update the DNS servers.

To simplify deployment at the end-user, the provider may provide a web-based dynamic DNS service which requires a standard username and password, and requires the user to first create an account at the dynamic DNS server website. The end-user's device is then configured to send updates to the dynamic DNS server whenever its IP address changes.

2.8.2.6.3 Reverse DNS Lookup

A reverse DNS lookup is the process of determining the domain name for a DNS query that provides a specific IP address (Figure 2.26). This reverse lookup process accomplishes the opposite of the DNS lookup process already discussed above in which the DNS resolution returns an IP address to a DNS query supplying the domain name. The latter is the more commonly used approach, and DNS servers are not required to implement the former as an additional feature. Although the relevant IETF standards suggest that compliant DNS servers should be capable of performing reverse DNS lookups, they are generally not implemented or adopted because these are not critical to the normal functioning of internetworks and the Internet.

In a reverse DNS lookup, a query carrying an IP address is sent to a DNS server to obtain a DNS PTR (pointer) record which provides the associated domain name. PTR records are stored in the top-level domain *arpa*. For IPv4 addresses, the domain for reverse lookup is *in-addr.arpa* while for IPv6, the domain is *ip6.arpa*. The IPv4 addresses are stored in PTR records with their dotted segments reversed, and then the string "*.in-addr.arpa*" is appended at the end.

For example, if the domain name *example.com* has an IPv4 address of 192.0.2.3, the PTR record will store that information as *3.2.0.192.in-addr.arpa*. When the DNS server receives a reverse lookup request for 192.0.2.3, it begins by sending a query to the root servers, which point to the name servers of ARIN (American Registry for Internet Numbers) for the *192.in-addr.arpa* zone. If we assume that the ARIN name servers have delegated *3.2.0.192.in-addr.arpa* to hold records for *example.com*, the

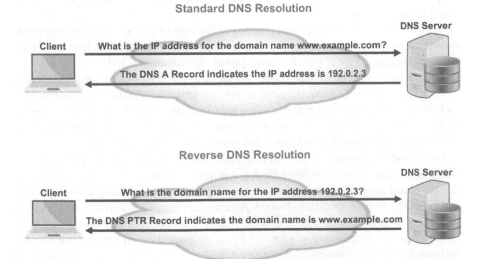

FIGURE 2.26 Standard versus reverse DNS.

DNS server sends another query to *3.2.0.192.in-addr.arpa*, resulting in an authorita-tive DNS response (supplying the domain name *example.com*).

Reverse DNS lookups are very commonly used by message transfer agents (i.e., mail servers). Most mail servers are designed in such a way that they will reject mes-sages sent from any mail server that does not support reverse DNS lookups. This is done to avoid receiving emails from spammers who typically do not use valid IP addresses. So, using reverse DNS lookups, a mail server will check to see if a received message comes from a valid mail server (having a valid IP address) before accepting it to be sent to receivers. Also, some logging software commonly uses reverse DNS lookups to convert IP addresses in their log data to the more human-readable domain names for users as opposed to letting them assimilate the less recognizable numeric IP addresses.

2.9   ENERGY-EFFICIENT DESIGNS

The consumerization of IT is continuously driving new devices and new applications into the home and workplace at an unprecedented rate. Mobility has enabled access to the network from a diverse set of devices. The growing footprint of devices in networks has spurred home users and organizations to take into account their envi-ronmental responsibilities and look for ways to optimize their energy consumption. Saving energy at the home, enterprise, and data centers requires enforcing different sets of energy policies (right from the chip level to the device level).

As microprocessors become more powerful, they also consume more power and generate more heat. The rapid growth in computing power and storage capacity is also creating the environment where the typical data center power consumption (kWh) and power density (kWh/sq. ft.) are both trending upward, placing a strain on many existing data center power distribution and cooling systems. Another aspect of the problem is the growing cost of electricity and the environmental impact of fossil-based energy systems.

With the demand for computing power and the cost of electrical power continuing to escalate, power can be expected to consume a larger share of home user and enter-prise IT budgets. Power provisioning, power consumption, and cooling are now among the top issues in the data center. As these trends continue to unfold, data cen-ter managers are now seeing the need to give careful consideration to the impact of each investment on the power and cooling profile of their facility. In addition, the life cycle costs of power and cooling are becoming increasingly important factors in TCO calculations used to guide selection among competing solutions.

Home users, enterprises, and service providers are increasingly making energy efficiency a critical criterion when buying equipment from vendors. The energy con-sumption and cooling costs for data centers, for example, can make up a significant portion of the overall operations budget. As a result, enterprises and service providers are facing unprecedented challenges to reduce energy, cooling, space, and cost. These pressures are driving vendors to focus on providing equipment with highly optimized energy consumption characteristics. The goal of this drive is to minimize both the cost of electricity and cooling in data centers, service provider, and enter-prise facilities.

To address their energy needs, some companies have even built their facilities in locations that provide access to power and cooling resources at the lowest cost. There is also now a global awareness, and in many cases, lots of national focus in many countries on green technology as a key driver of economic development. Some governments are encouraging purchasing decisions to be based on green technologies and provide some subsidies.

2.9.1 MAKING THE SWITCH/ROUTER ENERGY EFFICIENT

Vendors are increasingly adopting and incorporating certain design techniques into their equipment in order to minimize energy consumption. There are many techniques available for minimizing power in communications equipment, and in many cases, some of these techniques must be incorporated as a design objective from the outset of the product development cycle. The areas of development focus for network equipment are the power system design, ASIC design, functional integration and compaction, and density [FOR10POWMIN09].

2.9.1.1 Power System Design

Resistive and reactive power losses are key issues in the design of power distribution systems. The power and distribution subsystems in network devices such as switches, routers, and switch/routers have to be specifically designed to provide high efficiency with very low loss across a wide range of temperatures. This can be done by utilizing high-quality power supplies optimized for low voltage and high current. The overall power efficiency depends very much on a well-designed power distribution subsystem, which comprises the components between the power supplies and the loads, such as the connectors, backplanes, modules, cabling, and decoupling. The goal is to minimize both resistive and reactive (mostly, capacitive) losses between the power supply and the load(s). This reduces the amount of power losses in the distribution system.

For high-density, chassis-based switch/routers required for large data centers, power efficiency largely depends on the power characteristics of the device's backplane. In addition to providing the physical connectivity for the switch fabric carrying data between the line cards, the backplane serves as the grid that distributes power to the line cards and control modules of the switch/router. For passive copper backplanes [FOR10ESER05], power efficiency is primarily a function of the resistance of the copper traces.

2.9.1.2 ASIC Design

Many energy-efficient designs (e.g., line cards, switch fabric modules, etc.) are implemented using custom ASICs developed by the equipment vendors. The benefits of full-custom design usually include reduced area, performance improvements, and also the ability to integrate analog components and other pre-designed (and thus fully verified) components, such as microprocessor cores that form a system-on-chip (SoC). The custom ASICs facilitate a minimal number of devices on the I/O cards while providing a high degree of functionality with a low power consumption.

A number of techniques have been used to minimize the power consumption of systems based on ASIC architecture [FOR10POWMIN09].

- **Functional Integration**: By utilizing fewer chips, the system can be made to have lower overall power consumption. The level of integration designed into the ASICs can be made extremely high depending on the technology available. Integrating high levels of functionality (Ethernet MACs, serialization/deserialization (SerDes) components, etc.) minimizes the number of physical devices in the system and saves overall power consumption.
- **ASIC Libraries**: ASIC libraries are important tools when designing with ASICs. Such ASIC libraries typically include both high power (highest performance) and low power instances of logic functions. The energy-conscious designer preferably uses low power and high-performance functions in the ASICs.
- **Logic Structure**: Logic synthesis is a process by which an abstract form of the desired circuit behavior, typically register transfer level (RTL), is turned into a design implementation in terms of logic gates. Common examples of this process include synthesis of HDLs, including VHDL and Verilog. During compilation (translation from RTL to gates), the designer has the option to build either lower or higher power structures. Higher power structures tend to be more prolific of gates (usually flattened) as opposed to more hierarchical structures.
- **Floor Planning**: Floor planning is the process of choosing the best grouping and connectivity of logic in a design, and manually placing blocks of logic in the ASIC, with the goal of increasing density, routability, and performance. The intent is to reduce route delays for selected logic by looking for the optimum placement of logic blocks. Power dissipation in an ASIC is often a function of the amount of capacitance that a gate must drive. This can be minimized by careful floor planning.

2.9.1.3 Functional Integration and Compaction

To minimize overall power consumption, the system can be designed to allow better sharing of resources. The alternative, more power-consuming approach is to allow the system components to use resources on a dedicated basis which often leads to the case where more of these resources are needed in the system. For example, devices using large CAMs and RAMs tend to be very high power consuming. The system can be designed to minimize the number of CAMs and RAMs by compacting databases and incorporating multiple lookup tables into a single physical memory array. Typically, CAMs are used for lookup functions associated with forwarding decisions. They include Layer 2/Layer 3 table and ACL lookups. Some of the functions that use CAMs are Layer 2/Layer 3 lookups, ingress, and Layer 2/Layer 3 ACL lookups.

A system can implement all the above functions using a single well-designed CAM system. The shared use of the CAM requires additional software that intelligently manages the CAMs; however, there is a significant overall reduction in system power consumption by doing so. Furthermore, the system can utilize very compact data structures and efficient database implementation (variable length database) that minimizes the amount of RAM required and to reduce power consumption. Additionally, the designer may decide not to use daughter cards on the line cards as they also increase the amount of power loss in the system. Daughter cards require drivers to enable the motherboard and daughterboard to communicate. The drivers consume power, thus, eliminating them also lowers overall power consumption.

2.9.1.4 Port Density

Using system with high port density minimizes rack and wiring closet space but also has additional benefits. Port density is also a primary consideration in minimizing power in any network design. With high port density, networks can process more traffic per card, leaving more chassis space available when bandwidth demand increases. High port density also means very low power contribution per port. In addition, high density means that the (fixed) system power consumption is amortized over greater system capacity (ports). This enables customers to deploy more compact, high-performance, and scalable networks cost effectively. Enterprises and service providers now prefer solutions that provide high bandwidth per rack length.

Low power consumption and loss (and low cost, high port density) has become the design goal of network devices. Vendors continue to develop new techniques to minimize power consumption and loss. Low power consumption and loss translates to much lower operational costs for carriers and data center providers.

2.9.2 ADDRESSING ENERGY EFFICIENCY

To address network energy efficiency, the IEEE has developed a standard pertaining to Energy-Efficient Ethernet (EEE) – IEEE 802.3az. IEEE 802.3az is an amendment to the IEEE 802.3 Ethernet standard that defines mechanisms and protocols for reducing the energy consumption of network links during periods of low data transfer activity, by transitioning network interfaces into a low-power state. EEE defines a set of enhancements to the twisted-pair and backplane Ethernet family of standards that allow less power consumption on network interfaces during periods of low data activity.

IEEE 802.3az defines a method to reduce power consumption on network interfaces without interrupting the network connections they support. The goal of the standard is to ensure maximum efficiency on network interfaces under normal use scenarios, allow vendors to develop equipment designs for lower energy use when operating at lower utilization, and allow network devices to have minimum energy usage over their operational lifetime.

2.10   FUTURE-PROOFING THE NETWORK – SUPPORT FOR INDUSTRY STANDARDS AND IPv6

Given the significant investments that are being directed to enterprise and service provider networks, it makes sense to try to maximize the lifetime of any investment that is being made in the network infrastructure. One way of future-proofing the network infrastructure is deploying network platforms that have the functionality and flexibility to accommodate the application-rich environment of tomorrow. Investment protection is one of the most important factors in the selection of network platforms because the network represents a very large share of the overall investment of an organization. Investment protection involves protecting the network infrastructure against rapid obsolescence and possible "forklift upgrades".

2.10.1 SUPPORT FOR INDUSTRY STANDARDS

An important aspect of investment protection is designing the network based on industry standards rather than on proprietary ones. Network operators prefer network devices based on industry standards because they provide generally good backward compatibility with previous generations of standards. Standards also enable multivendor interoperability and also protect an organization from excessive reliance on a single vendor.

Some network equipment vendors also try to develop a competitive advantage by offering a number of proprietary features aimed at adding value and differentiating their products from the competition. The proprietary features provided are often quite attractive and, frequently, provide the stimulus for industry standardization of similar functionality. A good number of current industry standards evolved from vendor proprietary developments. However, there is a downside in deploying network devices with dominant features that are proprietary, because such features often involve significant hidden costs when there is a transition to equivalent (and often superior) industry standards, as they become available.

Thus, future-proofing the network with devices that are designed based on key industry standards is the best approach, as this allows the network operator to evolve the network to accommodate new standard-based features as they become available. From a product perspective, because of the availability of cheaper components in the market with standardized features (some of them off the shelf), focusing on industry standards significantly simplifies the design of hardware and software, reduces product costs, and improves operational stability.

2.10.2 FUTURE-PROOFING THE NETWORK WITH IPV6

The primary reason for IPv6 is the need to meet the demand for globally unique IP addresses. IPv6 quadruples the number of network address bits from 32 bits (in IPv4) to 128 bits, which provides more than enough globally unique IP addresses for every networked device on the planet. By being globally unique, IPv6 addresses inherently enable global reachability, functionality that is crucial to the applications and services that are driving the demand for the addresses. Additionally, the flexibility of the IPv6 address space reduces the need for private addresses; IPv6 enables new application protocols that do not require special processing by border devices at the edge of networks.

Migration to IPv6 may be inevitable in some networks, however, by starting with the deployment of IPv6-capable hardware, the transition can be more controlled and less disruptive to the network. Many organizations are increasingly deploying IPv6, and deployment in many countries is on the rise. In fact, some government agencies are mandating the purchase of IPv6-capable switches and routers. Therefore, it is important that enterprises and service providers plan to deploy IPv6-capable devices to capitalize on the trending change to IPv6.

Some networks are in the early stages of large-scale IPv6 production deployment. Although the success of IPv6 will ultimately depend on the new applications that run over it, a key part of the IPv6 design is the ability to integrate into and coexist with existing IPv4 devices within the network and across networks during the steady migration from IPv4 to IPv6.

With IPv6-capable network devices, customers can deploy the devices in their networks knowing they are IPv6-capable hardware today, and with future software upgrades, they can support IPv6 routing and advanced IPv6 features down the road. Most vendors are currently offering IPv6-capable network devices, which support the 128-bit addressing format, and also offer new software releases for older IPv4 platforms, offering an easy migration path for interworking IPv4 and IPv6 devices in the network.

2.11   EASE OF USE: PLUG AND PLAY

Switches and switch/routers, typically, those at the network access and directly connected end-user devices, may support the IEEE 802.1AB Link Layer Discovery Protocol (LLDP) [IEEE802.1AB] and ANSI TIA 1057 LLDP-Media Endpoint Discovery (LLDP-MED) [ANSI/TIA1057] standards. These standards enable organizations to deploy interoperable multi-vendor solutions for unified communications. Configuring IP endpoints such as VoIP stations can be a complex task requiring manual and time-consuming configuration. LLDP and LLDP-MED address these challenges, providing organizations with a standard and open method for configuring, discovering, and managing their network infrastructure.

2.11.1   LLDP

LLDP is a Link Layer protocol (specified in IEEE 802.1AB) that is used by network devices to advertise their identities, capabilities, and neighbors on Ethernet LANs. The information gathered with LLDP can be stored in a network device's MIB and queried using SNMP. By examining the hosts in an LLDP-enabled network and querying their MIBs, the topology of the network can be discovered (Figure 2.27). The information that may be retrieved from the device MIBs include system name and description, device port name and description, VLAN ID and name, IP management address, system capabilities (Layer 2 forwarding, Layer 3 forwarding, etc.), MAC and PHY information, Power-over-Ethernet (PoE) Medium Dependent Interface (MDI) power, IEEE 802.3ad Link Aggregation.

LLDP is used as a component in some network management and monitoring applications, allowing devices to pass capabilities and configuration information. For example, network devices use LLDP to advertise PoE capabilities and requirements, and negotiate power delivery.

LLDP is defined as an extensible Layer 2 mechanism that allows network elements to discover and exchange information with neighboring devices. The LLDP

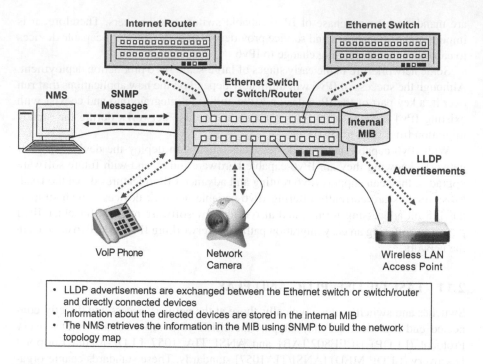

FIGURE 2.27 Example use of LLDP.

protocols help reduce operations costs by simplifying and automating network operations. LLDP greatly simplifies and enhances network management, asset management, and network troubleshooting. For example, it enables discovery of accurate physical network topologies, including those that have multiple VLANs/subnets where all VLANs/subnets may not be known.

LLDP can be used to discover and support "plug-and-play" operation of any LLDP-enabled device connected to an access port, including IP telephones, WLAN access points, network cameras, etc. Each device stores received information in a neighbor database, accessible via the CLI or an LLDP SNMP MIB. Management applications can access the database in each device to build topologies, locate devices (e.g., for Enhanced 911 services), take device inventories, etc. LLDP has the flexibility and extensibility to support a very wide range of management functions, including automated configuration management, fault diagnostics, and power management.

2.11.2 LLDP-MED

An enhancement of LLDP, known as LLDP-Media Endpoint Discovery (LLDP-MED), provides the following facilities:

- Auto-discovery of policies on an Ethernet LAN (such as VLAN, IEEE 802.1p priorities and IP DiffServ settings), enabling plug-and-play networking.

- Device location discovery in a network, allowing the creation of location databases for use with services such as VoIP and Enhanced 911 services.
- Extended and automated power management of PoE endpoints.
- Inventory management, allowing the tracking of network devices and determining their characteristics by a network administrator (e.g., manufacturer, serial or asset number software, hardware versions).

LLDP-MED addresses the unique needs that voice and video communications demand in a converged network by advertising media and IP telephony-specific messages that can be exchanged between the network and the endpoint devices. LLDP-MED defines a set of organizationally specific IEEE 802.1AB TLV extensions and a related MIB module. These allow improved deployment and multi-vendor interoperability between VoIP endpoint devices and IEEE 802 network infrastructure elements. LLDP-MED provides an extensive list of benefits including the following:

- Simplified troubleshooting of end-devices (e.g., IP telephony troubleshooting)
- Inventory management of end-devices (i.e., IP phones, etc.)
- Discovering and maintaining network topologies
- Multi-vendor interoperability and management
- Automatic deployment of policies
- Rapid startup and emergency call service location, identification, and discovery of endpoints to assist in E911 emergency call situations
- Managing unplanned user moves and security policy violations with endpoint move detection notification

For example, LLDP-MED provides an open protocol for configuring QoS, security policies, VLAN assignments, PoE power levels, and service priorities. Additionally, LLDP-MED provides for the discovery of device location and asset identity, information that is used for inventory management and by emergency response services, such as Enhanced 911 (E911). These sophisticated features make converged network services easier to deploy and operate while enabling new and critical services.

2.11.3 OTHER CONFIGURATION MECHANISMS

The switch/routers may support DHCP client-based auto-configuration features, simplifying user deployment and configuration (plug-and-play). Enterprises can use this feature to automate IP address and feature configuration without the presence of a highly trained network administrator on-site. Technicians can simply power up a switch/router and the unit will automatically get its IP address and configuration from DHCP and TFTP servers. These sophisticated features make converged network services easier to install, manage, and upgrade – and they significantly reduce operations costs.

2.12 POWER-OVER-ETHERNET

The IEEE has defined several PoE standards that describe how Ethernet interfaces can pass electric power along with data on twisted-pair Ethernet cabling to end

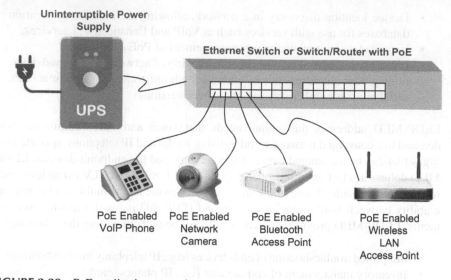

Uninterruptible Power
Supply

Ethernet Switch or Switch/Router with PoE

UPS

PoE Enabled   PoE Enabled   PoE Enabled   PoE Enabled
VoIP Phone   Network   Bluetooth   Wireless
                     Camera   Access Point   LAN
                                                               Access Point

FIGURE 2.28   PoE application example.

devices. POE allows a single Ethernet cable to carry both data and electric power from a switch or switch/router (the Power Sourcing Equipment (PSE)) to devices (the Powered Devices (PDs)) such as VoIP phones, IP cameras, and wireless access points (WAPs) as shown in Figure 2.28.

The PoE standards include the original IEEE 802.3af-2003 PoE standard, the updated IEEE 802.3at-2009 PoE standard also known as PoE+ or PoE plus, the IEEE 802.3bt-2018 standard also known as PoE++ or 4PPoE, and the IEEE 802.3bu-2016 amendment also known as Power over Data Lines (PoDL). PoE supports signaling that allows a PSE to detect the presence of a PoE conformant device, and for the device and the PSE to negotiate the amount of power required or available. In access networks, switches and switch/routers that serve end devices can be designed to support PoE capabilities; PoE is generally not needed on aggregation and backbone switch/routers.

As organizations progressively adopt collaboration as part of their business processes, the number of IP devices on their networks increases. While the productivity and cost-saving benefits of collaboration technologies are apparent, it is also clear that they place some new requirements on the network infrastructure. Customers must expect more from their network than simple packet forwarding. This provides an opportunity to provide flexible power management capabilities in the network devices. New energy cost-saving features on the network devices provide effective energy management and control for the network operator.

As noted above, PoE can provide inline Ethernet power to the IP video surveillance cameras (with higher resolution, plus pan/tilt/zoom capabilities), wireless access points (for higher bandwidth and broader area coverage), and higher-function IP phones. Using PoE eliminates the need for costly rewiring of dedicated power cables and provides centralized power management for these devices.

The primary concerns of IT managers deploying VoIP are: 1) network availability to meet expectations for "always on" communications applications, 2) delay and delay variations, and 3) sufficient bandwidth to support the new application mix. As VoIP and other emerging applications and edge devices are deployed, the edge of the network will need to be capable of evolving to provide additional security, manageability, serviceability, and power provisioning functionality.

PoE is increasingly becoming the default solution for connectivity for the converged desktop, often in combination with Gigabit Ethernet (GE). The capability of switches and switch/routers to deliver high-density, full-power PoE on all ports reduces the need to purchase additional hardware to support the higher power requirements. A switch may support the deployment of IP telephony, wireless LAN, and any third-party Ethernet line-powered device, by offering PoE support on 10/100 Mb/s and 10/100/1000 Mb/s Ethernet interface modules. Many switch vendors offer a wide range of switch models that include PoE-ready base models and PoE upgradeable base models.

PoE based on any of the IEEE PoE standards is a highly convenient means of providing power to VoIP phones, WiFi access points, and other small Ethernet-attached devices, such as surveillance cameras and office hubs. As a result, PoE line cards, together with switch-based utilities to manage the delivery of power to attached devices, have become basic requirements for enterprise applications. IEEE 802.3af-2009 or PoE Plus (PoE+) is an enhanced PoE standard that provides increased power (of at least 30 Watts per port) to Ethernet-powered devices. As the enterprise workspace evolves with more and more end devices for communication, collaboration, security, and productivity, the need for PoE is also evolving to support newer end devices with increased power requirements.

Power conversion efficiency and power efficiency measured in Gbps/Watt are additional aspects of managing power and cooling resources within the switching centers and wiring closet. Power efficiency is already a major concern in the data center and will become more of an issue as port densities increase and more ancillary devices are powered from the wiring closet via PoE.

Some high-end switches support the use of an intelligent power management system to monitor and control the provisioning of power both for the chassis and for inline PoE. The power management system monitors both allocated and consumed power for chassis subsystems, line cards, and PoE ports. A redundancy design ensures that full system power is maintained in the event of a chassis power supply failure. Using the PoE Management LLDP TLV, endpoints can advertise the required power level and the desired power priority.

2.13   IMPROVED PRICE/PERFORMANCE

Price/performance (or price-to-performance) has become one of the most important factors in switch selection for the network designer. Performance is closely linked to reliability with many of the design options for today's networks being a trade-off between performance needs and reliability requirements. Equipment vendors typically achieve improved price/performance by implementing normal routing and forwarding functions using hardware integration and advanced silicon (ASICs, FPGA, etc.).

As networks, the number of network devices, application data, storage requirements, and power consumption continue to rise, demand for higher port density and bandwidth grows. Fortunately, switch vendors favor higher density systems because density and scalability help to drive down the cost per port as well as maximize the deployable lifetime of the system. In particular, high scalability network devices are designed not only to fully exploit current technology but also to accommodate future technology developments and changes. Device scalability that addresses future technology developments allows enhancement of the device by a staged program of subsystem upgrades (e.g., line cards, switch fabric modules, and route processor modules), while protecting prior investments in the chassis, power, cooling, and backplane. Organizations looking to reduce TCO need solutions with higher scalability and density per rack unit that consume less power and dissipate less heat.

Network device vendors address those needs with state-of-the-art ASICs, front-to-back airflow, automatic fan speed control, and power-efficient optics to ensure the most efficient use of power and cooling. For low-cost, low-latency (0.25 µs), and low-energy-consuming (0.1 watts) cabling within and between the racks, the switch could support direct attached SFP+ copper (Twinax) cabling at up to 10 meters. For switch-to-switch connectivity, the switch could support low-power-consuming (1.0 watts) SFP+ optics at up to 300 meters. In high-port-density deployments, these features save significant operating costs.

Recent innovations in multi-gigabit Ethernet switching (10 GE, 25 GE, 40 GE, 50 GE, 100 GE, 200 GE, and higher) and routing have changed the networking landscape significantly. However, when 10 Gigabit Ethernet was first introduced, it was expensive and the latency of 10 Gigabit Ethernet switches was high. The situation has changed now, and the price of 10 Gigabit Ethernet network interface cards (NICs) has fallen dramatically and their latency performance has improved substantially. These innovations have driven latency to tens of nanoseconds, enabling Ethernet to compete both on throughput and performance. These continued improvements are removing the price and latency advantage of legacy technologies like those based on SONET/SDH, Infiniband, and Fibre Channel.

The above developments are clearing the way for simplifying the technology makeup of enterprise and service provider networks (including data centers), and leveraging the cost-effectiveness of Ethernet to minimize TCO without any compromise in performance. As data centers move toward virtualized applications and infrastructure, the combination of lower port prices and lower latency has become crucial drivers of the adoption of multi-gigabit Ethernet as the converged data center switching technology.

2.14 PORTABILITY OF SWITCH/ROUTER OPERATING SYSTEM ACROSS MULTIPLE PLATFORMS

One of the most effective ways to simplify vendor product development, and customer equipment deployment and management, is reducing the number of different types of operating systems and software versions that are deployed in a network. Definitely, network operators always want to simplify the management of their network infrastructures by reducing the number of different types of devices and software versions that are deployed. They prefer to consolidate their networks by reducing the number of equipment vendors and the number of different models of devices from each vendor.

Network operators do not like dealing with multiple, divergent network operating systems, even from the same equipment vendor. They seek to minimize the number of operating systems and operating system releases in use for each switch or router vendor. The reality is, no single family of devices can scale or has the features to meet the diverse needs of the wiring closets, data centers, and network cores across the range of sites typically found in large organizations. The hardware features across a single device family would be very different but the operating system on the other hand could be, beneficially, streamlined to run across the multiple platforms.

Vendors offering multiple platform products have addressed the problem of network operating system complexity by leveraging the concept of hardware abstraction to port their operating system to the different platforms [FOR10HAL08]. This allows network operators the option of running a single operating system across all the switches and routers in the network. They can run the same operating system end-to-end in their Layer 2/Layer 3 network infrastructure from the access/aggregation tier to the network core and the data center. With a single operating system deployed end-to-end across the network, network operators have the flexibility to simplify operation and management of the network.

To support portability of an operating system across multiple platforms, vendors typically implement a hardware abstraction layer in the operating system. The hardware abstraction layer provides abstraction of the capabilities of the different hardware platforms. With hardware abstraction, porting the operating system to a new hardware platform is accomplished without rewriting the core of operating system (i.e., the modular processes and kernel) that comprises the majority of the codebase. New code that is required for porting is decoupled from the core code by the abstraction layer and is restricted to hardware-specific interfaces on the hardware abstraction layer. With operating system portability across the entire vendor switch/router product line, a single switch/router operating system can span the entire network.

A hardware abstraction layer is a layer of software that decouples the kernel of an operating system from the specific details of the underlying hardware. By hiding the differences in the underlying hardware, the hardware abstraction layer isolates the operating system kernel source code from a major rewrite whenever there are changes in the hardware platform. Most major operating systems like Windows, Mac OS X, Linux, BSD Unix, Solaris, and other portable operating systems employ the concept of hardware abstraction. Hardware abstraction allows these operating systems to be readily adapted to employ a wide range of hardware subsystems (e.g., storage, sound, or video), run on different generations of the same microprocessor architecture, and even be ported to completely different microprocessor architectures.

Offering a single operating system across multiple platforms reduces the complexity of the network operating system environment and has a number of advantages for the vendor and end-user [FOR10HAL08].

2.14.1 Vendor Advantages

The advantages of using one operating system across different platforms from the vendor perspective are described here:

- **Product Development and Roll-out**: One operating system with a single code base provides a single accelerated path for rolling out new products and software features. Software designers can work on cross product development with a single control plane architecture for all switch/router platform. A single operating system allows all software testing resources to a directed to a single operating system release, greatly enhancing the efficiency and speed of the entire test cycle. A single code base also results in fewer discontinuities in features/functions across platforms. Through this, end users benefit from a more stable, reliable operating system with fewer software errors and system restarts.
- **Cross Platform Protocol Interoperability**: A single operating system with only one Layer 2/Layer 3 protocol stack across all switch/router models and platforms, simplifies interoperability testing and verification – no need for extensive testing between product operating systems and protocols. The result is better interoperability among devices.
- **Technical Support**: A codebase based on a single operating system streamlines diagnostics, troubleshooting, product end-user training, and software management and upgrades. It simplifies operating system software patch management, patch processes, and version management because only one patch process and one bug-fix process are required to support multiple product families. Also, a single codebase simplifies tracking new releases, features/feature changes, and security alerts.
- **Solving Security Vulnerabilities**: With a single operating system across platforms, security vulnerabilities and holes are easier to resolve within a single code base rather than in a plethora of parallel operating system releases.

2.14.2 End-User Advantages

In this section, we describe the advantages using one operating system across different platforms from the end-user perspective:
- **Network Management and Administration**: Deploying multiple network operating systems across the network adds considerable complexity to a wide range of network management tasks. It is much better to have a single operating system across multiple products (cross-platform operating system) to simplify network design and facilitate consistent implementation of end-to-end QoS and security policies in the network. With this, the exact same access control, ACL, and roles-based security frameworks can be utilized across all platforms, allowing the deployment of a single,

comprehensive, network-wide security model. This allows for a simpler recovery from security vulnerabilities and intrusions.

A single operating system also eliminates the need for interfacing with diverse CLIs, MIBs, SNMP traps, diagnostics, and instrumentation features. It allows network management with a single interface and management framework spanning the entire network. A common CLI and management model across all models of switch/router simplifies network management and provides the opportunity to automate management functions. This allows consistent troubleshooting procedures and hardware diagnostics to be performed across all platforms. This also results in simplified trouble-shooting and fault management, contributing to a significant reduction in network operational expense, which is a major component of the TCO for a network.

A common management system and operating system across all products is essential for simplifying configuration management and operational support across the network. A high percent of network downtime is related to human error, much of this coming from operational errors. Management and operating system consistency greatly reduces these errors and their associated costs.

- **Feature Upgrade**: Deploying a single operating system allows newly developed common features to be released and installed on all platforms very quickly. Also, a single operating system greatly simplifies the planning, staging, and deploying of operating system upgrades to take advantage of new features. In addition, patching and upgrade processes are greatly simplified because the network operator does not need to deal with the complexity of multiple operating systems versions in the network.
- **Product Deployment and Protocol Interoperability**: In a multiple platform environment, a single common Layer 2/Layer 3 protocol stack ensures interoperability across all platforms. A switch/router software environment that features higher reliability and more consistent functionality and management capabilities, contributes significantly to faster interoperability testing and network operational uptime. System downtime due to network operator errors is reduced because of the commonality of configuration management functionality and user interfaces across the switch/router product spectrum.

REVIEW QUESTIONS

1. What is a managed device in SNMP?
2. Explain briefly the main functions of an SNMP manager, SNMP agent, and a MIB in SNMP.
3. Why is SNMPv3 considered the more secure version of SNMP?
4. What are the main functions of an SNMPv2 proxy agent?
5. What is a bilingual Network Management System (NMS) in SNMP?

6. What is the difference between an SNMP *Trap* message and an SNMP *InformRequest* message?
7. What is authentication, authorization, and accounting in networking?
8. Explain briefly what RADIUS and TACACS+ are used for and what their main differences are.
9. What is the purpose of IEEE 802.1X in networking?
10. What are the three main entities involved in IEEE 802.1X authentication? Describe briefly their roles.
11. What is MAC Authentication Bypass (MAB)?
12. What is Secure Shell (SSH) and why is it preferred over protocols such as TELNET and rlogin?
13. What are the uses of sFlow, NetFlow, and RMON in networking?
14. What are the primary functions of DHCP in networking?
15. Describe briefly the four main phases or stages in DHCP operation.
16. Why is it not possible for a DHCP server in one subnet to serve DHCP clients in different subnets?
17. What is a DHCP relay agent used for in networking?
18. Describe two methods for DHCP IP address conflict detection and resolution.
19. What is the difference between standard (forward) DNS and Reverse DNS?
20. What is a DNS record?
21. Explain the differences between a DNS resolver, DNS root name server, DNS top-level domain name server, and a DNS authoritative name server.
22. What is a DNS zone transfer?
23. What is the difference between a non-recursive, recursive, and iterative DNS query?
24. What are the advantages of caching in DNS name resolution?
25. Why is a DNS record stored in a DNS caching system (host, resolver, or server) assigned a time-to-live (TTL)?
26. What is Dynamic DNS (DDNS)?
27. What four techniques are used to minimize the power consumption of systems based on ASIC architecture?
28. What is IEEE 802.1AB Link Layer Discovery Protocol (LLDP) used for in networking?
29. What is Power-over-Ethernet (PoE)?
30. What are the advantages of using one operating system across different platforms from the network equipment vendor perspective?
31. What are the advantages using one operating system across different platforms from the end user perspective?

REFERENCES

[ANSI/TIA1057]. ANSI/TIA-1057-2006, Link Layer Discovery Protocol for Media Endpoint Devices.
[BARSILSSH01]. Daniel J. Barrett and Richard E. Silverman, *SSH, The Secure Shell: The Definitive Guide*, ISBN: 0-596-00011-1, O'Reilly, 2001.

[CISC2TMANM11]. Cisco Catalyst 6500 Series Supervisor Engine 2T: Management and Monitoring, Cisco Systems, *White Paper*, 2011.

[CISC2TMUL11]. Building Next-Generation Multicast Networks with Supervisor 2T, Cisco Systems, *White Paper*, April 13, 2011.

[CISC2TNETF11]. Cisco Catalyst 6500 Supervisor Engine 2T: NetFlow Enhancements, Cisco Systems, *White Paper*, March 5, 2011.

[CISC2TQOS11]. Catalyst 6500 Sup2T System QOS Architecture, Cisco Systems, *White Paper*, April 13, 2011.

[CISCID13838]. TACACS+ and RADIUS Comparison, Cisco Systems, *Document ID: 13838*, January 14, 2008.

[DRAFTTACACS+]. T. Dahm, A. Ota, D. Medway Gash, D. Carrel, L. Grant, "The TACACS+ Protocol", *Internet-Draft, draft-ietf-opsawg-tacacs-05*, August 20, 2016,

[FOR10ESER05]. Force10 Networks, "The Force10 E-Series Architecture", *White Paper*, 2005.

[FOR10HAL08]. Force10 Networks, "The Hardware Abstraction Layer: Enabling FTOS to Span the Switching and Routing Infrastructure with a Consistent Feature Set and Unified Management", *White Paper*, 2008.

[FOR10POWMIN09]. Force10 Networks, "Building Scalable, High Performance Cluster and Grid Networks: The Role of Ethernet", *White Paper*, 2006.

[IEEE802.1AB]. IEEE Std 802.1AB-2005 - Station and Media Access Control Connectivity Discovery.

[IEEE 802.1X]. IEEE Std 802.1X-2004 - Port-Based Network Access Control.

[RFC792]. J. Postel, "Internet Control Message Protocol", *IETF RFC 792*, September 1981.

[RFC826]. David C. Plummer, "An Ethernet Address Resolution Protocol", *IETF RFC 826*, November 1982.

[RFC1013]. Robert W. Scheifler, "X Windows Systems Protocol, Version 11", *IETF RFC 1013*, June 1987.

[RFC1034]. P. Mockapetris, "Domain Names - Concepts and Facilities", *IETF RFC 1034*, November 1987.

[RFC1035]. P. Mockapetris, "Domain Names - Implementation and Specification", *IETF RFC 1035*, November 1987.

[RFC1155]. M. Rose and K. McCloghrie, "Structure and Identification of Management Information for the TCP/IP-based Internets", *IETF RFC 1155*, May 1990.

[RFC1157]. J. Case, M. Fedor, M. Schoffstall, and J. Davin, "A Simple Network Management Protocol (SNMP)", *IETF RFC 1157*, May 1990.

[RFC1213]. K. McCloghrie and M. Rose, "Management Information Base for Network Management of TCP/IP-based Internets: MIB-II", *IETF RFC 1213*, March 1991.

[RFC1441]. J. Case, K. McCloghrie, M. Rose, and S. Waldbusser, "Introduction to version 2 of the Internet-standard Network Management Framework", *IETF RFC 1441*, April 1993.

[RFC1546]. C. Partridge, T. Mendez, and W. Milliken, "Host Anycasting Service", *IETF RFC 1546*, November 1993.

[RFC1901]. J. Case, K. McCloghrie, M. Rose, and S. Waldbusser, "Introduction to Community-based SNMPv2", *IETF RFC 1901*, January 1996.

[RFC1909]. K. McCloghrie, Ed., "An Administrative Infrastructure for SNMPv2", *IETF RFC 1909*, February 1996.

[RFC1910]. G. Waters, Ed., "User-based Security Model for SNMPv2", *IETF RFC 1910*, February 1996.

[RFC1928]. M. Leech, M. Ganis, Y. Lee, R. Kuris, D. Koblas, and L. Jones, "SOCKS Protocol Version 5", *IETF RFC 1928*, March 1996.

[RFC2131]. R. Droms, "Dynamic Host Configuration Protocol", *IETF RFC 2131*, March 1997.

[RFC2132]. S. Alexander and R. Droms, "DHCP Options and BOOTP Vendor Extensions", *IETF RFC 2132*, March 1997.

[RFC2136]. P. Vixie, S. Thomson, Y. Rekhter, and J. Bound, "Dynamic Updates in the Domain Name System (DNS UPDATE)", *IETF RFC 2136*, April 1997.

[RFC2151]. G. Kessler and S. Shepard, "A Primer On Internet and TCP/IP Tools and Utilities", *IETF RFC 2151*, June 1997.

[RFC2526]. D. Johnson and S. Deering, "Reserved IPv6 Subnet Anycast Addresses", *IETF RFC 2526*, March 1999.

[RFC2572]. J. Case, D. Harrington, R. Presuhn, and B. Wijnen, "Message Processing and Dispatching for the Simple Network Management Protocol (SNMP)", *IETF RFC 2572*, April 1999.

[RFC2578]. K. McCloghrie, D. Perkins, and J. Schoenwaelder, "Structure of Management Information Version 2 (SMIv2)", *IETF RFC 2578*, April 1999.

[RFC2613]. R. Waterman, B. Lahaye, D. Romascanu, S. Waldbusser, "Remote Network Monitoring MIB Extensions for Switched Networks Version 1.0", *IETF RFC 2613*, June 1999.

[RFC2743]. J. Linn, "Generic Security Service Application Program Interface Version 2, Update 1", *IETF RFC 2743*, January 2000.

[RFC2819]. S. Waldbusser, "Remote Network Monitoring Management Information Base", *IETF RFC 2819*, May 2000.

[RFC2845]. P. Vixie, O. Gudmundsson, D. Eastlake 3rd, and B. Wellington, "Secret Key Transaction Authentication for DNS (TSIG)", *IETF RFC 2845*, May 2000.

[RFC2865]. C. Rigney, S. Willens, A. Rubens, W. Simpson "Remote Authentication Dial In User Service (RADIUS)", *IETF RFC 2865*, June 2000.

[RFC2866]. C. Rigney, "RADIUS Accounting", *IETF RFC 2866*, June 2000.

[RFC2882]. D. Mitton, "Network Access Servers Requirements: Extended RADIUS Practices", *IETF RFC 2882*, July 2000.

[RFC3176]. InMon Corporation's sFlow, "A Method for Monitoring Traffic in Switched and Routed Networks", *IETF RFC 3176*, September 2001.

[RFC3411]. D. Harrington, R. Presuhn, and B. Wijnen "An Architecture for Describing Simple Network Management Protocol (SNMP) Management Frameworks", *IETF RFC 3411*, December 2002.

[RFC3412]. J. Case, D. Harrington, R. Presuhn, and B. Wijnen, "Message Processing and Dispatching for the Simple Network Management Protocol (SNMP)", *IETF RFC 3412*, December 2002.

[RFC3414]. U. Blumenthal and B. Wijnen, "User-based Security Model (USM) for version 3 of the Simple Network Management Protocol (SNMPv3)", *IETF RFC 3414*, December 2002.

[RFC3415]. R. Presuhn, B. Wijnen, and K. McCloghrie, "View-based Access Control Model (VACM) for the Simple Network Management Protocol (SNMP)", *IETF RFC 3415*, December 2002.

[RFC3416]. R. Presuhn, "Version 2 of the Protocol Operations for the Simple Network Management Protocol (SNMP)", *IETF RFC 3416*, December 2002.

[RFC3417]. R. Presuhn, "Transport Mappings for the Simple Network Management Protocol (SNMP)", *IETF RFC 3417*, December 2002.

[RFC3418]. R. Presuhn, "Management Information Base (MIB) for the Simple Network Management Protocol (SNMP)", *IETF RFC 3418*, December 2002.

[RFC3576]. M. Chiba, M. Eklund, D. Mitton, and B. Aboba, "Dynamic Authorization Extensions to Remote Authentication Dial In User Service (RADIUS)", *IETF RFC 3576*, July 2003.

[RFC3577]. S. Waldbusser, R. Cole, C. Kalbfleisch, and D. Romascanu, "Introduction to the Remote Monitoring (RMON) Family of MIB Modules", *IETF RFC 3577*, August 2003.

[RFC3584]. R. Frye, D. Levi, S. Routhier, and B. Wijnen, "Coexistence between Version 1, Version 2, and Version 3 of the Internet-standard Network Management Framework", *IETF RFC 3584*, August 2003.

[RFC3748]. B. Aboba, L. Blunk, J. Vollbrecht, J. Carlson, and H. Levkowetz, Ed., "Extensible Authentication Protocol (EAP)", *IETF RFC 3748*, June 2004.

[RFC3954]. B. Claise, Ed., "Cisco Systems NetFlow Services Export Version 9", *IETF RFC 3954*, October 2004.

[RFC4033]. R. Arends, R. Austein, M. Larson, D. Massey, and S. Rose, "DNS Security Introduction and Requirements", *IETF RFC 4033*, March 2005.

[RFC4034]. R. Arends, R. Austein, M. Larson, D. Massey, and S. Rose, "Resource Records for the DNS Security Extensions", *IETF RFC 4034*, March 2005.

[RFC4035]. R. Arends, R. Austein, M. Larson, D. Massey, and S. Rose, "Protocol Modifications for the DNS Security Extensions", *IETF RFC 4035*, March 2005.

[RFC4120]. C. Neuman, T. Yu, S. Hartman, and K. Raeburn, "The Kerberos Network Authentication Service (V5)", *IETF RFC 4120*, July 2005.

[RFC4121]. L. Zhu, K. Jaganathan, and S. Hartman, "The Kerberos Version 5 Generic Security Service Application Program Interface (GSS-API) Mechanism: Version 2", *IETF RFC 4121*, July 2005.

[RFC4251]. T. Ylonen, and C. Lonvick, Ed., "The Secure Shell (SSH) Protocol Architecture", *IETF RFC 4251*, January 2006.

[RFC4252]. T. Ylonen, and C. Lonvick, Ed., "The Secure Shell (SSH) Authentication Protocol", *IETF RFC 4252*, January 2006.

[RFC4253]. T. Ylonen, and C. Lonvick, Ed., "The Secure Shell (SSH) Transport Layer Protocol", *IETF RFC 4253*, January 2006.

[RFC4254]. T. Ylonen, and C. Lonvick, Ed., "The Secure Shell (SSH) Connection Protocol", *IETF RFC 4254*, January 2006.

[RFC4256]. F. Cusack and M. Forssen, "Generic Message Exchange Authentication for the Secure Shell Protocol (SSH)", *IETF RFC 4256*, January 2006.

[RFC4291]. R. Hinden and S. Deering, "IP Version 6 Addressing Architecture", *IETF RFC 4291*, February 2006.

[RFC4419]. M. Friedl, N. Provos, and W. Simpson, "Diffie-Hellman Group Exchange for the Secure Shell (SSH) Transport Layer Protocol", *IETF RFC 4419*, March 2006.

[RFC4462]. J. Hutzelman, J. Salowey, J. Galbraith, and V. Welch, "Generic Security Service Application Program Interface (GSS-API) Authentication and Key Exchange for the Secure Shell (SSH) Protocol", *IETF RFC 4462*, May 2006.

[RFC4786]. J. Abley and K. Lindqvist, "Operation of Anycast Services", *IETF RFC 4786*, December 2006.

[RFC5227]. S. Cheshire, "IPv4 Address Conflict Detection", IETF RFC 5227, July 2008.

[RFC5646]. A. Phillips, and M. Davis, Ed., "Tags for Identifying Languages", *IETF RFC 5646*, September 2009.

[RFC6353]. W. Hardaker, "Transport Layer Security (TLS) Transport Model for the Simple Network Management Protocol (SNMP)", *IETF RFC 6353*, July 2011.

[RFC7015]. B. Trammell, A. Wagner, and B. Claise, "Flow Aggregation for the IP Flow Information Export (IPFIX) Protocol", *IETF RFC 7015*, September 2013.

[RFC8415]. T. Mrugalski, M. Siodelski, B. Volz, A. Yourtchenko, M. Richardson, S. Jiang, T. Lemon, and T. Winters, "Dynamic Host Configuration Protocol for IPv6 (DHCPv6)", *IETF RFC 8415*, November 2018.

[SCHEIXWIN96]. Robert W. Scheifler and James Gettys, *X Window System: Core and Extension Protocols*, ISBN 1-55558-148-X, X version 11, Releases 6 and 6.1, Digital Press 1996.

[STEINXWIN88]. Jennifer G. Steiner and Daniel E. Geer, Jr., "Network Services in the Athena Environment", *Proceedings of the Winter 1988 Usenix Conference*, 21 July 1988.

3 Designing the Switch/Router
A Design Example and What to Consider When Designing a Switch/Router

3.1 INTRODUCTION

High-performance switch/routers have substantially higher packet processing capabilities that enable the support of high-speed packet forwarding, and advanced quality of service (QoS) and security processing features. Some of the key functions in high-performance switch/routers include performing Layer 2 and 3 header modifications, packet classification, policing and shaping of flows, and queuing and scheduling, all at line-speed rates. In this chapter, we discuss the design and architectural considerations, as well as the typical processes and steps used to build practical switch/routers. We describe how a switch/router can be built by coalescing all the concepts and methods discussed in the previous chapters and Volume 1 of this two-part book.

This chapter reviews the common building blocks of the typical high-performance switch/router, and identifies the various components within the switch/router that require special attention during the design stage. To understand the design of a switch/router, we examine the basic functionality and components of both the route processor and the packet forwarding engine. The discussion also includes the components that are used in configuring, managing, and monitoring a switch/router.

3.2 HIGH-LEVEL ARCHITECTURE OF A HIGH-PERFORMANCE SWITCH/ROUTER

Designing a switch/router requires a clear understanding of the various building blocks that make up the system and their interaction. This allows the designer to determine the various stress points and weakest links in the packet processing chain within the device, and to identify the best system components and sub-components, as well as determine the best design approach for the overall system.

DOI: 10.1201/9781003311256-3

The following are the key concerns a designer must address when designing a high-performance switch/router:

- Forward Layer 2 and 3 unicast and multicast traffic at line-speed with maximum throughput and low latency. A key concern here is the ability to forward at line-speed Layer 2 and 3 traffic with a minimum packet size of 64 bytes (for IPv4) or 84 bytes (for IPv6 due to the larger IPv6 header).
- Ensure that the forwarding of multicast traffic does not degrade the forwarding performance of unicast traffic.
- Filter packets at line speed based on MAC addresses, IP addresses, TCP or UDP port numbers, or a combination of the source and destination IP addresses and Transport Layer port numbers (i.e., the N-tuple). This includes filtering based on access control lists (ACLs).
- Perform traffic classification and prioritization based on class-of-service/quality-of-service (CoS/QoS) markings. This includes handling CoS/QoS markings based on IEEE 802.1p or Differentiated Service Code Point (DSCP) (see discussion in Chapter 8 of Volume 1).
- Police and shape traffic based on user-defined rate limits (see discussion in Chapter 9 of Volume 1).
- Priority scheduling based on scheduling algorithms such as strict priority, weighted round-robin (WRR), and weighted fair queuing (WFQ) (see discussion in Chapter 8 of Volume 1).

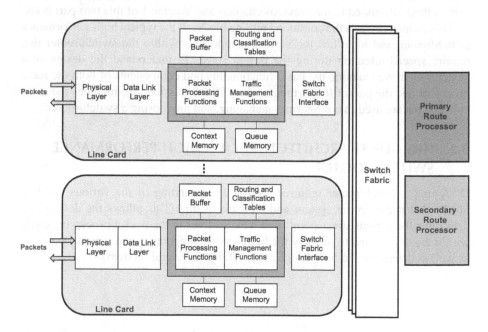

FIGURE 3.1 Components of a high-performance switch/router.

- Prevent Head-of-Line (HOL) blocking when using switch fabrics such as crossbar switch fabrics (see discussion in Chapter 1 of this volume).

Figure 3.1 shows a high-level architecture and the major components of the typical high-performance switch/router, including the line card, the switch fabric module (SFM), and the route processor module (RPM), also called the controller card. We provide in the following sections and subsections, descriptions of the main components that are crucial to the correct functioning of such a switch/router.

3.3 THE LINE CARD

The line card of the switch/router supports circuitry and processors that provide one or more transmit/receive interfaces (or ports) for attachment to an external LAN, MAN, or WAN. An interface may support copper or optical fiber media (see "Ethernet Physical Layer Types" section in Chapter 6 of this volume). In some cases, an interface may support a wireless medium (e.g., microwave interface as in mobile backhaul networks). Figure 3.1 shows the major components in the line card of the typical high-performance switch/router. The line card communicates with other line cards and also with the RPM via the SFM.

3.3.1 PACKET BUFFER

A packet that arrives at a line card is stored in the packet buffer memory shown in Figure 3.1. This is after the line card has performed the relevant Layer 2 packet sanity checks to determine if the packet should be received and processed further (see "Data Plane" section in Chapter 5 of Volume 1). The packet buffer in Figure 3.1 serves as temporary storage for packets arriving at the line card while they wait to be processed. Because the packet processor performs many tasks and also, several input ports may be contending for a particular destination port, when a packet arrives on a line card interface, it is held in the packet buffer until the packet processor is ready to process it. The packet header is copied into a processor memory for processing as soon as the packet processor is ready for a new packet.

3.3.2 PACKET PROCESSING FUNCTIONS – PACKET PROCESSOR

The packet processing functions (simply called, the packet processor) can be implemented as a programmable device such as Network Processing Unit (NPU) or an optimized special-purpose ASIC. The packet processor (which operates in the fast path of the data plane) is responsible for receiving packets from an interface, processing, and forwarding them over the switch fabric to other line cards or the route processor.

The packet processor performs specific key tasks such as parsing the header of a received packet, performing destination IP address lookup in the forwarding table (or FIB), performing packet classification (or pattern matching), performing the modification of certain fields in the packet, and forwarding the packet to other modules in the switch/router via the switch fabric.

3.3.3 Routing and Classification Tables

Typically, the switch/router would support a special memory that holds the routing and classification tables used by the packet processor. This memory may contain the following databases:

- **IP Forwarding Table (or FIB)**: This database contains the information used to route incoming packets to their correct outgoing interfaces and next-hop nodes.
- **Access Control Lists (ACLs)**: These databases contain information used to filter packets (i.e., grant or deny a packet access to a specific network).
- **Flow Classification Table**: This database contains information used to identify a particular stream of packets (or flow) which can be, a particular user, group of users, protocols, or application. Flow identification information is used for QoS or CoS control, traffic policing, traffic shaping, per-flow queuing, and billing.
- **Label Table**: This database contains information about VLANs (as defined in IEEE 802.1Q), stacked VLANs (also called Q-in-Q as defined in IEEE 802.1ad), Multiprotocol Label Switching (MPLS) labels, etc.

When the header of a packet in the packet buffer is copied into the processor memory, the packet processor (NPU or specialized ASIC) parses the packet header and then performs IP destination address lookup in the FIB, in addition to packet and flow classification. The packet processor does this to determine whether the packet should be forwarded or filtered (i.e., discarded). The information stored in the local tables, as described above, is maintained by the route processor, which operates in the control plane. The profiles or rules in the ACLs pertain to a given flow or packet and are used for packet filtering, while those in the flow classification table are for QoS and CoS control.

3.3.4 Context Memory – Route/Flow Cache Memory

The context memory contains instructions that indicate whether a received packet should be forwarded or denied, the system module (line card or route processor) to which a packet should be forwarded, the internal system headers (i.e., internal packet tags) needed to move a packet from its ingress interface to the correct egress interface, and the external packet header information for an outgoing packet (i.e., new source MAC address, new destination MAC address, outer IP address if tunneling is to be performed, new VLAN tags, MPLS stacks, etc.).

The context memory also holds information relevant to metering packets in a flow, and other miscellaneous information needed to process a packet. During FIB lookup and classification, the process typically points to many pieces of information usually stored in the context memory about the packet being processed. The context memory can be viewed as a form of a cache memory (i.e., a route cache) that holds instructions for routing packets that have the same forwarding characteristics as discussed in section "Second Generation Routing Devices" of Chapter 2 of Volume 1, and section "Architectures Using Route Cache Forwarding" of Chapter 6 of Volume 1.

3.3.5 Traffic Management Functions

The traffic management functions are responsible for controlling the flow of traffic, and forwarding traffic according to user-defined rules (pertaining to latency and bandwidth guarantees, traffic priority levels, and system and network congestion levels). These functions also allocate the buffering required for any queuing mechanism configured to manage traffic flow across the switch fabric. The traffic management functions may also coordinate with the switch fabric interface (on the line card) that connects the line card to the switch fabric. Figure 3.2 shows the major traffic management functions in the line card of the typical high-performance switch/router.

The traffic management functions include functions for advanced queuing, congestion management, and hierarchical scheduling of traffic, possibly, handling a large number of flows. Some functions are responsible for forwarding traffic according to a set of user-defined rules as mentioned above.

The traffic management functions interface with the switch fabric through the switch fabric interface which supports a Serializer/Deserializer (SerDes) function. Depending on the architecture, communication between the line card and the switch fabric may require flow control (or backpressure), which means, additional flow control information may have to be exchanged between the line card and switch fabric, creating some overhead and increasing switch fabric bandwidth requirements.

3.3.5.1 Metering and Statistics Collection

The packet metering functions (traffic policing and shaping) are responsible for checking if a specific flow is conformant to a subscribed traffic policy or contract [RFC2475]. The information obtained from metering a flow is used to decide whether to forward or discard a packet belonging to that flow. A contract may establish bandwidth allocations that can include specifying per-flow, steady-state, and short-term

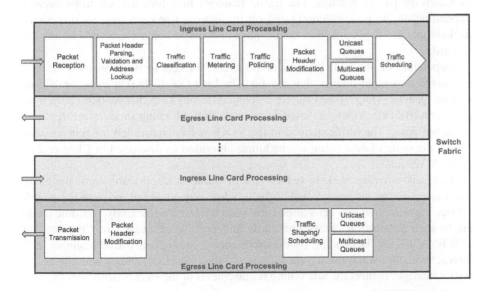

FIGURE 3.2 Traffic management functions in a high-performance switch/router.

peak (traffic burst) rates. As discussed in Chapter 9 of Volume 1, typically, a Single-Rate Two-Color Marker (or Token Bucket Algorithm), Single-Rate Three-Color Marker (srTCM) [RFC2697], or Two-Rate Three-Color Marker (trTCM) [RFC2698] is used at the ingress interface to check if a flow is compliant to its contract.

During metering, packets that violate the assigned contract can be marked for treatment in various ways including the following:

- Unconditionally drop the packet
- Drop packet if there are no internal resources available to handle the packet
- Pass on the packet to the downstream device to make the dropping decision.

The egress line card can also mark packets to allow downstream devices to decide whether to drop such packets when they experience local resource oversubscription, or when local traffic metering indicates that the packet stream does not conform to some defined traffic profile. The egress line card may also perform traffic shaping as discussed in Chapter 9 of Volume 1. Traffic shaping is a mechanism used to smooth out the flow of packets; it is used to regulate the rate and volume of traffic sent into the external network.

3.3.5.2 Traffic Queuing and Scheduling

For each new packet that is received on an interface, the packet processor will parse the packet header and perform IP destination address lookup and other classification tasks. The packet processor then appends an internal forwarding header or tag to each packet before forwarding the packet to the traffic management functions (referred to here as the traffic manager). The internal forwarding tag contains information that indicates to the ingress and egress traffic managers, the egress line card and inter-face (port) to which the packet should be sent to, and the characteristics of the flow to which the packet belongs. The traffic manager then forwards or drops packets depending on the policies configured on the ingress line card (ingress direction); packets that are flagged as non-conformant to a traffic contact may be eligible for discard.

Depending on the switch/router architecture, packets to be forwarded may be organized into input queues that are First-In First-Out (FIFO) queues, or Virtual Output Queues (VOQs) (see Chapter 2 "Understanding Crossbar Switch Fabrics" in [AWEYA2BK19]). VOQs are used to prevent HOL blocking in architectures with input buffering. The traffic queued in the VOQs is then scheduled for transmission across the switch fabric using a scheduling algorithm as discussed in Chapter 2 of [AWEYA2BK19].

Typically, arriving packets (which are of variable length) are segmented into fixed-size units (sometimes called "cells") before being transmitted onto the switch fabric. Segmentation into cells allows the switch fabric to efficiently schedule pack-ets to their destination egress line cards and ports as discussed in Chapter 2 of [AWEYA2BK19]. Depending on the architecture, the ingress traffic manager may interact with the switch fabric (via the switch fabric interface) and/or with the egress traffic manager to meet the scheduling requirements of the switch fabric and to avoid internal congestion.

On the egress line card, the traffic manager first performs reassembly of the cells of a given packet if segmentation was performed on the ingress line card. The packets received on the egress line card are then placed into an output queue based on the information that is carried by the internal forwarding tag (that was appended at ingress line card). If per-class or per-flow queuing is supported on the egress line card, the egress traffic manager maintains additional output queues for each class or flow. Traffic shaping may be applied to each output queue to maintain conformance to configured CoS or QoS policies.

The queuing architecture of the switch/router affects the QoS experienced by the traffic flows in times of heavy traffic load. An important concern is, how packets are queued at the ingress and egress sides of the switching fabric, that is, how traffic is prioritized and queued at the ingress and egress line cards. For switch fabrics that support internal buffering (e.g., internally buffered crossbar switch fabrics [AWEYA2BK19]), the way in which traffic flows are queued in the switch fabric is equally important.

When strict priority queuing (also called strict priority scheduling) is used in an architecture, all high-priority packets are forwarded before any lower-priority packets. A device may use scheduling mechanisms such as WFQ to statistically schedule packets into the system. Unlike strict priority queuing, WFQ allows packets from lower priority queues to be scheduled and interleaved with higher priority traffic. WFQ prevents the lower priority traffic from being completely blocked or starved by the higher priority traffic, since each traffic class is guaranteed service for a pre-defined proportion of the time.

Some switch/routers are designed with a small number of queues for traffic prioritization and scheduling. Using a small number of queues is common in architectures that support class-based queuing. These architectures limit the ability of the device to perform fine-grain prioritization and to provide fairness to different traffic flows. A device that performs queuing based on traffic classes typically prioritizes traffic based on IEEE 802.1p priority levels in the Layer 2 header (see section "IEEE 802.1p/Q" in Chapter 8 of Volume 1), or DSCP (see section "IETF Differentiated Services (DiffServ)" in Chapter 8 of Volume 1), rather than on higher-level information such as application or protocol type.

Architectures that support large number of queues are more capable of granular prioritization and providing greater fairness to traffic flows. A system that establishes queues on a per-flow basis would provide each user session with its own queue. However, implementing a large number of per-flow queues and QoS policies requires the device to use large amounts of memory and, possibly, hierarchical schedulers in order to be able to handle hundreds of thousands of flows. In practice, switch/routers use per-class queuing to avoid the complications and architectural complexities associated with per-flow queuing and scheduling.

The CoS/QoS markings in each packet header can be used in the classification of the traffic passing through the switch/router. On the ingress line card, the CoS/QoS markings are also used for assigning a specific priority queue to a packet, packet metering, and policing. On the egress line card, the CoS/QoS values can be used for packet shaping (see Figure 3.2). For Layer 2 forwarding, the switch/router may perform a Layer 2 destination address lookup, check the IEEE 802.1p priority value on the ingress line card, queue the packet in the designated queue, and then possibly modify the CoS/QoS setting in the packet header, before forwarding the packet on the egress interface.

3.3.5.3 Internal Congestion Control

As discussed above, the packet buffer on the ingress side of a line card is a temporary repository for packets arriving at an interface while waiting to be processed by the packet processor. If the processing speed of the packet processor cannot keep up with the packet arrival rate, data in the packet buffer can build up, resulting in high intermittent or short-term packet loss, latency, or packet delay variation (PDV).

Generally, data buildup in the buffer in any part of the system can occur for the following reasons:

- The device is running out of processing resources as traffic load increases.
- Contention for resources in the device exists, for instance, multiple ingress ports on the device are contending for a given egress port at the same time.

In a device with multiple buffering stages, packet buildup in a buffer can create a chain reaction, leading to unpredictable behavior in the device. Another challenge in packet buffering in network devices with high-speed interfaces (e.g., 10 Gigabit Ethernet (GE), 25 GE, 40 GE, 50 GE) is, dealing with small packets that arrive back to back. For example, on a 10 Gb/s interface, arriving 64-byte packets must be written into the buffer memory every 67 nanoseconds (ns), and read every 67 ns. Thus, to process a stream of 64-byte packets that arrive back to back, the packet buffer memory subsystem must be able to write and read packets at intervals not greater than every 67 ns. Writing and reading back-to-back 64-byte packets slower than this rate will lead to packet losses.

In modern architectures, when the packet buffer begins to fill beyond a preset threshold, the packet processor initiates congestion control using a mechanism such as Random Early Detection (RED) (or any of its variants such as Weighted RED (WRED)) as a way of signaling upstream TCP sources to throttle their packet transmission rates (see the "Random Early Detection (RED)" section in Chapter 8 of Volume 1). Packet discard is an important part of traffic management as discussed in Chapter 8 of Volume 1. During times of heavy traffic and internal congestion, the traffic manager may need to make packet discard decisions based on the availability of buffer space, or traffic priority, using packet discard algorithms like RED or WRED, mechanisms that are very effective for controlling TCP traffic flow rates.

3.3.6 PACKET HEADER MODIFICATION

Upon completion of the IP destination address lookup in the FIB and the classification process in the ingress line card, if the packet is to be forwarded to a particular egress line card, an internal forwarding tag is attached to the packet indicating the packet's egress interface plus other relevant information required to route the packet. The ingress line card may also modify the packet header with new QoS/CoS markings if necessary. The context memory (in Figure 3.1) contains the instructions that describe all the internal system header information needed to transfer the packet to its egress line card and interface.

On the egress line card, packet header modification (i.e., header rewrites) also takes place where the egress packet processing function inserts a new external packet header (new source and destination MAC addresses, outer IP header when tunneling is required, new VLAN tags, MPLS stacks, etc.). This means the switch/router must perform packet header parsing and IP destination address lookups (as well as IP header updates) on the ingress line card, and then modify the Layer 2 header on the egress line card before forwarding the packet on the egress interface.

3.3.7 Packet Classification Using Ternary Content Addressable Memory

During IP destination address lookup and packet classification of an arriving packet, the packet processor parses the IP destination address and various fields in the packet header to construct search keys. These search keys are then used to match entries in various tables described in the "Routing and Classification Tables" section above. Most high-end architectures use Ternary Content Addressable Memories (TCAMs) or equivalent technologies to store the FIB and the classification tables. The search keys are applied to the table entries maintained in the TCAMs (see section "Lookup Tables used in Layer 3/4 Forwarding Operations, QoS and Security ACLs" in Chapter 5 of Volume 1 for more discussion on TCAMs).

Large TCAMs are capable of storing millions of entries and can perform key searches in a matter of few internal clock cycles. However, searches in an FIB (using longest prefix matching (LPM)) or a packet/flow classification table may require multiple key searches to produce a look-up result. Furthermore, the packet processor may be required to perform separate searches in different classification tables (related to ACLs, QoS, etc.) to determine how to handle an arriving packet. For example, the packet processor may perform a look-up in an ACL first to decide whether to deny or forward a packet and then perform a look-up in the FIB to determine the outbound interface to which the packet should be forwarded (assuming ACL lookup is required before FIB lookup).

The packet processor may also perform look-ups in a flow classification table in order to provide enhanced QoS services and to enforce policies (e.g., priority queuing, CoS tagging or remarking, policy-based routing (PBR)). Flow classification may be performed to provide a finer level of granularity that allows the switch/router to establish policies based on application type. The packet processor may perform flow classification using any number of combinations of Layer 3 and Layer 4 information in order to define the QoS or security policies that are to be enforced.

In some switch/routers that support, for example, firewall and other deep packet processing functions in addition to packet classification, flow classification may include performing stateful analysis of packets within packet streams. The flow classification functions will track the protocol state of each flow as the connection develops. This allows the switch/router to track control connections on well-known Transport Layer ports that generate data connections on ephemeral ports. Keeping such protocol state is important, because protocols such as TCP, establish connections and negotiate services on well-known ports, and then establish another ephemeral port to transfer data for the TCP session.

Other than a router which forwards packets at Layer 3 only, a switch/router can forward packets at both Layer 2 and 3 as discussed in the section "How a Switch/Router Decides When to Use Layer 2 or Layer 3 Forwarding for an Arriving Packet" of Chapter 5 of Volume 1. An advanced switch/router design may be required to parse and process more fields in a packet, the number and type fields depending on the services provided by the switch/router. A simple switch/router may be required to inspect only the Layer 2 header (i.e., MAC header fields including VLAN tags), and the Layer 3 header (i.e., IPv4 or IPv6 header fields), and, in some cases, performs ACL lookups and limited flow classification.

Figure 3.3 shows a flowchart that describes the typical Layer 2 forwarding (with QoS and Security ACL processing) in a switch/router. Figure 3.4, on the other hand, shows the typical Layer 3 forwarding (with QoS and security ACL processing) in a switch/router. Figure 3.5 shows as an example, how an ACL TCAM is populated, while Figure 3.6 describes the lookup process in an ACL TCAM.

A switch/router may be called upon to perform more advanced searches and packet classification. Complex classification may require the packet processor to perform searches based on multiple Layer 2 and Layer 3 headers for a given packet. Also, the packet processor may be required to handle packets of a given protocol that are encapsulated within one or more tunneling protocols. For example, when the switch/router supports IPv6 over Generic Routing Encapsulation (GRE), the packet processor is required to examine and process two Layer 3 headers (IPv4 and IPv6) in

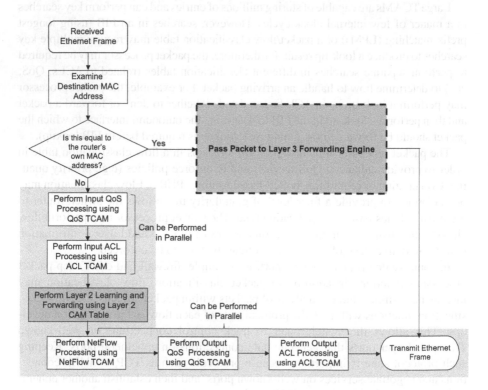

FIGURE 3.3   Layer 2 forwarding with QoS and security ACL processing.

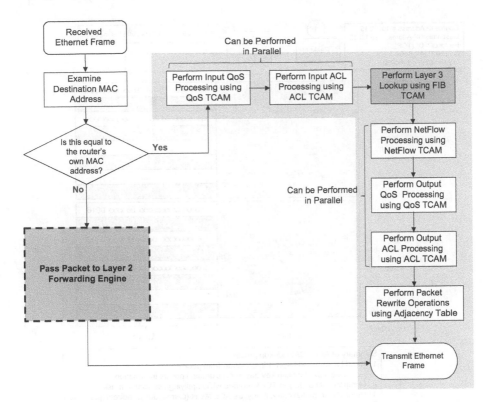

FIGURE 3.4 Layer 3 forwarding with QoS and security ACL processing.

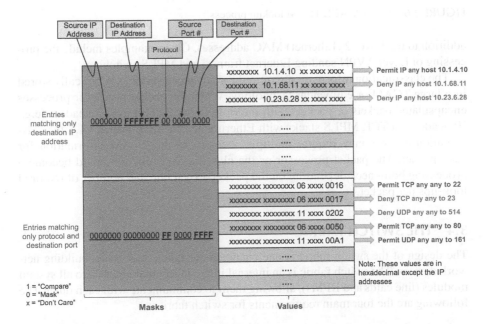

FIGURE 3.5 Example ACL TCAM population.

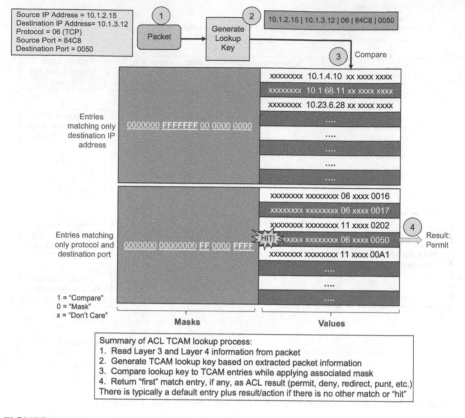

FIGURE 3.6   Example ACL TCAM lookup process.

addition to the Layer 2 (Ethernet) MAC addresses. Other examples include the processing of Layer 2 VPN sending Ethernet frames, and MPLS switching.

The classification information required for advanced services is typically stored in a large TCAM or equivalent technology. When the packet processor processes encapsulated packets or packets with multiple Layer 2 and Layer 3 headers (e.g., IP header and TCP, MPLS stacks with Ethernet header, IPv6 over IPv4,) the classification process may require multiple accesses to the TCAM information for each packet. The packet processor or the classification function could become a processing bottleneck depending on the packet arrival rate and number of required look-ups per packet.

3.4   THE SWITCH FABRIC

The design of the switch fabric is one challenge designers face when building network devices. The switch fabric is an internal interconnect that attaches to all system modules (line cards and RPMs), allowing them to communicate with each other. The following are the four main requirements for switch fabrics:

- It must provide a method for transferring packets between system modules and from input ports to outputs ports.
- It must provide a traffic arbitration mechanism when more than one packet arrives concurrently at input ports and are destined for the same output port.
- It must provide sufficient buffering to handle situations where the input traffic rate is greater than the switch fabric's data transfer capacity. Buffering may be provided at the input of the switch fabric (input buffering), internally to the switch fabric, or at the output of the switch fabric (a shared-memory switch fabric is a special case of output buffering).
- It must manage traffic flow (via some form of flow control) on packets arriving at the output of the switch fabric (output buffering).

The main types of switch fabrics are shared-bus, shared-memory, and crossbar switch fabric [AWEYA1BK18] [AWEYA2BK19]. High-performance switch/routers generally use crossbar switch fabrics which may also have redundancy and fault-tolerance built into them (see Chapter 2 "Understanding Crossbar Switch Fabrics" in [AWEYA2BK19]).

The switch fabric functions as a data plane interconnect linking all ingress and egress interfaces in the system. The switch fabric operates in the data plane and interconnects all line cards and the route processor in the system. An $N{\times}N$ crossbar switch fabric allows N line cards including the route processor to be interconnected in the switch/router. This allows each slot in the system (which may house a line card or the route processor) to send and receive traffic over the switch fabric simultaneously. Typically, high-end switch/routers employ single-stage or multi-stage crossbar switch fabrics [AWEYA2BK19] to move packets between system modules, while mid-end and low-end switch routers use various forms of shared-bus or shared-memory switch fabrics [AWEYA1BK18].

3.4.1 Avoiding Head-of-Line Blocking

Modern network devices that employ crossbar switch fabrics use VOQs and special scheduling algorithms to avoid HOL blocking [AWEYA2BK19]. HOL blocking occurs with FIFO input queuing, when a packet at the head of the queue is held up because its output ports are busy (and not ready to accept packets), thereby blocking the transmission of any packets behind it even if their output ports are available to receive packets. In a crossbar switch with FIFO input queuing, HOL blocking allows the switch fabric to achieve not more than 58.6% of its total throughput assuming random traffic arrival at the FIFO input queues.

The scheduling algorithms used in modern crossbar switch fabrics require a traffic scheduler and the input VOQs to exchange information, including requests for permission to transmit (from the VOQs to the scheduler), grants giving permissions to transmit (sent from the scheduler to VOQs), and other information. Figure 3.7 shows a high-level view of a crossbar switch with VOQs and a scheduler. Reference [AWEYA2BK19] describes in greater detail the use of VOQs and the different types of scheduling algorithms employed in crossbar switch fabrics.

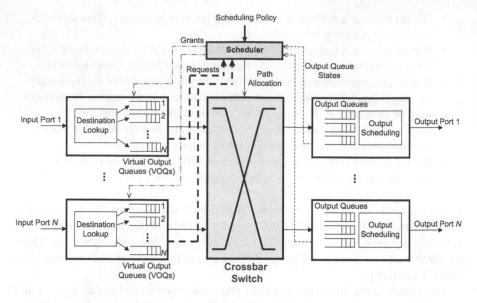

FIGURE 3.7 Crossbar switch fabric with Virtual Output Queues (VOQ) and scheduler.

3.4.2 HANDLING MULTICAST TRAFFIC

Typically, a device using a crossbar switch fabric maintains separate queues and scheduling disciplines for unicast and multicast traffic on the ingress line cards. Some architectures use multicast packet replication to transfer multicast traffic across the crossbar switch fabric to the destination output ports [AWEYA2BK19]. To get around packet replication issues, some architectures simply broadcast an incoming multicast packet across the switch fabric to any number of egress ports, allowing any port that is not currently supporting multicast traffic to simply filter such broadcast packets. This approach, however, can significantly degrade the overall throughput of the switch fabric, particularly, for unicast traffic.

Implementing packet replication is one of the biggest challenges when designing the switch fabric to handle multicast packets. Generally, packet replication in the crossbar switch fabric is accomplished in two stages.

- **Branch Replications**: This first stage handles packet replications from a given ingress line card across the switch fabric to multiple egress line cards.
- **Leaf Replications**: This second stage is typically accomplished on the egress line card and handles replications to the ports/interfaces on the egress line card.

Depending on how the packet replication function is implemented on the switch fabric, it could cause starvation of processing and memory resources, as well as contention with unicast packets. Replicated packets will have to compete with unicast packets for system processing and memory resources.

The preferred method of handling multicast traffic in crossbar switch fabrics is to use scheduling algorithms that transfer multicast traffic efficiently from an input port to multiple output ports without resorting strictly to packet replication as discussed in [AWEYA2BK19]. This method (which is implemented in newer generation network devices) uses traffic scheduling algorithms that take advantage of the natural multicast properties of a crossbar switch fabric. In this method, packet replication is performed within the switch fabric by closing multiple internal cross-points simultaneously. Using this method, the ingress line card is relieved from performing packet replication. However, the leaf replication on the egress line card may still have to be carried out (when necessary), which can still cause resource starvation and congestion on the line card.

3.4.3 SWITCH FABRIC SPEEDUP

To prevent the accumulation of data at the input queues and to minimize contention for the switch fabric, the crossbar switch fabric is typically run at a speed higher than that of the highest speed interface (say, by a factor 1.5 to 2). This additional bandwidth over the highest speed interface is called the *speedup* of the switch fabric. The speedup is the additional bandwidth required to support all input queues (VOQ) without significant input blocking and data loss.

For example, a crossbar switch with all 10 GE line cards may support 15 Gb/s to the switch fabric, thereby offering 50 percent speedup. Speedup, which means running the crossbar switch fabric faster than the external line rate, has become a common feature in crossbar switch fabric design and is an effective method for reducing input and output blocking. For example, if the crossbar switch fabric is two times faster than the input line rate, the traffic scheduler can transfer (during each cell time) two cells from each input port across the switch fabric, and two cells from the switch fabric to each output port.

The advantage of supporting speedup in a crossbar switch fabric is, it allows the switch fabric to offer more predictable delay and delay variation across the device by transferring more cells per cell time. In the absence of HOL blocking, speedup reduces the delay experienced by each cell through the switch fabric. With sufficient speedup, the switch fabric can guarantee that every cell at the head of an input queue will be immediately transferred to its output port, where its departure time is more predictable.

3.5 SYSTEM BACKPLANE

A *backplane* is the circuitry in a network device (e.g., a switch, router, or switch/router) that provides connectivity between the different modules on the device (Figures 3.8 and 3.9). A backplane supports a group of parallel electrical traces and connectors to which the system modules are attached. The primary function of a backplane is to provide data communication channels between the different system modules as well as distribute electric power to them. In some cases, the backplane may support timing buses to distribute clock signals from a reference clock source and a control bus to distribute control messages to various system modules such as line cards, switch fabric, and the route processor.

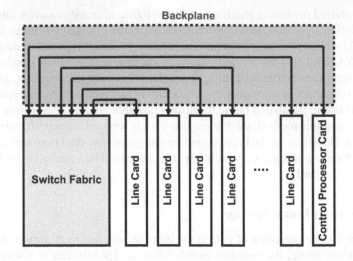

FIGURE 3.8 Line card-switch fabric connectivity (centralized switch fabric system).

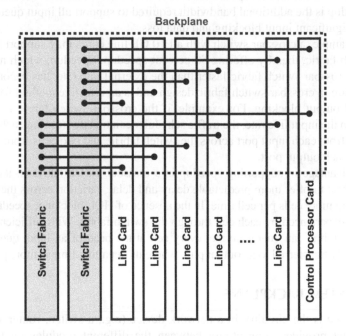

FIGURE 3.9 Line card-switch fabric connectivity (distributed switch fabric system).

There are two main types of backplanes, *passive backplanes* and *active backplanes*. Passive backplanes contain buses and bus connectors but do not contain active components (e.g., bus driving circuitry for driving signals to the connectors or slots). Any desired components required for communication over the backplane (e.g., arbitration logic) are placed on the modules or cards that attach (via connectors/slots) to the backplane. Active backplanes contain the slots (to which the modules/cards are attached) and all the necessary circuitry to manage and control all the communication between the slots (e.g., bus control and circuitry which buffer the various signals sent to the slots). Given that passive backplanes do not carry active components, they usually do not present a single point of failure in the system. Active backplanes are more complex given that they have active components and, thus, have a non-zero risk of failure or malfunction.

Note that a *midplane* (which provides similar functions as a backplane) provides connectivity between back-side system components and the front-end components. A midplane has slots on both sides for connecting to devices; cards can be plugged into either side of a midplane. In network devices, one side of the midplane may accept the RPMs and system peripheral components such as timing cards, power supplies, and cooling fans, and the other side may accept network interface cards. *Orthogonal midplanes* connect horizontal cards plugged in on one side to vertical cards on the other side.

Backplane design and backplane interfacing are some of the additional challenges designers face when implementing network devices. The backplane connects the SFM, all the line cards, and the RPM in the system. Figure 3.8 shows a backplane architecture that connects a single centralized SFM to all the line cards in the system. The SFM connects to the line cards through high-speed serial links on the backplane. In Figure 3.9, the system has two switch fabrics, and the backplane connects each switch fabric to every other line card in the system using high-speed serial links.

Figure 3.10 shows the placement and relationship of the backplane with respect to other components in a typical switch or routing device using a switch fabric. A typical system consists of a switch fabric, backplane, backplane transceivers, and backplane interfaces, and multiple line cards (each containing a number of Physical Layer devices (PHYs), a MAC or Framer, packet processing and forwarding functions, traffic manager, and queue and buffer management system).

The PHYs and MACs/Framers process packets received from the network, and then pass the data on to the packet processor (i.e., processing and forwarding functions) for classification, prioritization, and forwarding. The line card sends data over the backplane to the switch fabric, which passes the data to other line cards or the

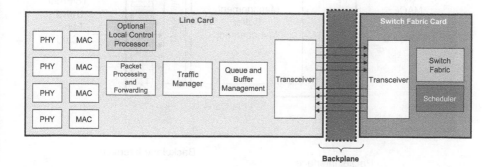

FIGURE 3.10 Line card, backplane interface, and switch fabric.

route processor. A *backplane interface* is a set of connectors through which a system module connects to the backplane.

The backplane transceiver functions are implemented on both the line card and the SFM (which also contains a scheduler). The line cards and SFM in the system chassis are electrically or optically interconnected via the backplane. The SFM interconnects all network interfaces or ports on all the line cards in the system. In high-end systems, the actual devices that make up the switch fabric are typically implemented on a separate card and then connected to the line cards using high-speed transceiver devices.

Figures 3.11 and 3.12 show details of the backplane interface on a line card and on an SFM, respectively. In Figure 3.11, the designer implements the traffic management, buffer management, backplane transceiver, and backplane interface on the line card. In Figure 3.12, the SFM has an integrated backplane transceiver and interface, in addition to its switch fabric components (e.g., VOQs, scheduler, high-speed external memory interfaces). The line card and SFM each connect to the system backplane via their respective backplane interface.

3.5.1 BACKPLANE ETHERNET

Other than the Ethernet standards used for designing interfaces on traditional computing devices and embedded systems (e.g., 1000BASE-T, see also Chapter 6 of this volume), a variety of other Ethernet standards have been defined that provide numerous alternatives to Ethernet Physical Layers (PHYs) suitable for different applications such as Ethernet backplanes (see Figures 3.13–3.16). *Backplane Ethernet* standards are relatively new and define standards for using high-speed Ethernet links

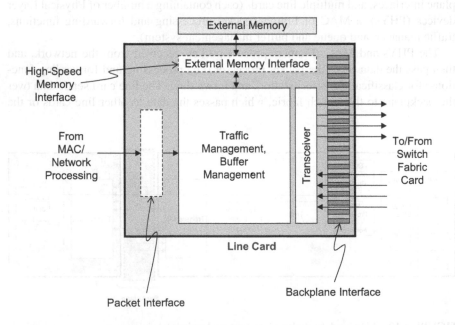

FIGURE 3.11   Backplane interface on line card.

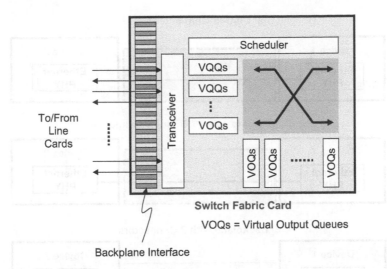

Switch Fabric Card

VOQs = Virtual Output Queues

Backplane Interface

FIGURE 3.12   Backplane interface on switch fabric card.

over backplanes and connectors for interconnecting components and sub-systems (boards such as line cards, route processors, switch fabrics, etc.) in network and computing devices. The backplane Ethernet standards provide a number of options for connecting together multiple system boards/cards, particularly, in network devices such as switches, routers, and switch/routers.

Ethernet owes its popularity to its proven effectiveness in implementing device connectivity. Ethernet is known for its low cost, versatility, robustness, high performance, and broad interoperability. Ethernet has long been implemented in computing devices and on embedded systems to provide connectivity between modules, and connectivity to external devices. Ethernet PHY speeds range from 10 Mb/s to several hundreds of multi-gigabits, and work over media such as twisted pair cabling, or fiber optics.

Ethernet also works well over backplanes as defined by the various backplane Ethernet standards. Backplane Ethernet standards promote multivendor device interoperability, broadening the number of components a designer can select for a design, and allowing the optimization of the design by leveraging the strengths of different device suppliers. Backplane Ethernet interface (PHY) solutions are now readily available and cheap to integrate into a system architecture; there are a lot of off-the-shelf components available that can be used to build Ethernet backplanes.

One of the key philosophies of Ethernet (standards) is interoperability; Ethernet devices should be able to interwork whenever possible. To enable interoperability, many Ethernet PHY specifications define auto-negotiation, link training, and other techniques to ensure that compatible devices will interoperate, even if they have different underlying technologies and capabilities.

- Auto-negotiation allows the end devices of a link to share information on various capabilities they support that is relevant for establishing communication between them (i.e., the highest common denominator among their capabilities).

a) Chip-to-Chip over PCB

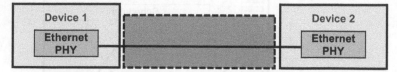

b) Chip-to-Chip over PCB plus 1 Connector

c) Backplane with 2 Connectors

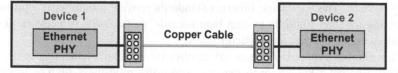

d) 2 Connectors with Copper Cable

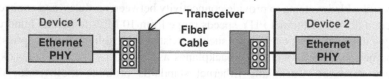

e) 2 Connectors with Optical Fiber Cable plus Transceivers

FIGURE 3.13 Examples of device interconnection using Ethernet.

- Link training allows the characteristics of the signal transmitted by an end-point to be optimally tuned to be carried over the intervening transmission medium. Essentially, link training enables the finite impulse response (FIR) filter for each transmission channel (i.e., channel equalization settings) to be automatically tuned so that the desired bit error rate (BER) can be achieved.

Backplane Ethernet PHY standards have been defined for 1 Gb/s, 2.5 Gb/s, 5 Gb/s, 10 Gb/s, 25 Gb/s, 40 Gb/s, 50 Gb/s, 100 Gb/s, 200 Gb/s, and 400 Gb/s data rates over one to four SerDes lanes (e.g., 1000BASE-KX, 2.5GBASE-KX, 5GBASE-KR, 10GBASE-KR, 10GBASE-KX4, 40GBASE-KR4, 100GBASE-KR, 100GBASE-KR2, 100GBASE-KR4). All of the backplane Ethernet PHY standards include auto-negotiation, making it possible for two devices that have different native

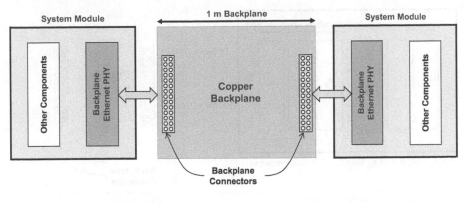

Examples:
* 10GBASE-KR: 1 m over a single lane of backplane
* 10GBASE-KX4: 1 m over 4 lanes of backplane
* 40GBASE-KR4: 1 m over 4 lanes of backplane

FIGURE 3.14   Reach of Backplane Ethernet.

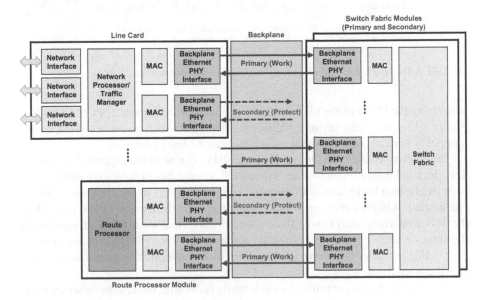

FIGURE 3.15   Using Backplane Ethernet in a routing device.

Ethernet PHY types to connect. Based on the capabilities of each device, the two dives will negotiate to use the best capabilities supported by both. Ethernet backplanes are becoming cost-effective and attractive for implementing all sorts of backplanes including mesh backplanes as discussed next.

Figure 3.13 shows various options for interconnecting devices in a system using Ethernet. Initially, a great deal of attention was paid to connecting chips on a single printed circuit board (PCB). Today, there is equal interest in connecting boards on network devices such as switches and routing devices. The Ethernet medium

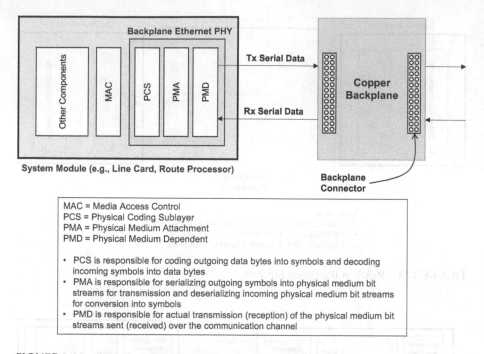

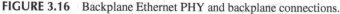

MAC = Media Access Control
PCS = Physical Coding Sublayer
PMA = Physical Medium Attachment
PMD = Physical Medium Dependent

- PCS is responsible for coding outgoing data bytes into symbols and decoding incoming symbols into data bytes
- PMA is responsible for serializing outgoing symbols into physical medium bit streams for transmission and deserializing incoming physical medium bit streams for conversion into symbols
- PMD is responsible for actual transmission (reception) of the physical medium bit streams sent (received) over the communication channel

FIGURE 3.16   Backplane Ethernet PHY and backplane connections.

connecting the PHYs of the two devices may consist of only pairs of PCB traces (see Figure 3.13a), or it may include additional components such as connector(s), cables (copper or optical), and transceivers (see Figures 3.13b–3.13e).

The medium connecting the two Ethernet PHYs can be either electrical or optical. The electrical medium can be traced on a PCB, copper-based cables (twisted pair or twin axial), or a backplane with multiple connectors. Due to the differences in the characteristics of the cable, connectors, and backplane, the Ethernet PHY for each of the device interfaces may have different specifications that are medium-dependent. If the interconnecting medium is optical (single-mode fiber (SMF) and multi-mode fiber (MMF)), transceivers have to be used for electrical/optical and optical/electrical signal conversion as shown in Figure 3.13e.

In Figure 3.15, two centralized switch fabric cards with backplane Ethernet interfaces (one card for operation and the other for backup) are implemented on the chassis. All the line cards and RPM in the system then connect to the switch fabrics via their backplane Ethernet interfaces; the SFM manages the movement of data between the different system modules. Each board supports primary and secondary backplane Ethernet interfaces, providing built-in reliability. In practice, the Ethernet backplane in Figure 3.15 will be composed of switch fabric slots, line card and RPM slots, and the links that interconnect them. To provide high availability, the backplane will additionally support hot swap capabilities, allowing cards to be removed or inserted without powering down the system.

3.6 MESH INTERCONNECTS

Switch fabric design has always taken center stage in the design of high-performance, high-capacity network devices. With growing demands for high-speed networks and tremendous increases in line speeds, designers have always evaluated different methods for increasing switch fabric capacity (including backplane architectures) in order to support the throughput necessary for handling multi-gigabit line rates. Fortunately, advances in high-speed serial interconnects, including short-range high-speed Ethernet backplanes, have enabled designers to develop high-capacity switch fabrics as well as mesh fabric backplanes. Using such high-speed interconnects and backplanes allows a switch fabric to offer advantages in scalability, availability, cost, and interoperability (particularly, using Ethernet backplanes).

3.6.1 Most Common System Interconnect Topologies

The three most common topologies for interconnecting the various boards and components of a network device are the bus, star, and mesh topologies. Each of these topologies has its own advantages, limitations, and implementation complexity in terms of speed, pin count, additional logic, reliability, and overall complexity [EET04.2002] [EET07.2002]:

- **Bus Topologies**: A bus topology uses a multi-drop configuration to interconnect the various components of a system. Busses are typically wider in bit length but are slower in transmission speed. In most cases, a bus requires numerous pins to interconnect system components, and this high pin count makes it impractical to use a bus interconnect to connect lots of individual components to the shared bus. The resulting sharing and slower speed (albeit using a wider bus) make busses rapidly reach their maximum performance limits. A bus being a shared resource makes reliability a very big issue; failure of the bus can compromise the integrity of the whole system.
- **Star Topologies**: In a star topology, a dedicated point-to-point link is used to connect each system component to a central resource, allowing the connected component to send/receive data to/from the central resource. The central resource serves as a data distribution point for the system. Reliance on a single central resource presents a single point of failure. Thus, to provide reliability, a star topology requires redundancy. In most cases, particularly, in high-availability applications, a dual-star topology is used.
- **Mesh Topologies**: In a mesh topology, the various system components are interconnected using point-to-point connections in such a way that each component has connections to all other components as shown in Figure 3.17. Each component connects to every other component in the system using high-speed serial links, creating a high-speed mesh backplane. A mesh topology is a form of a distributed switch fabric since each node handles its own traffic, and does not depend on a central resource as in the star topology; all nodes are equal peers in the system.

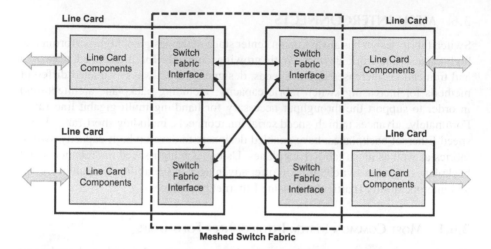

FIGURE 3.17   Fully meshed switch fabric.

As discussed above, advances in high-speed serial interconnects and Ethernet back-planes have enabled mesh interconnects to replace traditional bus and star-based architectures. In high-speed serial mesh interconnects and backplanes, the PHY consists of transmit and receive pairs per channel, each channel using high-speed SerDes and appropriate encoding at speeds that can be up to 40 Gb/s and more as in Ethernet backplanes.

The advances made in SerDes technology have allowed a higher number of high-speed connections to be used in mesh interconnects, and system implementations to be very cost-effective. There are now a wide range of SerDes available from many vendors, many of which are incorporated in FPGAs. Advances in Ethernet backplane technology have also made practical and cost-effective multi-gigabit high-speed mesh backplanes possible. The availability of cost-effective SerDes, Ethernet back-planes, and connectors makes it easier for designers to justify the adoption of mesh backplanes; these devices are becoming more cost-competitive.

3.6.2   EXAMPLE MESH INTERCONNECT ARCHITECTURE

Unlike the traditional bus and star topologies, a mesh topology provides a distributed system interconnect. In a mesh topology, each node has a direct path to each other node and each node constitutes a portion of the overall system fabric (Figure 3.17). Other than using a system interconnect with an $N{\times}N$ crossbar switch fabric, a designer can use a fabric architecture that consists of M, $1{\times}N$ switches distributed among the line cards as shown in Figures 3.18 and 3.19 [EET04.2002] [EET07.2002]. Figure 3.19 shows details of a $1{\times}N$ switch with an integrated traffic management and buffer management system.

In Figure 3.17, the mesh (switch) fabric interface accepts traffic from the line card (that includes a traffic manager or network processor) over one of many available standard-based parallel interfaces (i.e., a standard-based interface between a traffic

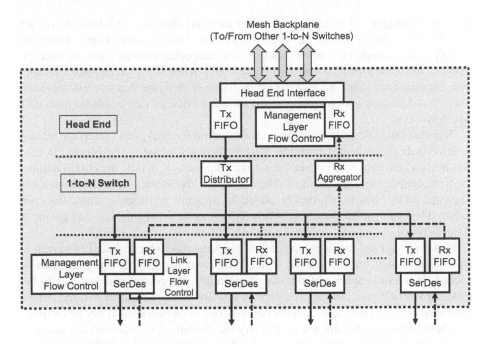

SerDes = Serializer/Deserializer

FIGURE 3.18 Example block diagram of a 1-to-N switch.

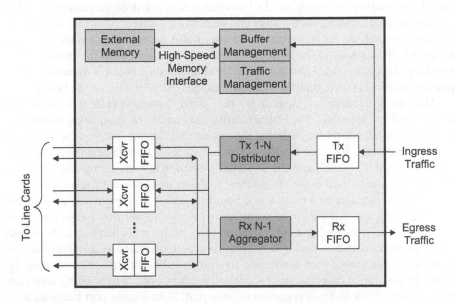

FIGURE 3.19 Distributor/aggregator with traffic and buffer management.

processor/manager and switch fabric). The fabric interface then serializes the traffic over the links that make up the mesh backplane. The *Common Switch Interface* (*CSIX*) is one example of a packet protocol that encapsulates traffic, allowing interoperability (between a traffic processor/manager and switch fabric) in an open architecture environment. Using a CSIX-based interface, a designer can use off-the-shelf traffic managers, network processors, and switch fabrics, as well as blades from different vendors.

The distributed design of the mesh switch fabric (or backplane) interface allows nodes/boards (equivalently, costs) to be incrementally added to the system. In this architecture, the only fixed cost in the mesh backplane is generally the cost of adding the fixed copper mesh interconnect. The designer adds capacity (and cost) with each line card added, which can also be added in different increments. Thus, the cost incurred by adding a 2.5 Gb/s system board does not have to be the same as adding a 10 Gb/s board.

However, in a star topology, the central resource has to be designed to have full anticipated system capacity, even if the additional capacity is not immediately needed. If the full capacity of the star is exceeded, the central resource has to be replaced with a higher capacity resource. A star architecture requires two redundant central resources of the same capacity but only one of them is used at any given time. Star-based fabrics, generally, use proprietary mechanisms for synchronizing and mirroring data through the central resource. Fault containment is also an advantage in a mesh architecture.

In Figure 3.18, the 1-to-N switch [EET04.2002] [EET07.2002] is responsible for transferring all traffic sent between the serial mesh interfaces and the head end interface. In the transmit direction, the Tx Distributor is responsible for outbound packet forwarding by examining each packet and selecting its correct outbound port. In the receive direction, the Rx Aggregator performs traffic aggregation into the head end Rx FIFO. The capacity of the head end determines the capacity of the (mesh) fabric interface. All traffic sent in and out of the local mesh fabric board is intended for the ports on this local board, thus no additional capacity is necessary on the board.

Also, the serial mesh interfaces do not necessarily have to operate at the same rate as the head-end interface. The overall traffic only has to be adequately distributed among the serial interfaces such that the average traffic rate matches the head-end interface capacity; the serial interfaces can individually operate at different speeds. References [EET04.2002] and [EET07.2002] describe how buffering and link-level flow control can be used in the mesh fabric architecture to prevent packet losses.

The traffic manager on a mesh fabric board (Figure 3.19) is primarily responsible for scheduling and shaping traffic according to its own local rules. Each egress port receives traffic from multiple ingress ports. The traffic schedulers at each ingress traffic manager schedule traffic according to preset contracts. If an egress device starts to be overloaded with lower priority traffic, it has the option of shaping the traffic by discarding excess data, or it can initiate flow control by sending flow control messages back to the appropriate ingress port. If an ingress port becomes congested, it has the option of performing traffic shaping or further backpressure the ingress channel.

3.7 THE ROUTE PROCESSOR MODULE

We discussed in Volume 1 and the previous chapters that a routing device has two primary processing components, the route processor (also called the control engine or control processor) and the packet forwarding engine. The forwarding engine is the architectural portion of the routing device that receives packets from input interfaces/ports, processes, and forwards them to their correct output interfaces. The forwarding engine may apply filtering and routing policies, and other processing functions when forwarding a packet to its next hop (which lies on the route to the final destination). The packet forwarding engine may be a hardware or software entity in the routing device dedicated to packet forwarding.

The route processor is the architectural portion or module in a routing device that runs all the routing protocols (e.g., RIP, EIGRP, OSPF, IS-IS, BGP), control and management protocols (ICMP, IGMP, SNMP, etc.), as well as all device/system access, configuration and management tasks (Command-Line Interface (CLI), secure logins and system access, file copying, etc.) (Figure 3.1). The route processor supports all the tools for system configuration, management, monitoring, and debugging/troubleshooting.

As discussed in Chapters 2, 5, 6, and 7 of Volume 1, the route processor operates in the control plane and is responsible for running not only the routing protocols and network management, but also many system management functions such as statistics and alarm collection and reporting, background diagnostics, and control of interfaces and chassis components. Some architectures support a control subsystem separate from the route processor. This control subsystem normally works with its companion route processor to provide control and monitoring functions for system components. In our discussion in this chapter, we consider the control subsystem as a logical part of the RPM.

3.7.1 ROUTE PROCESSOR MODULE FUNCTIONS

As discussed above, the route processor is responsible for performing several tasks associated with routing, packet forwarding, and system maintenance. The main functionality of the RPM includes tasks such as determining the topology of networks and internetworks, creating and maintaining routing tables, responding to routing update requests, controlling device configuration, maintaining and reporting network interface statistics, and processing and responding to network management messages (e.g., SNMP requests, traps).

System maintenance tasks include maintaining configuration registers, configuring packet memories, and environmental monitoring. Current routing devices typically use software-controlled configuration registers that eliminate the need to remove/move jumpers in order to configure the device. This section discusses in detail the various system tasks.

3.7.1.1 Running Routing Protocols, Sending, and Receiving Protocol Updates

All routing protocol packets (including routing updates) received from other routing devices are directed to the route processor. The route processor is also responsible for generating and sending routing updates to other routing devices. The route processor is the key component responsible for providing routing protocol intelligence in the routing device. It is responsible for exchanging routing updates with other routing devices and creating the routing table, which contains all the best routes to all known network destinations.

The route processor may run interior gateway protocols (IGPs) such as RIP, OSPF, and IS-IS to determine the network topology within a routing domain [AWEYA2BK21V1] [AWEYA2BK21V2]. The route processor may also run external gateway protocols such as BGP to determine routes across autonomous systems or routing domains [AWEYA2BK21V2].

3.7.1.2 Creating and Maintaining the Routing Table

A fundamental task of the route processor is to compute the best path that a packet should take through the network from the source to its destination. This computation normally takes into account various network constraints (or routing metrics) and policies. For example, a user policy may dictate that the network path taken by traffic should minimize bandwidth usage, minimize latency and deliver the fastest possible response times to users, maximize network efficiencies, or meet some other set of criteria.

Other than system management and related tasks, the RPM performs route calculations, routing and forwarding tables construction and maintenance, and related statistics processing and reporting. The route processor constructs and maintains the routing table which contains the information for routing packets through the network to their destinations. The route processor examines the routing table, generates a smaller set of routing information (a subset of the routing table information that is mainly used for packet forwarding), and stores this information in the forwarding table. In some architectures, in addition to creating and managing the routing tables, the route processor creates and manages the route caches (see Chapters 2 and 6 of Volume 1).

3.7.1.3 Monitoring System Interfaces and Environmental Status

The RPM handles general system maintenance functions such as diagnostics, line card monitoring, and console support. The RPM also provides control and monitoring functions for the routing device, including controlling and monitoring the cooling fans, electric power supply, and system status (which involves monitoring ambient temperature, voltage levels, alarm conditions of chassis components and installed cards, and activating and shutting down of cooling fans when necessary). Other than alarm handling, the route processor may also perform statistics gathering and reporting.

For example, the routing device may allow statistics information to be gathered on every port. The route processor may also collect statistics from all environmental sensors in the system. When it detects a failure or alarm condition, it may send a signal to the network operator's management stations, which in turn generates control messages, or sets an alarm to draw the attention of the network administrator to the event. The route processor may also relay control messages originating from the network administrator to the routing device's components.

The route processor may control the power-up sequence of the routing device's components such as the line cards as they start, and also power down these components when necessary. The route processor may also monitor the system clock and line card clocks (e.g., SONET/SDH line card clocks) to verify that they are providing the required signal quality. It may generate an alarm if a clock signal is incorrect or out of specifications.

3.7.1.4 Providing Management Interfaces

The RPM provides the required hardware and software interfaces for device and network configuration, management, and monitoring. Typically, the hardware interfaces include console, auxiliary, and Ethernet ports (see "System Physical Management Interfaces" section below). The route processor is also responsible for providing user interaction functions, such as the CLI, SNMP management, and craft interface. The route processor allows configuration files, system images, and microcode to be stored and maintained in primary and secondary storage systems, permitting local or remote system upgrades.

3.7.1.5 Building and Distributing Forwarding Tables to Line Cards

In distributed forwarding architectures and systems in which the forwarding engines are decoupled from the route processing functions, the route processor is responsible for distributing and updating the forwarding tables used by the distributed forwarding engines. A line card with a local forwarding engine will have its own forwarding table. The route processor is responsible for populating and updating these forwarding tables.

After a line card is powered up and properly initialized, the route processor sends a copy of its main forwarding table to the line card. The route processor also updates the distributed forwarding tables when network changes occur and the main routing table is updated. The contents of the forwarding tables are continuously synchronized with the routing information maintained by the route processor.

3.7.1.6 Communicating with Line Cards

In distributed forwarding architectures, an in-band or out-of-band control path exists that links the route processor to the line cards. Using the control path, the route processor distributes copies of its forwarding table to the line card processors, allowing them to forward packets locally. The control path also allows the route processor to perform system monitoring and statistics collection functions on each line card. The control path also allows the route processor to download the operating system (OS) software required by a line card at card power up. Each line card has an onboard processor and memory, to which the route processor downloads the OS software image.

The route processor communicates with the line cards either through the switch fabric (i.e., in-band control path) or through a control and maintenance bus (CMBus) (i.e., out-of-band control path) [AWEYA2BK19]. The route processor also performs general maintenance functions, such as line card diagnostics and monitoring via the control path. In some architectures, the route processor communicates with the line cards through the switch fabric just as the line cards do. In these architectures, the control traffic consumes part of the overall available switch fabric bandwidth, overhead that has to be accounted for during the switch fabric design stage.

In most high-performance architectures, the route processor and the line cards (as well as other chassis components) communicate via the out-of-band control path (i.e., the CMBus). In this case, the switch fabric serves as the main data path for routing and controlling protocol messages to and from the route processor, and for the movement of packets between the line cards. The CMBus provides a channel for the route processor to download software images and forwarding tables to the line cards, collect or load diagnostic information, and perform general system maintenance operations.

3.7.1.7   Supporting Redundancy and High-Availability Features

To provide redundancy and high-availability, a routing device may support two route processors as shown in Figure 3.1 (see also Chapter 1 of this volume). One route processor works as the primary processor (i.e., system master), and the other as the secondary (i.e., standby processor). Using route processor redundancy increases system availability during planned and unplanned network outages. On some systems with two installed RPMs, when the standby processor stops receiving keepalive signals from the master processor, it automatically assumes mastership (see Chapter 1 of this volume).

3.7.2   Implementing the Route Processor Module

In high-end routing devices, the route processor is typically a separate module (card) in the device (supporting various system management interfaces). For redundancy, fault-tolerance, and high-availability, a high-end router supports dual-route processors, primary and secondary route processors as shown in Figure 3.1 (see also the "Node Redundancy and Resiliency" section in Chapter 1 of this volume). Particularly, Chapter 1 explains that, to enhance overall system and network uptime, routing devices are often configured with redundant route processors. This involves plugging a second (often identical route processor card) into a second slot in the chassis, and configuring the system so that, in the event of a failure on the primary route processor card, a switchover to the backup or secondary route processor is performed (see Chapter 1 of this volume).

In low-end devices, the route processor is typically a software (or logical) component sharing the same system resources with other components such as the forwarding engine (which is another software (or logical) component) in the device. In some mid-end devices, the route processor is a separate module integrated into the system, or implemented on a daughter card that also supports the system management interfaces.

The route processor runs the device's OS (e.g., the Internetwork Operating System (IOS) in Cisco devices, Junos Operating System (JUNOS) in Juniper Networks devices) which is generally packaged with a wide range of network functions such as routing protocols, control and network management protocols, software diagnostics tools, CLI, etc. The router OS software also provides comprehensive debugging and logging capabilities. Typically, the router OS software contains a complete software documentation of the OS software package including help files, and user and configuration guides (which can be accessed via user CLI or Web browser). In the distributed architecture, where each line card has its own forwarding engine and FIB, the route processor typically supports software that is responsible for generating and distributing the master FIB to the line cards for local packet forwarding operations. The route processor also supports software that is responsible for synchronizing the line card FIBs to the master FIB it maintains.

As discussed above, in the distributed architecture, the control path that connects the route processor(s) to the various line card forwarding engines allows the route processor to initialize, configure, perform diagnostics, and most importantly, to set up or update the line card FIBs, Layer 2 forwarding tables, classification tables, and routing policies. Generally, the route processor is able to update any location in the FIBs, route caches, and other databases when the need arises for route addition and removal, table flushing during route flaps, and policy updates for a given flow. Typically, the destination address lookup operations and FIB management is asynchronous; the route processor may update the FIB while the forwarding engine is performing a lookup.

3.7.3 ROUTE PROCESSOR MODULE COMPONENTS

The RPM contains a processor (CPU), various types of memories, and runs the routing device's OS. The OS supports all the various processes and programs needed for running the device and the network. The OS supports various mechanisms for installing routes in the routing table (e.g., directly connected networks and static routes), as well as dynamic routing protocols, and with each mechanism implemented as a software module running in the router.

For example, RIP, EIGRP, OSPF, IS-IS, and BGP will each be implemented as routing protocol modules that exchange routing information with other routers. A mechanism, generally, referred to as a routing table manager, will be responsible for distilling the routing information and installing routes in the routing table. Generally, the routing device supports various commands for starting an instance of a routing protocol, specifying which interfaces will participate in the started routing protocol, and which directly connected networks will be announced to other routing devices. The type of route processor used depends on the route processing requirements of the routing device. This section describes the main components of the RPM.

3.7.3.1 CPU

The router OS software runs on a CPU and requires several types of memory to store the OS images, system configuration information, running applications, and data. The CPU runs the OS software, the routing protocols (that maintain the routing tables), and other control and system management protocols and processes. In the distributed forwarding architectures, the CPU is mainly responsible for processing all the non-time-critical key functions in the system. In these architectures, the CPU is responsible primarily for running the routing protocols and for maintaining a master copy of the forwarding table which is downloaded to the line cards for local packet forwarding. It also executes the management functions that configure and control the overall system.

The CPU runs at a specified internal clock speed or frequency (usually measured in megahertz (MHz) or gigahertz (GHz)) which represents the number of cycles per second it can execute instructions. This determines the speed at which the CPU executes instructions of various types. The motherboard on which the CPU resides provides an external clock which the CPU uses to determine its own operational speeds. The external clock is different from the CPU clock speed (its internal frequency). The system uses a clock multiplier (CPU multiplier or CPU Clock Ratio) to set the ratio of the internal CPU clock speed to the externally supplied clock. The internal CPU speed is obtained by multiplying the externally supplied clock speed by the clock multiplier.

3.7.3.2 Memory Components

Generally, an RPM uses different types of memory components (some of which are now obsolete and superseded by newer technologies) such as random-access memory (RAM) of various types, read-only memory (ROM) of various types, onboard Flash memory, and memory cards that fit into system card slots. These memory types are described in this section (see Figure 3.20). The specific memory types and their sizes depend on the router architecture, sizes of the different databases supported (e.g., routing table, FIB, Layer 2 forwarding table, QoS/CoS tables), packet memories and forwarding capacity, and the target network application area (i.e., access, aggregation, or core portions of a network).

Most routing devices use error correction code (ECC) mechanisms to protect their memories. A system may use ECC for error detection of single-bit and double-bit errors, and correction of single-bit errors as follows:

- If a single-bit error is detected by ECC, the error is automatically corrected, and the system continues operation as it was.
- If a double-bit error is detected by ECC, the system logs the error, stops the main processor on the RPM, and takes the module offline.

If a double-bit error is detected by ECC in a system that contains a redundant RPM, the redundant module becomes active and the system continues to operate. However,

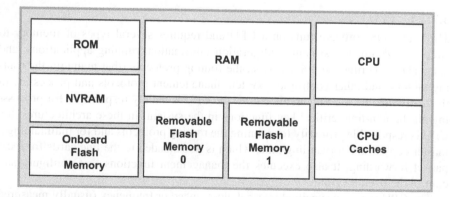

FIGURE 3.20 Main memory components.

it is still prudent for the network administrator to troubleshoot the RPM with the double-bit error to uncover the circumstances surrounding the error.

3.7.3.2.1 Main System Memory (RAM)

In most architectures, the main system memory (a RAM of some type) provides storage for the routing table (or RIB), forwarding table (or FIB), routing protocol applications, network management applications, and the router OS software images. This memory serves as the storage for the RIB, FIB, buffer headers, and other data-bases and data structures required for processing and forwarding routing and control packets. The main memory may support ECC, ensuring very high levels of system availability and service.

Generally, a *startup* or *bootstrap* code loads the OS into a buffer space in the main memory (RAM) of the routing device, after which, the OS then takes care of loading other system software as needed. The startup or bootstrap code first checks the system hardware and loads the OS into the RAM. After the OS has been loaded into RAM and is handed control, it copies the system configuration (which is stored in the NVRAM (see below)) into a buffer space in the RAM. The OS then passes the system configuration to a parser for processing, which then proceeds to dynamically process corresponding configuration commands.

3.7.3.2.2 Packet Memory (RAM)

Some architectures support a separate packet memory (a RAM) that is used to buffer incoming packets. In other architectures, the packet memory is simply a partition of the main system memory (described above) that provides storage for the routing and forwarding tables, the OS executable, and other route processor applications (see discussion in the "Centralized Processor or CPU Based Architectures" section of Chapter 6 of Volume 1). Packet buffers are used in a network device to compensate for differences in input and output processing speeds; for example, during times when a forwarding processor cannot keep up with the aggregate packet arrival rate from all interfaces. Bursts of data can be stored in packet buffers until they can be handled by a processing module in the routing device.

3.7.3.2.3 CPU Cache Memory

The CPU also has several levels of memory for CPU cache memory functions. The CPU normally uses a hierarchy of multiple cache memory levels (typically, Level 1 (L1), Level 2 (L2), and Level 3 (L3) caches) to temporarily hold instructions and data that it is likely to reuse. A cache memory stores copies of the instructions and data from frequently used main memory locations.

A cache memory, which is a smaller, faster memory, located closer to the CPU, reduces the average cost (time or energy) it takes for the CPU to access data from the main memory. Most CPUs support L1, L2 (often plus L3) caches, with separate data-specific and instruction-specific caches at L1. The principal function of the CPU cache memory in the RPM is to act as a staging area (fast memory) for routing protocol updates, and also for holding control and management information data received from and sent to the network interfaces.

3.7.3.2.4 Non-Volatile RAM (NVRAM)

An NVRAM is a type of RAM whose contents are not lost when the system is powered down or reloaded. Typically, electronic devices use NVRAMs to store certain systems settings and information about local components and devices so that they can be accessed quickly when the system powers up or reloads. Unlike a static RAM (SRAM) and a dynamic RAM (DRAM) which both maintain data only for as long as the system power is applied, the information in an NVRAM is retained from one use to the next even if the system is powered off.

The NVRAM provides memory storage for system configuration files, software configuration register settings, and environmental monitoring logs (see discussion in the "Centralized Processor or CPU-Based Architectures" section of Chapter 6 of Volume 1). A system configuration file in NVRAM allows the router OS software to control several system variables upon system initialization. The NVRAM may also contain system hardware configuration registers for setting default boot instructions. The NVRAM may also contain a pool of hardware addresses (Ethernet MAC addresses) for the network interfaces/ports. When the RPM stores the system configuration in NVRAM, configuration information read from the NVRAM is buffered in RAM following system initialization, and is written to the NVRAM device when the user saves the configuration.

3.7.3.2.5 Erasable Programmable Read-Only Memory

An erasable programmable read-only memory (EPROM) is a type of memory that can be erased and reused (i.e., written into), and does not lose its contents when the system is powered off; it holds its content without power. Electronic devices generally employ EPROMs for programs designed for repeated use, but these programs can be replaced or upgraded with newer versions when the need arises. As a rewritable storage chip, the content of an EPROM chip is written using an external programming device before being placed on the circuit board of the routing device. When reprogramming is required, the chip is extracted from the circuit board and placed under an intense ultraviolet (UV) light (for approximately 20 minutes) before reprogramming.

The EPROM stores the startup or bootstrap code that bootstraps the routing device. The bootstrap code initializes the OS after power-on or general reset. Typically, the default system software resides on EPROM components, which are often also called *system ROMs*, *software ROMs*, or *boot ROMs*. The boot ROM is a read-only memory that stores the boot image for bringing up the OS image. In most cases, the EPROM also stores the serial number of the route processor. Although these EPROMs are replaceable, most system software upgrades are distributed via flash drives (or compact disks (CDs) in older systems), or via web downloads from a remote server. The latter enables easy download and boot from upgraded software images stored remotely, without the need to remove the RPM and replace the EPROMs. For instance, when a router boots an image from Flash memory, it first loads the boot ROM image, and then the OS image into the main memory.

EPROMs have largely been superseded by the newer electrically erasable programmable read-only memory (EEPROM) and the Flash memory devices (and are generally no longer used except in older systems). The main difference between an EEPROM and an EPROM is that the contents of an EEPROM are erased by using electric signals, while the contents of an EPROM are erased by using UV rays. Both the EEPROM and Flash memory can be erased while still on the circuit board.

3.7.3.2.6 Flash Memory

Flash memory retains its contents when the system is powered down or restarted, and provides storage for system software images and microcode. A system *restart* on a router is also often referred to as a *reload*. In most designs, the OS image is typically stored in a Flash memory. The Flash memory is used as the primary storage of software images, configuration files, and microcode. Multiple router OS software and microcode images can be remotely loaded and stored in Flash memory. For example, a new image can be downloaded from a remote or local server over the network, and added to Flash memory or the new image can replace existing images. The router can then boot either manually or configured to boot automatically from any of the multiple stored images.

The Flash memory may also function as a Trivial File Transfer Protocol (TFTP) server to allow other routers to boot remotely from the stored images in the TFTP server or to copy the stored images into their own Flash memory. On some platforms, the Flash memory may store configuration files and boot images. An onboard Flash memory (called a *bootflash*) normally contains the router OS boot image, and a PC slot card memory (see below) contains the router OS software image. Basically, the bootflash is a Flash memory that holds the boot image used to do the initial booting of the router. Depending on the platform, the Flash memory is available in the following forms:

- **Internal (or Onboard) Flash memory**: The internal Flash memory is often used to store the system software image.
- **Bootflash**: The bootflash often stores the boot image.
- **Flash memory PC (or PCMCIA) cards**: Some systems support one or more PCMCIA (Personal Computer Memory Card International Association) slots into which Flash memory cards can be inserted (see discussion below). In this case, the Flash memory card can be used to store system software images, boot images, and configuration files.

3.7.3.2.7 PC Card Slots

The PC Card (or Flash Disk) slot, developed by the PCMCIA, provides expansion capabilities for computing devices (desktops, laptops) as well as network devices such as switches, routers, and switch/routers. A Flash memory card or disk can be inserted into a PC Card slot (usually located on front side of the network device), providing additional memory. The Flash memory card is keyed for proper insertion into the slot. Each PC Card slot has an ejector button for ejecting a Flash memory card from the slot. The PC Card slot is designed such that the ejector button will not pop out unless the Flash memory card is inserted correctly into the slot. A routing device may support a number of PC Card slots (e.g., two slots).

In a routing device, for example, the PC Card slot may accept a removable Flash card which stores software images, used possibly, for system upgrades. The removable memory card may also be used for loading and storing multiple router OS software and microcode images. In some architectures, the PC Card slot may hold a memory card which is loaded with the system's software images and configuration files.

A routing device may be designed such that, the OS software images that run the system can be stored in PC Card memory. In this case, the Flash memory may hold the main OS software image (see discussion in the "Centralized Processor or CPU-Based Architectures" section of Chapter 6 of Volume 1). In general, storing the router OS software images in removable Flash memory provides a simple but efficient way to download and boot from upgraded router OS software images remotely, or from software images located in the route processor Flash memory, without having to remove and replace ROM devices. Most vendors design routing devices that support downloadable system software for most OS software upgrades. This enables a net-work administrator to remotely download, store, and boot from a new router OS software image in Flash memory.

The PC Card slot and its successors like the CardBus were originally defined by the PCMCIA as a standard for memory-expansion cards in computing devices. The PC Card port has been superseded by the ExpressCard interface (also a specification from the PCMCIA), although some equipment vendors continued to offer PC Card ports. ExpressCard is built around the PCI Express and USB 2.0 standards, and is intended as a replacement for the PC Card.

Despite ExpressCard being much faster in speed/bandwidth, it has not been as popular as the PC Card, partly due to the wide availability and ubiquity of USB ports on modern computing devices. In today's market, however, most of the functionality provided by ExpressCard or PC Card devices is now available and provided by exter-nal USB devices. These USB devices have the advantage of being compatible with many different computing devices from desktop computers to portable devices. Because of the ubiquity of USB devices, newer computing devices like laptops and desktop computers are now rarely fitted with a PC Card or ExpressCard slots. The needed expansion that formerly required a PC Card or ExpressCard is catered for by USB, reducing the requirement in computing and network devices for internal expan-sion slots (based on ExpressCard or PC Card).

3.7.3.2.8   Hard Drives

Some routing devices support the traditional internal hard drive (as seen in desktop and laptop computers). Usually, such an architecture will support a Flash memory and an internal hard drive storage. The hard drive is used to store a backup copy of the router OS software, user files, log files, and debug outputs. The internal Flash memory stores the router OS software and configuration files for the router. A RAM memory in the route processor stores routing tables, forwarding tables, link-state databases (used in link-state protocols such as OSPF or IS-IS), and also provides operational memory space for the router OS software. However, given the great advances made in data storage devices, many newer architectures do not support hard disks anymore.

If no external Flash memory card is installed in the system, the hard disk typically provides primary storage for software images, configuration files, and microcode.

The router OS software resides on the internal Flash card, with an alternate copy residing on the system hard drive. If an external Flash memory card is installed, the hard drive provides secondary storage for log files and memory dumps, and can reboot the system if the external Flash disk fails. In some router architectures, a hard drive is provided for onboard data collection and for system and error logging.

3.7.3.3 Memory Types and Router Boot Sequence

Typically, when a router leaves the factory or is upgraded in the field, it is configured to store the bootable copies of the router OS software in a number of locations: the internal Flash memory card, internal hard drive, or removable memory card. The router is configured to be able to load the OS software into main memory from any of the storage locations and boot the router. Usually, the primary boot location is the internal Flash memory. In this configuration, the internal hard drive is designated the secondary boot location, while the removable memory card is used for disaster-recovery purposes.

Anytime the router boots up, it will first run a power-on self-test (POST) as a way of verifying if the basic system components are operating correctly. The router then locates a copy of the router OS software in one of the memory locations and loads it into the main memory. Typically, when the router begins to boot, it will first check the removable memory card, followed by the internal Flash memory, and finally the internal hard drive. An example boot sequence is described as follows [JNCIAGUIDE]:

1. The first boot location examined is the removable memory card. If the router finds a copy of the router OS software in this location, it will load the software in the main memory. It should be noted, however, that booting from the removable media presents some risks since all existing file systems and files on the router will be erased during this process. This method of booting returns the router to its default factory settings and should be used only for disaster recovery.
2. If no removable memory card is present, the router will load the OS software from the internal Flash memory. This method is the normal boot operation and should occur each time the router starts.
3. There is the possibility that the internal Flash memory may become corrupted or otherwise unusable when it is needed. When this happens, the hard drive is used as the router's boot location. In this case, when the router boots from this location, it will display a message to alert the network administrator about this event.

3.7.3.4 System Configuration, Management, and Monitoring

This section describes the components and tools used in switch/routers for system configuration, management, and monitoring.

3.7.3.4.1 System Software Management Interfaces

Typically, routing devices are designed to allow a user (network administrator or network engineer) to interact with the system via a Graphical User Interface (GUI) which may also be a web-based interface or a CLI. A GUI displays objects that communicate useful information about the device to the user. These objects may also represent actions that can be performed by the user to configure, manage, and monitor the device.

Unlike a GUI, a CLI is a text-based interface that allows a user to interact with a system (configure, manage, and monitor) via commands statements. The user enters commands into the interface and receive replies the same way via visual prompts. The processing entity or program that handles the CLI is called a *Command-Line Interpreter* or *Command-Line Processor*. The CLI of network devices typically includes a collection of commands for configuring, monitoring, and debugging/troubleshooting the system. The CLI is the most commonly used method for configuring, monitoring, and troubleshooting a network device.

For example, configuration commands include those used for modifying, storing, and downloading device (e.g., switch, router, switch/router) setting from an internal or external storage facility. Generally, the OS used in routing devices is multitasking and provides the user (network administrator) with a CLI and various software utilities and tools. Some of the in-built tools include commands that can be used to display the contents of various data structures such as the contents of the routing tables, or the internal data structures of the routing protocols running in the routing device.

Other software tools include Ping [RFC792] and Traceroute [RFC1393]. Ping is a software tool used for testing end-to-end connectivity between two devices running the Ping software (e.g., two routers). Traceroute is a software tool used to determine the path packets travel over from one point on a network to another.

3.7.3.4.2 System Physical Management Interfaces

The management ports on a routing device provide the network administrator, the necessary management access to the entire system. Routing devices typically support out-of-band management through one or more dedicated management Ethernet interfaces, as well as EIA-232 console and auxiliary ports. The management ports allow the routing device (via RPM) to be connected to one or more external devices through which the system administrators can issue commands (instructions) to manage the device.

In most architectures, the management facility on the RPM is a single dedicated module that interfaces with the rest of the modules in the routing device via the system backplane or midplane. Typically, all routing devices (more particularly, network devices in general) have the following basic system management interfaces (or ports) as shown in Figure 3.21.

3.7.3.4.2.1 Console Port

The console (CON) port is a port with a simple asynchronous serial controller that allows a user (e.g., network administrator) to access the routing device via a terminal. The console port connects the RPM to an external management terminal/console through an RS-232 (EIA-232) serial cable. This asynchronous EIA/TIA-232 (RS-232) serial port is used to connect an external terminal to the routing device, allowing local administrative access. Generally, all routers support a console port that provides an EIA/TIA-232 asynchronous serial connection.

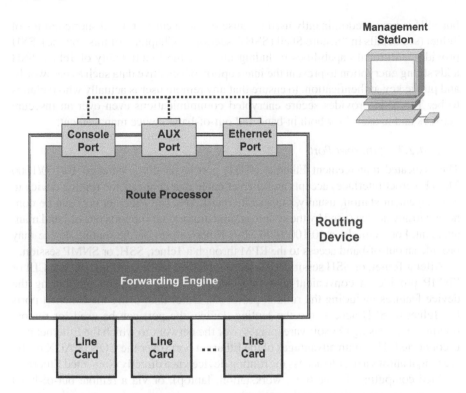

FIGURE 3.21 Route processor and system management interfaces.

A network administrator can connect a management terminal to the router and then to use a CLI running on the terminal to manage and monitor the device and the network. The serial connection used for configuring and monitoring the system's hardware/software configuration can be established from the traditional desktop, or laptop computer running terminal emulation and system access software such as Telnet and Secure Shell (SSH) as discussed in Chapter 2 of this volume.

3.7.3.4.2.2 Auxiliary Port

The auxiliary (AUX) port is an additional asynchronous serial port that can be used to access the routing device via either a direct connection to an external management terminal, or a remote out-of-band management connection using a modem. Most routing devices typically have two asynchronous serial ports, the console and auxiliary ports. Before the era of high-speed networking, the AUX port (also an asynchronous EIA/TIA-232 (RS-232) serial port) was used to connect a modem to the routing device for remote administrative access. The AUX port is very similar to the CON port but was typically used for remote router management via a modem connection. Today, it is mostly used to connect to a desktop, laptop, or other optional equipment for Telnet and SSH management.

For management access, there are several ways one can communicate with a network device from a remote console. The most common methods are Telnet and SSH, but SSH is now predominantly used because of the security-related shortcomings of Telnet (see details in "Secure Shell (SSH)" section in Chapter 2 of this volume). SSH provides additional capabilities including much of the functionality of Telnet. SSH adds strong encryption to prevent the interception of sensitive data such as passwords, and public key authentication, to ensure that any remote user is actually who it claims to be. The SSH provides secure encrypted communications even over an insecure network and is useful for both in-band and out-of-band device management.

3.7.3.4.2.3 Ethernet Port

The dedicated management Ethernet (ETH) port (typically a standard 10/100/1000 Mb/s Ethernet interface) accepts an Ethernet cable that connects the routing device to a management station, usually, via an Ethernet LAN. The Ethernet port can be connected to any device (used by the system administrator) that supports out-of-band management. For example, a 10/100/1000 Mb/s Ethernet port on the routing device may provide an out-of-band access to the RPM through a Telnet, SSH, or SNMP session.

After a Telnet, or SSH session has been established to the routing device, a CLI or SNMP provides a convenient user interface for configuring and monitoring the device features including the routing protocols. Other than using the Ethernet ports for Telnet or SSH access into the routing device, the port can be used for remote booting or accessing OS software images over the network to which the Ethernet port is connected. The main advantages of the Ethernet port (over the CON or AUX ports) are that, it allows a user to access the routing device via a directly connected Ethernet-enabled computing device (e.g., workstation, laptop), or via a remote out-of-band management connection over the network.

By providing the different management ports, the CLI of a routing device can be accessed via a direct connection from a management terminal to the CON port, or via a remote TELNET [RFC854] or Secure Shell (SSH) connection [RFC4251] over a network to the Ethernet port (see the "Secure Shell (SSH)" section in Chapter 2 of this volume).

3.7.3.4.3 Air-Temperature Sensors for Environmental Monitoring

Most high-end routers have an environmental monitoring system that monitors the physical conditions and health of the router. Most of the logic for the environmental monitoring functions normally reside on the RPM. A separate dedicated control and monitoring card may be present in the router, but for our discussion, we consider this as a logical extension of the RPM. In more sophisticated designs, environmental sensors and monitors located on the router chassis are connected to a chassis interface card which in turn interfaces with the RPM.

The RPM and it associated monitoring devices ensure that supply voltage(s) and the chassis temperature are maintained within the normal operating ranges. Temperature sensors are appropriately located in the system, and send temperature information to the associated monitoring card. The RPM would typically report all voltage and temperature readings, and make these readings available to the network administrator through standard software commands provided for environmental monitoring.

In some systems, if a measurement exceeds acceptable margins, a warning message is printed on the system management console. The system software may collect measurements once every 60 seconds, for example, but warnings for a given test point may be printed at most once every 4 hours. If the temperature measurements are out of specification (exceeding values set for system shutdown), the software shuts the router down (but generally, the cooling fans are allowed to remain on). In such a case, the router must be manually turned on after such a shutdown.

Typically, on routers with an environmental monitor, if the software detects that any of its temperature test points have exceeded maximum margins, it performs the following steps:

1. Saves the last measured values from each of the test points to internal non-volatile memory (e.g., the system may be configured to save the last six test points).
2. Interrupts the system software, and causes a shutdown message to be printed on the system management console.
3. Shuts off the power supplies after a few milliseconds of delay.

The system may display specific diagnostic message if temperatures exceed maximum margins, along with a message indicating the reason for the shutdown.

3.7.3.5 Other System Components

The RPM may support other components that include the following:

- *Craft Interface*: Typically, the craft interface, which is a visible interface on the front panel or plate of the device, provides short at-a-glance useful information (e.g., status and troubleshooting information) that allows the network administrator to perform several system control functions. The craft interface may provide the following information:
  o Alarm light-emitting diodes (LED) and alarm cutoff/lamp test button
  o Route processor management ports and status indicators
  o Link and activity status lights
  o Line card online/offline buttons
  o Alarm relay contacts
  o LCD display and navigation buttons
  o Management ports for management and service operations. This allows the network administrator can connect the route processor to an external console or management station through ports on the craft interface.
- *Alphanumeric Display (or LCD)*: This is usually a small alphanumeric display on the front panel that provides information on the state of the route processor and the system as a whole.
- *Alarm contacts*: Alarm relay contacts on the RPM provides a connection from the routing device to a site alarm maintenance system.
  o If this feature is used, all critical, major, and minor alarms generated by the router can activate the network administrator's site external visual or audible alarms as well as the LEDs on the router's front panel.

- o In some architectures, a network administrator can also connect the router to an external alarm-reporting devices through the alarm relay contacts on a craft interface, to provide for remote indication of critical, major, and minor router conditions.
- *External timing inputs*: These inputs provide timing signal inputs, for example, from an external timing source such as Global Navigation Satellite System (GNSS timing). This ensures that the clock signals used by the router remain synchronized with an accurate external reference or network clock.
  - o There may be two input clock ports (a primary clock and a secondary (or redundant) clock) on the RPM for external clock sources.
  - o The two input clock ports provide system clock redundancy, allowing the internal system clocks to remain synchronized with the external reference clock.
- *Reset button*: This button reboots the RPM (and possibly, the whole system) when pressed.
- *Offline button*: This button powers down the RPM (and possibly, the whole system) when pressed.
- *I2C controller*: This is used to monitor the status of various router components. The I2C (Inter-Integrated Circuit (I^2C or IIC)) controller is a bi-directional serial bus that provides a simple and efficient method of transmitting data between different devices in a device (intra-board communication) over short distances.
- *LEDs*: These indicate RPM (and system) status. Generally, the router vendor provides support instructions that describe the LED states. For example, a system may support a blue LED that may be labeled MASTER, a green one labeled OK, and an amber one labeled FAIL.

3.7.4 IMPORTANT ROUTE PROCESSOR MODULE SOFTWARE COMPONENTS

The software processes that run in the route processor run on top of a kernel that also interacts with other components such as the packet forwarding engines, the line cards, and the network interfaces (Figure 3.22). These software processes include the processes that manage the routing protocols and maintain the routing tables (i.e., *routing protocol process*), control the router's interfaces (i.e., *interface process*), control chassis components (i.e., *chassis process*), provide the interface for system management and control user access to the router (i.e., *management process*), and several others, depending on the architecture and functions performed by the router in the network.

The routing protocol process is responsible for handling all routing protocol tasks (e.g., RIP, EIGRP, OSPF, IS-IS, BGP). It runs on top of the kernel (that coordinates the communication among the various routing protocols) and the software process that interfaces with and controls the packet forwarding engine. Through the software routing protocol process, a network administrator can configure the routing protocols that run on the system and the properties of the network interfaces participating in a routing protocol. The software control processes control the router's interfaces and the chassis itself. These processes include the interfaces process and the chassis process.

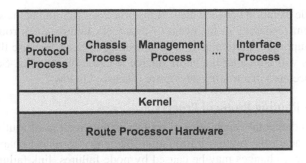

FIGURE 3.22 Routing Operating System (ROS) software.

The software management process provides the interface for system management (i.e., all the software processes for the management of network interfaces, the network itself, the Management Information Bases (MIBs), and the router chassis). For example, after a software configuration has been activated in the system, the management processes can be used to monitor the protocol traffic passing through the router, and to troubleshoot protocol and network connectivity problems.

This section describes the various software processes in the typical routing device.

3.7.4.1 Route Processor Kernel

The *kernel* is the central part of the router OS that manages the system's processing and memory resources, and the communication between hardware and software components. The kernel is the basic component of an OS that also provides abstraction layers for hardware (especially for the memory and processors), and the I/Os that allows hardware and software to communicate. The kernel also makes these facilities available to user-level applications through inter-process communication mechanisms and system calls.

Different kernels handle these tasks differently, depending on their design and implementation. Monolithic kernels are designed such that, all the code is executed in the same address space (kernel space), with the goal of increasing the performance of the system. Microkernels, on the other hand, are designed to run most of their services in user space, with the goal of improving maintainability and modularity of the codebase.

The route processor's kernel provides the underlying software infrastructure for all the router software processes. It also provides the link between the routing tables (maintained by the routing protocol process running on the route processor) and the master forwarding table maintained by the route processor itself. Additionally, the kernel coordinates communication with the forwarding engine(s), which involves, primarily, synchronizing the forwarding table of the forwarding engine with the master forwarding table maintained by the route processor.

The kernel is responsible for running a number of daemons that perform important tasks in the router. A *daemon* is a program that runs continuously as a background process with the sole purpose of handling periodic service requests that the system receives. Typically, in high-performance routing systems, each daemon operates in its own protected memory space. This protected memory space and the communication among the processes is also controlled by the kernel. This separation is done to provide isolation between the various processes running in the system, which leads to system resiliency in the event of a process failure. Such router resiliency features are important, particularly, in core routing devices, because the failure of a single process will not cause the entire device to cease functioning. Some common daemons (processes) in a routing device are discussed below.

3.7.4.2 The Routing Protocol Process

One of the fundamental capabilities of routers is the creation of routing tables that automatically adapt the routing information they carry to reflect changing network topologies. These changes may be caused by node failures, link failures, and additions and deletions of nodes and links to the network. The dynamic routing protocols detect such network changes and provide the routing information to populate the routing tables. It is important to note here that routing protocol processing (i.e., exchanging routing updates and maintaining the routing table) is a software-based activity that operates independently from the packet forwarding process (i.e., forwarding table lookups, packet rewrites and forwarding).

As noted above, the *routing protocol process* in the route processor runs the routing protocols that control the routing behavior in the network (i.e., the routing behavior of the router in relationship to other routers in the network). The routing protocol process is responsible for starting all configured routing protocols on the interfaces on which they are enabled as well as handling all routing messages sent and received. This process consolidates the routing information learned by the routing protocols and all information sources (i.e., directly connected networks and static routes) into a common routing table.

From the routing table, the routing protocol process extracts the most relevant information for packet forwarding (e.g., destination IP network prefix, next-hop IP address, and outgoing interface) for each network destination and installs this in the route processor's forwarding table (see discussion in Chapter 5 of Volume 1). The routing protocol process also controls the communication between the route processor and the packet forwarding engine(s). An additional function (depending on router architecture) may be to retrieve the packet forwarding statistics such as per-interface input/output statistics from the packet forwarding engine(s).

Also, other than exchanging routing protocol messages and performing routing table updates, the routing protocol process implements all the routing policies that the network administrator specifies. The routing policies determine how routing information is transferred between the routing protocols, and the route import and export policies of the routing table (see Figures 3.23 and 3.24). Note that the packet filter policies (which are applied to network interfaces) are configured and managed via the management process (see Figure 3.25).

3.7.4.3 The Interface Process

The *interface process* manages the router's physical interfaces as well as logical interfaces. This process configures and controls both the physical and logical properties of the interfaces on the router. It is responsible for implementing the commands and configuration statements that the network administrator uses to specify interface properties such as device location (e.g., the slot in which a line card is located and

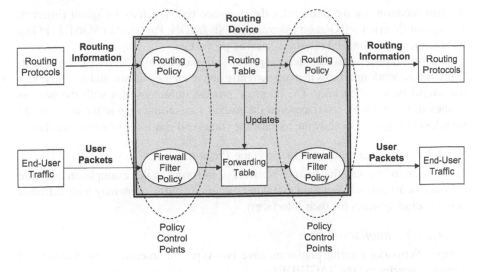

FIGURE 3.23 Policy control points in a routing device.

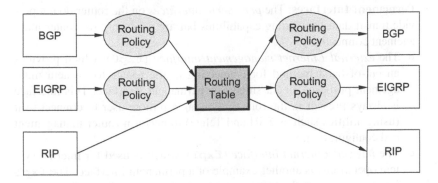

FIGURE 3.24 Controlling routing information flow using routing policies.

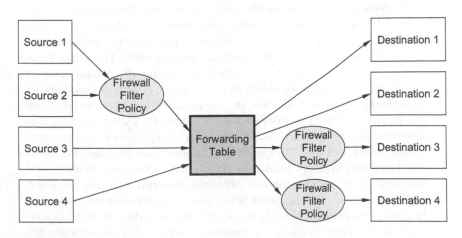

FIGURE 3.25 Controlling packet flow using firewall filters.

the port locations on the line card,), the interface type (such as 1 Gigabit Ethernet, 10 Gigabit Ethernet, 40 Gigabit Ethernet, SONET/SDH, Packet over SONET (POS), ATM), encapsulation, and interface-specific properties.

In some architectures, the interface process allows a network administrator to configure the network interfaces that are currently active on the chassis, and the interfaces that might be installed later. The interface process communicates with the process entities in the network interfaces and the packet forwarding engine through the OS kernel, enabling the OS software to track the status and condition of router interfaces.

3.7.4.3.1   Physical and Logical Interface Properties

In this section, we use the Juniper Networks platforms as the example architecture for our discussion, while noting that other routers manufacturers may have different classification systems for their interfaces.

3.7.4.3.1.1   Interface Types

Juniper Networks routing platforms have two types of interfaces: *permanent* and *transient* interfaces [JNCIAGUIDE].

- **Permanent Interfaces**: The *permanent interfaces* on the router do not provide transit data forwarding capabilities but are used only for router management connectivity.
  - The *external Ethernet management interface* (**fxp0**), which provides an out-of-band method for connecting to the router for system management purposes, is classified as a permanent interface. This interface is always present in the router, allowing a remote user to connect to it (using utilities such as SSH and Telnet) to perform router management and configuration.
  - The *internal Ethernet interface* (**fxp1**), which is used for internal system operations, is another example of a permanent interface. The **fxp1** interface connects the route processor to the packet forwarding engine. This interface provides a path for sending routing protocol packets to the route processor for routing table calculations. This interface is also used for updating the forwarding table used by the packet forwarding engine. When the JUNOS software boots, it configures, addresses, and enables the internal Ethernet interface automatically. The **fxp1** interface is never disabled or reconfigured, as altering the default behavior can seriously impair the ability of the router to perform its functions.
- **Transient Interfaces**: These are physical interfaces (cards or modules) that can be removed from or inserted into the router by a user. The *transient interfaces* are the interfaces that receive user data packets to be processed by the router, and then forwarded toward their final destinations. These interfaces are physically located on the line card modules called Physical Interface Cards (PICs) in Juniper Networks platforms. The user can insert and remove them from the router at any time. Each transient interface must be configured before it is ready for operational use. Also, the JUNOS software allows a user to configure transient interfaces that are not currently

installed in the router chassis (i.e., preconfigure a chassis slot to accept an interface with a specific configuration). Then, when a user installs a new transient interface in the chassis (for which the router has some existing configuration), the JUNOS software will activate the parameters for that interface [JNCIAGUIDE]. When the JUNOS software is in the process of activating the router's configuration, it detects which transient interfaces are actually present in the chassis and activates only those interfaces.

3.7.4.3.1.2   Interface Naming and Numbering

The physical placement of interfaces on a router is important. The router references all active interfaces within its running configuration and also when forwarding packets to their destinations. Each chassis slot in which a line card (including the RPM) is inserted, as well as each interface on the line card (if the line card supports multiple interfaces or ports) receives a unique name (identifier) based on its location on the router chassis. Also, each router manufacturer has an *interface naming convention* that is usually easy to understand (see, for example, [JNCIAGUIDE] for a description of the naming convention in Juniper Network routing platforms).

Usually, the media type of a physical interface (including its speed) is uniquely identified by a two-character designator (e.g., ge for Gigabit Ethernet interface). Typically, each router model supports a specific maximum number of slots on the chassis for the line cards and the RPM(s). The physical media cable from the user's network (e.g., Ethernet or SONET/SDH) connects to a port on the line card and these data ports are represented in the interface naming structure as numbers. Note that routing platforms like those from Juniper Networks have line modules that go into chassis slots (called Flexible PIC Concentrators (FPCs)) which also contain sub-slots (called PICs). The PIC is the module on which the actual network interfaces or ports are located. In this architecture, the chassis slots (FPCs), the sub-slots (PICs), and the actual data ports are each uniquely identified using the manufacturer's naming convention [JNCIAGUIDE].

3.7.4.3.1.3   Logical or Virtual Interfaces

A router may also support *logical* or *virtual interfaces* that are located on a physical interface or port. An Ethernet port may be handling several *sub-interfaces*, and an ATM port may be supporting several *virtual circuits* (VCs). *Sub-interfaces* are logical interfaces that are created by dividing a single physical interface into multiple virtual interfaces. The sub-interface uses the parent physical interface for receiving and sending packets. Sub-interfaces can be configured just like physical interfaces. A Layer 3 sub-interface can have its own properties such as IP addressing and routing, packet forwarding policies, maximum transmission unit (MTU) and IPv4 fragmentation, ACLs, and QoS policies.

An IEEE 802.1Q VLAN sub-interface is a virtual interface on a routed physical interface that is associated with a VLAN ID. A sub-interface can be created only on a Layer 3 physical interface. The parent interface is the physical port on which it is created. Other than being assigned unique Layer 3 parameters such as IP addresses, sub-interfaces can be assigned dynamic routing protocols. The IP address assigned to each sub-interface must be in a different IP subnet from any other sub-interface created on the parent physical interface.

For example, we may want a router with one physical interface to be connected to two IP subnets and to route traffic between the two subnets. In this case, we can create two sub-interfaces within the physical interface, each assigned an IP address within each subnet. The router can then route packets between the two subnets. Sub-interfaces can be used for inter-VLAN traffic routing using a one-armed router (or router-on-a-stick) configuration as discussed in Chapter 6 of Volume 1. We discussed in Chapter 6 of Volume 1 that sub-interfaces can be used to create unique Layer 3 interfaces to VLANs that are supported on the parent physical interface. In this case, the parent physical interface located on a routing device connects to a Layer 2 trunking port attached to another device. The user configures the sub-interface and associates it to a VLAN ID using IEEE 802.1Q trunking.

In addition to the placement of an interface (or port) on a line card located in a chassis slot, any logical or virtual sub-interfaces and channels within that physical interface are also numbered since they have to be represented within the router configurations, and also known (and identified) for packet forwarding purposes. For example, an Ethernet port may contain a number of sub-interfaces or logical channels. A channelized (i.e., non-concatenated) SONET/SDH OC-48 interface has four OC-12 channels which may be numbered 0 to 3. A channelized OC-12 interface has 12 DS-3 channels which may be numbered 0 to 11.

3.7.4.3.1.4 Interface Properties

When created, interfaces have both *physical* and *logical properties*. The type of media on which the interface is created (e.g., Ethernet, SONET/SDH) often determines the physical properties of the interface. The logical properties of an interface represent the Layer 2 transmission and Layer 3 routing parameters required to operate the interface in a network.

- **Physical Properties**: When an interface is activated on a router, it inherits certain default values for its physical properties that are placed in the router configuration file. Some of the possible physical property options include the following [JNCIAGUIDE]:
  - **Description**: This is a user-defined text that describes the interface's purpose.
  - **Diagnostic characteristics**: These properties are user-configured on a per-physical interface basis and include information such as circuit-testing capabilities, such as loopback settings or Bit Error Rate Test (BERT) tests. A loopback test can be used to verify the connectivity of a circuit, while BERT is used to track poor signal quality due to noise on a line.
  - **Encapsulation**: This describes the type of encapsulation used on the interface.
  - **Frame check sequence (FCS)**: This value is used for error-checking packets that are received on the interface. In [JNCIAGUIDE], the default value can be from a 16-bit value to a 32-bit value.

o **Interface clock source**: Point-to-point interfaces such as those based on SONET/SDH and PDH (Plesiochronous Digital Hierarchy) technologies require a clock signal source for synchronization purposes. The clock signal options include using an internal clock signal (i.e., **internal** which is the default), or an external signal source such as a GNSS reference (i.e., **external**).

o **Interface MTU size**: This represents the MTU of the physical interface, a value that can be changed. Each interface has a different default MTU value, ranging from 256 to 9192 bytes.

o **Keepalives**: This is a Physical Layer mechanism used to determine whether the interface is still up and functioning correctly.

o **Payload scrambling**: This is a mechanism used in long-haul communications to help make transmissions error-free.

- **Logical Properties**: Each interface in a Juniper Network routing device running the JUNOS software requires at least one logical interface (called a *logical unit*) which holds all configured addressing and protocol information [JNCIAGUIDE]. Some physical interface encapsulation types such as Point-to-Point Protocol (PPP) use only a single logical unit. The loopback and non-VLAN Ethernet interfaces also use only one logical unit which is assigned a *unit value* of 0. Interfaces based on ATM or those that support VLAN-tagged Ethernet networks have multiple logical units. Each logical unit in an ATM interface (called a virtual circuit (VC)) is assigned a *virtual circuit identifier* (VCI), while a logical unit in an interface that supports VLAN-tagged networks is assigned a VLAN number or ID. These mechanisms allow the user to create and map multiple logical interfaces onto a single physical interface. Each logical interface is viewed as a separate entity in the JUNOS software. Common logical interface properties in the JUNOS software include a protocol family, logical Layer 3 protocol addressing, protocol MTU, and VC (Layer 2) addressing information:

  o **Protocol Families**: Each logical interface is capable of supporting one or more protocol families which enable the interface to receive and process packets for the router [JNCIAGUIDE]. An interface has to be configured with a protocol family for it to accept and process packets for that family. Some of the possible protocol families are as follows:
    - **inet**: This protocol family supports IPv4 packets.
    - **inet6**: This protocol family supports IPv6 packets.
    - **iso**: This protocol family supports IS-IS packets.
    - **mpls**: This protocol family supports MPLS packets.

  o **Layer 3 Protocol Addresses**: When any of the protocol families **inet**, **inet6**, or **iso** is configured on an interface, a corresponding logical Layer 3 address (IPv4, IPv6, or IS-IS) must also be configured to be used for routing packets in the network.

    Juniper Networks JUNOS software uses the concepts of a *primary address* and the *preferred address* for addressing interfaces. JUNOS assigns a single *primary address* to each interface, which by default, is the lowest numerical IPv4 address configured (e.g., the IPv4 address 10.10.20.2 /24 is lower than 172.16.1.2 /24). The *primary address* is used as the source address of a packet (originated by the router) when the destination address is not part of the interface's IP subnets (i.e., not local to a configured subnet on the interface). For example, let us assume the interface fe-0/0/0.0 on a Juniper Networks router has both 10.10.20.2/24 and 172.16.1.2/24 configured on it. We also assume the ping command is used on the router and an IPv4 packet is sent with a destination address of 192.168.200.20. Since the destination address is not part of the IP subnets on the interface, the router uses the primary interface address of 10.10.20.2 as the source IP address within the ping packet.

A logical unit may have multiple *preferred addresses* assigned to it at the same time unlike the *primary address*. The router uses the *preferred address* when an interface has two addresses configured within the same IP subnet. Similar to the *primary address*, the default selection of the *preferred address* is the lowest numerical IPv4 address. The *preferred address* can also be used as the source IP address of a packet similar to the *primary address*. The JUNOS software allows multiple IPv4 addresses to be configured on a logical unit. The inet family allows multiple addresses to be assigned to each logical unit, with each IPv4 address equally represented on the interface.

Cisco router allows multiple IP addresses to be assigned to an interface using the concept of *primary* and *secondary addresses*. However, the router uses only the *primary address* for all interface functions. Juniper Networks routers, on the other hand, do not see any functional differences between the addresses assigned on an interface; all the addresses are equal to the JUNOS software. Each of these multiple addresses forms routing protocol neighbor relationships separately, and each is advertised into the corresponding IGP.

o **Protocol MTU**: The difference between the *interface MTU* and the *protocol MTU* is highlighted in [JNCIAGUIDE]. The *interface MTU* is the largest size packet an interface can send on the physical media. This MTU value includes all Layer 2 overhead information (such as the Ethernet sources and destination MAC addresses in Ethernet transmission, or MPLS labels in MPLS transmission), but excludes the Cyclic Redundancy Check (CRC) information. All encapsulation types contain a payload field where higher-layer information is stored. The size of the *protocol MTU* corresponds to the size of this payload field. Thus, the protocol MTU is the largest amount of logical protocol data, including the protocol header, that a particular interface is able to send.

Each logical unit in the router can be configured with an MTU value. The default values that can be configured for a logical unit vary for each physical media type and the protocol family configured on the interface:

■ **Broadcast interfaces**: The different Ethernet interface types (i.e., 100 Mb/s, 1 Gb/s Ethernet) have the same properties for MTU sizes. The MTU size for inet family is 1500 bytes, iso family is 1497 bytes, and the mpls family is 1488 bytes.

- ■ **Point-to-point interfaces**: For encapsulation type PPP or ATM, for example, the default MTU for the `inet` and `iso` protocol families is 4470 bytes. The `mpls` protocol family uses an MTU value of 4458 bytes.
- o **Virtual Circuit Addressing**: An ATM interface, or an interface that supports Ethernet VLANs also requires Layer 2 virtual circuit or channel addresses. These options are examined here:
  - ■ **ATM VPI and VCI**: ATM uses the concept of a *virtual path* (VP) and a *virtual circuit* (VC) to connect any two devices in an ATM network. The VP is represented by a *virtual path identifier* (VPI), which represents a larger logical conduit between the devices. Each VP (with VPI) in a network may contain multiple VCs each represented by a *virtual circuit identifier* (VCI), which is the actual connection between the devices. Each logical unit in the router is assigned a VPI/VCI Layer 2 number. The VPI values range from 0 to 255, while the VCs on that VP can have VCIs with values from 0 to 4089. These values have local significance only, meaning the two endpoints can have different values, allowing for greater overall scalability. For example, let us assume that router Router1 is connected to Router2 through an ATM interface. We assume also that 10 logical units are created on this ATM interface, each assigned a unique IPv4 address and VCI value. When the two routers exchange data packets, the VCI value at Layer 2 helps each router determine which logical unit should receive and process the packet.
  - ■ **Ethernet VLAN Tags**: Ethernet has inherent broadcast capabilities and IEEE 802.1Q allows an Ethernet interface to be channelized into multiple logical interfaces, referred to as VLANs. As discussed in Chapter 6 of Volume 1, a VLAN is a logical subnet (or Layer 2 broadcast domain) that allows many different hosts to connect to an Ethernet switch while still staying in their separate logical subnets (or broadcast domains). IEEE 802.1Q can support up to 4094 VLANs as discussed in Chapter 6 of Volume 1.

3.7.4.3.2 *Software User Interfaces*

As discussed earlier, the OS software user interfaces allow the network administrator to interact with and configure the router. Typically, routers provide two main types of user interfaces. The first is using a management station running the Hypertext Transfer Protocol (HTTP) and a standard Web browser and a connection to any of the management ports (Console, Auxiliary, or Ethernet (see Figure 3.26)). This GUI-based method allows the user to configure, monitor, troubleshoot, and manage the router from a client host running a Web browser. The second method is using a CLI over an SSH (or Telnet) connection to any of the management ports. In the second case, the SSH or Telnet session through the management port is the access method, while the CLI is the user interface for interacting with the router. There is a third (mostly limited) option, that is, using scripts as shown in Figure 3.27. For example,

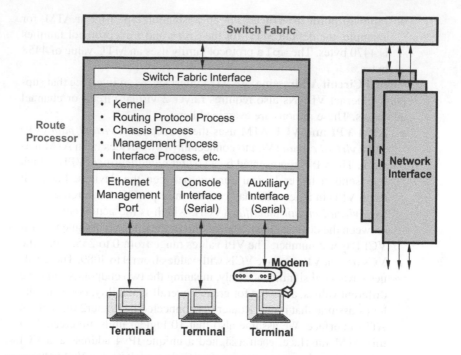

FIGURE 3.26 Management interfaces.

the user can develop custom scripts for configuring and monitoring the router as follows [JUNCLIGUID21]:

- Scripts to enforce custom configuration rules.
- Scripts to provide simplified aliases for configuration statements that are frequently used.
- Scripts that configure diagnostic event policies on the router and the actions associated with each policy.

In many cases, only authorized personnel (i.e., users with the right access privilege levels) are permitted access to the router for administration and configuration. These users can easily edit the router configuration using the Web browser interface or the CLI. Figure 3.27 shows the relationship of the user interfaces with respect to the rest of the processes involved in system/chassis management. In addition to configuring the interfaces, the user interfaces allow authorized personnel to access the router to monitor the status of physical and logical interfaces, verify their operation, and perform troubleshooting.

Other than the software interfaces in Figure 3.27, the user can use the Network Configuration Protocol (NETCONF) Application Programming Interface (API) to configure and monitor the router. NETCONF is an IETF-defined network management protocol that provides mechanisms for installing, manipulating, and deleting the configuration of network devices [RFC6241]. Data encoding for the

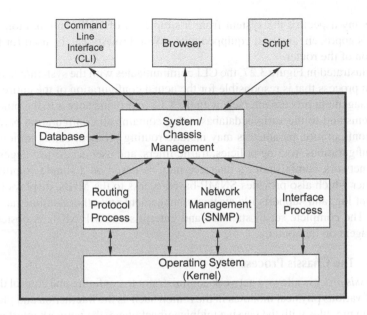

FIGURE 3.27 User interfaces for system/chassis management.

configuration data and protocol messages in NETCONF are based on the Extensible Markup Language (XML). XML is a markup language that defines a set of rules that allow a user to encode documents in both human-readable and machine-readable format [XMLW3C1.0]. NETCONF operates on top of a simple Remote Procedure Call (RPC) layer. The protocol message exchanges are done on top of a secure transport protocol (e.g., NETCONF Protocol over Secure Shell (SSH) [RFC6242] or NETCONF Protocol over Transport Layer Security (TLS) with Mutual X.509 Authentication [RFC7589]). The user can also use standard and enterprise-specific MIBS to retrieve information about the router hardware and software components.

Typically, a CLI provides full router configuration and status functionality through a local serial port, or over an SSH (or Telnet) connection from any reachable network. As discussed above, the CLI allows the user to perform various tasks including configuring the router, restarting system processes, and monitoring the operation of the routing protocols.

Most routers also support device management application systems (mostly Web browser-based GUIs) that provide a global method of managing all router components such as the route processor, line cards, and individual router ports; a single GUI for managing the entire system (see also discussion in Chapter 5 of Volume 1). The router OS software typically includes a Web browser, which is password protected, from which a network administrator can issue router commands. Typically, this router OS feature is accessible from the router's home page and can be customized for a particular business environment.

For example, for a simple GUI-based configuration, the user may maintain a set of Java-based applets that can be uploaded to any Web browser. Using these Java applets, the user can configure the router's ports, set up routing protocols, and manage any aspect of the system from a simple set of GUI configuration screens. With this approach, any host equipped with a Web browser can be used for the configuration of the router.

As illustrated in Figure 3.27, the CLI communicates with the system/chassis management process that is responsible for the actual configuration of the entire system. The management process interacts with the CLI to provide access to the other router subsystems and to the various databases that contains all configuration parameters. These configuration parameters may include routing protocol configurations, interface configurations, routing policies, traffic filters, and user access privileges.

The network administrator's interface may be based on a simple command line interpreter which also provides SNMP-based access to the MIBs, displays the router's event log, and supports file system management and other administrative commands. The complete set of standard and enterprise SNMP MIBs is hosted by an SNMP agent on the router.

3.7.4.4 The Chassis Process

The *chassis process* allows a network administrator to configure and control the properties of various physical modules in the router, such as the interaction and placement of system modules with the passive midplane/backplane, the network interface modules (or line cards), and the different system control boards. The chassis process also controls the power supplies, cooling fans, clock signal sources, the conditions that trigger alarms, and the sending of SNMP traps. The chassis process in the individual system components communicates directly with a chassis process in the router OS kernel.

3.7.4.5 The SNMP and MIB Processes

As described in Chapter 2 of this volume, SNMP is an Application Layer protocol for transferring information about managed network devices in an IP network. SNMP operates over UDP as also discussed in Chapter 3 of Volume 1. A network administrator can use SNMP to access information about a managed device such as network packet error rates. This allows the administrator to have a better view of network performance and also carry out troubleshooting/debugging tasks when network problems occur. As discussed in Chapters 5 of Volume 1 and Chapter 2 of Volume 2, an SNMP system consists of three main parts, the *SNMP manager*, the *SNMP agent*, and the *MIB*. The SNMP agent resides on a network device such as a switch, router, or switch/router, while the SNMP manager runs software that is typically implemented as part of a *network management system* (NMS).

An MIB is a group of managed objects that are stored in a virtual database in a managed device. The MIB is a repository for information about network performance and device parameters. The value of an MIB object can be retrieved or changed using SNMP commands, usually through a GUI-based NMS. The SNMP agent can receive MIB-related queries from the NMS, and then respond with the appropriate information. The SNMP manager can request or change the value of any of the MIB variables the SNMP agent contains. The SNMP manager can get a value from the SNMP agent or store a value in the agent's MIB. The SNMP agent gathers data from the MIB, and can also respond to requests sent by the SNMP manager to get or set data. The SNMP agent can also send unsolicited SNMP traps to the SNMP manager. SNMP traps are messages sent by the agent to alert the SNMP manager to a condition on the device and network. An SNMP trap can indicate link status (up or down), improper user authentication, restarts, loss of connection to a neighbor router, or closing of a BGP TCP connection.

Typically, all router OS software supports SNMP which provides a mechanism for monitoring the state of the router. This software is controlled by the *SNMP and MIB processes* (Figure 3.27), which consist of an SNMP master agent and an MIB agent. Routing devices typically offer a complete SNMP interface for device configuration, system status, and alarm reporting. A system would typically support both standard and enterprise MIBs.

As explained in Chapter 2 of this volume, the SNMP agent runs on the router, exchanging network management information with the NMS software running on SNMP manager (which is an IP host in the network). By polling the router, the SNMP manager collects information about network activity, events, and connectivity. The SNMP agent responds to requests for information and actions from the SNMP manager. The SNMP agent also controls access to the MIBs the agent maintains. Usually, an OS software chassis MIB represents each component in the router and contains information about the status of those components. The SNMP agent sends a trap to the SNMP manager software when an event occurs on the router. Generally, the agent sends SNMP traps to report significant events occurring on a router, most often failures or errors.

3.7.4.6 The Management Process

The *chassis process*, the *SNMP and MIB processes*, and the *management process* all form part of a bigger *system management process*. The *management process* starts all the other router OS software processes and the CLI when the router boots. It monitors the running OS processes and makes all reasonable attempts to restart any process that terminates. The management process also controls all the tools for accessing and configuring the router software. These include CLI, SNMP, XML, Web, and other management tools.

The CLI is the primary tool for accessing and controlling the OS software. A network administrator can use it when accessing the router via a console or an out-of-band connection from a remote management station. The CLI includes commands for configuring router hardware, the router OS software, and network connectivity. Most router CLIs are straightforward command interfaces. A network administrator can type commands on a single line and enter the commands by pressing the **Enter** key on the keyboard. The CLI also provides *command completion, command help facilities*, as well as *Emacs (Editor macros) style keyboard sequences* that allow the user to move around on a command line and scroll through a buffer that contains recently executed commands.

The management process also supports tools for monitoring the router software. Other than the commands for configuring router software and hardware, the CLI includes commands for monitoring and troubleshooting network connectivity, router software and hardware, and routing protocols. The CLI could support commands to display information from routing tables, information specific to routing protocols, and information about network connectivity derived from the `ping` and `tracer-out` utilities. After connecting to the router, the network administrator is prompted to supply authentication information (name and password) to the route processor. After successful authentication, the network administrator is able to perform configuration and management operations in the router, as well as carry out troubleshooting functions using tools like `ping` and `traceroute`.

A network administrator can also use the OS software's implementation of SNMP to monitor the router and the network. The software may also support tracing and logging operations, which a network administrator can use to track normal router operations, error conditions, and the packets that the router generates or forwards. A mechanism like the UNIX syslog command can be used by the logging operations in the system to record high-level, systemwide events such as user logins on the router, and network interfaces that are going down or up. Tracing operations provide the network administrator with more detailed information about the operation of routing protocols, such as the packets sent and received by the various routing protocols running on the system, and routing policy actions taken.

The management process may also support software upgrades on the router. Typically, the router is delivered with the router OS software preinstalled. To upgrade the software, a network administrator may use CLI commands to copy a set of software images over the network to appropriate memory storage on the route processor. The router OS software set may consist of several images provided in individual packages or as a bundle.

3.7.4.6.1   Juniper Networks JUNOS OS CLI Use Modes

As discussed above, a CLI is a software text-based interface that is used for accessing most network devices. After a router has successfully booted and appropriate software loaded, the CLI can be used to monitor and configure the router. A network administrator, using a station connected to the console port or a network connection through the management Ethernet port, can use the CLI to configure the device, monitor its operations, and adjust its configuration when needed. Generally, the CLI of a Juniper Network router has two main use modes: *operational* and *configuration* modes [JNCIAGUIDE] [JUNCLIGUID21].

In this section, we use the CLI of the Juniper Networks JUNOS OS software (as the reference example) to explain the main features and inner workings of the typical router CLI [JNCIAGUIDE]. The Junos OS CLI is a command shell that runs on a FreeBSD UNIX-based OS kernel. It uses industry-standard tools and utilities to provide a set of commands that can be used to configure and monitor network devices that run the JUNOS OS software.

We discuss this CLI by examining the operational and configuration modes of a Juniper router. We discuss the *context-sensitive help feature* that assists a user in navigating through the CLI, and the *command completion process*. We also discuss the differences between a *candidate* and *active configurations*, how configurations are *committed*, as well as the *rollback* functionality in the router. The discussion here is not meant to generalize to other router OS software CLIs like the Cisco IOS CLI, but only to highlight the role of the CLI in router configuration and management. The discussion aims to explain the role of the CLI, as well as provide an understanding of the main concepts related to router configuration and management: using configuration commands to configure routers; managing router configurations; and using configuration commands to monitor routers.

3.7.4.6.1.1 Operational Mode

The *operational mode* displays the current status of the router and is the mode through which the network administrator monitors and troubleshoots the router. The operational mode is entered after a successful login, and the router *prompt* displays the user's status graphically. Typically, the *default prompt* for the CLI is a combination of the username (e.g., **userjames**) and the router hostname (e.g., **Router1**). The character ">" indicates to the user that the system is in the operational mode [JNCIAGUIDE]:

```
userjames@Router1>
```

Most router CLIs use a *command hierarchy* structure within the operational mode. This allows the network administrator to find only the relevant information requested in a timely manner. For example, accessing the *top-level hierarchy* in the operational mode on router **Router1** may give the following:

```
userjames@Router1> ?
Possible completions:
  clear        Clear information in the system
  configure    Manipulate software configuration
               information
  file         Perform file operations
  help         Provide help information
  monitor      Real-time debugging
  mtrace       Trace multicast path from a source to a
               receiver
  ping         Ping a remote target
  quit         Exit the management session
  request      Make system-level requests
  restart      Restart a software process
  set          Set CLI properties, date, time, craft
               display text
  show         Show information about the system
  ssh          Open a secure shell to another host
  start        Start a software process
  telnet       Telnet to another host
  test         Diagnostic debugging commands
  traceroute   Trace the route to a remote host
```

This hierarchy level can be used for several different purposes. For example, when the **ping**, **telnet**, **traceroute**, and **ssh** commands are used, the router behaves as if it is an IP end host, generating corresponding IP packets on the route processor and

sending them into the network through a particular router interface. The user controls
the router's operations using commands such as **request**, **restart**, and **start**.
When upgrading the OS software, the **request** command can be used to load a new
version of the OS software.

The **show** command can be used to view the router's current status. The hierarchy
located in this directory allows the user to access routing protocol information, the
current routing table, and interface statistics. For example, the **show** hierarchy directory on the router gives the following information (listing here just a small number of
possible completions):

```
userjames@Router1> show ?
Possible completions:
    accounting      Show accounting profiles and records
    arp             Show system ARP table entries
    bgp             Show information about BGP
    chassis         Show chassis information
    cli             Show command-line interface settings
    configuration   Show configuration file contents
    interfaces      Show interface information
    isis            Show information about IS-IS
    log             Show contents of a log file
    ospf            Show information about OSPF
    policer         Show interface policer counters and
                    information
    policy          Show policy information
    rip             Show information about RIP
    route           Show routing table information
    snmp            Show SNMP information
    system          Show system information
    version         Show software process revision levels
```

Other features of the CLI include [JNCIAGUIDE] [JUNCLIGUID21]:

- **Context-Sensitive Help**: The question mark character **"?"** gives a *context-
  sensitive help* that assists the user in navigating the command hierarchy. The
  help function can be used at a specific hierarchy level, but it also assists the
  user in locating specific options within a particular level. For example, to
  see information about the OSPF routing protocol, the question mark can be
  used within the next level of the command hierarchy:

```
userjames@Router1> show isis ?
Possible completions:
    adjacency       Show the IS-IS adjacency database
    database        Show the IS-IS link-state database
    hostname        Show IS-IS hostname database
    interface       Show IS-IS interface information
```

```
route        Show the IS-IS routing table
spf          Show information about IS-IS SPF
             calculations
statistics   Show IS-IS performance statistics
```

- **Command Completion**: This CLI functionality allows each unique combination of characters that are entered at a particular hierarchy level to expand into the full command when either the spacebar or the Tab key is used. When the user enters a command partially and presses the spacebar, the entry is expanded to complete the command. The router OS parses the CLI when the user is typing a command which provides the user with an immediate syntax check. With this CLI feature, the user does not have to type out a long command and then press Enter only to be informed that an error has been made at the beginning of the command. The *command completion feature* is available at each level of the command hierarchy and also applies to other strings, such as usernames, filenames, configuration statements, and interface names.
- **Editing Command Lines**: This CLI feature allows the router to store operational mode commands in a history buffer as the user types them. This allows a user to repeat a command by simply accessing the previously entered version and then pressing the Enter key. For example, when the CLI and the user's terminal emulator are configured to use vt100 as the character output, the user can use the keystrokes left, right, up, and down arrows to access the CLI history, and to easily edit previously entered commands. The Backspace key in vt100 mode can be used to delete characters to the left of the cursor.

  The JUNOS software CLI is designed to respond to common *Emacs keystrokes*, allowing a user to edit the command line on all terminal types [JNCIAGUIDE]. The following are some of the common Emacs keystrokes:
  - **Ctrl+A**: Moves the cursor to the beginning of the current command line.
  - **Crtl+E**: Moves the cursor to the end of the current command line.
  - **Ctrl+B**: Moves the cursor back one character and is equivalent to the Left arrow key.
  - **Ctrl+F**: Moves the cursor forward one character and is equivalent to the Right arrow key.
  - **Ctrl+N**: Displays the next line in the CLI history buffer and is equivalent to the Down arrow key.
  - **Ctrl+P**: Displays the previous line in the CLI history buffer and is equivalent to the Up arrow key.
  - **Ctrl+W**: Deletes the word to the left of the cursor.
  - **Ctrl+X**: Deletes the entire current command line.
  - **Ctrl+L**: Redraws the current command line.
  - **Esc+B**: Moves the cursor back one word at a time. The Esc key must be released and re-pressed for each keystroke.
  - **Esc+F**: Moves the cursor forward one word at a time. The Esc key must be released and re-pressed for each keystroke.

- **Operational Command Variables**: The CLI provides the user with additional useful capabilities such as the ability to attach variables to valid commands through the use of the pipe key "|". For example, the help system can be used with the pipe functionality as follows:

```
userjames@Router1> show cli | ?
Possible completions:
    count      Count occurrences
    display    Display additional information
    except     Show only text that does not match a
               pattern
    find       Search for the first occurrence of a
               pattern
    hold       Hold text without exiting the
               --More-- prompt
    match      Show only text that matches a pattern
    no-more    Don't paginate output
    resolve    Resolve IP addresses
    save       Save output text to a file
    trim       Trim specified number of columns from
               start of line
```

The CLI also provides the user with the flexibility of combining multiple pipe options together.

- **Modifying the Command Output**: Each time a command is entered in the CLI by a user, the router generates the entire output information (which is stored in a buffer) before displaying any characters on the user's screen. When the router determines that the display output is longer than the user's terminal length, it paginates the output by displaying a prompt of, say, `--(more 18%)---`. This informs the user that more information is to follow, and how much of the output buffer the router has displayed (i.e., the user has seen); In this case, the user has seen 18 percent of the total output. Each time the terminal displays information and stops at a page break, the user has the option of modifying and manipulating the output. The user can access these features by pressing the **h** key at the **more** prompt. The following are some commonly used options and keystrokes:
  - The user can use **Ctrl+E** keystroke to access the bottom (or end) of the output buffer. This is useful when the user is examining log files where new information is placed at the end of the file.
  - The user can use the **Ctrl+B** sequence to move backward through any CLI output. This is useful for viewing information presented earlier in the display output without having to retype the command over again.
  - The user can exit from the display output and return to the command line at any time by using the **q** key.

○ The user can use the forward slash (/) key to search for a particular string in the display output. This moves the user prompt to the first occurrence of the supplied string and paginates the display output at that point. This use of the slash key is similar to the **find** pipe variable.

3.7.4.6.1.2 Configuration Mode

The *configuration mode* provides the network administrator with a method for altering the router's current environment. When a new router is first started in the field, the router OS typically runs an *autoinstall* process which prompts the user to answer a few questions. The router OS then configures the system based on the input provided by the user. After this initial setup, most commonly, the user modifies the configuration using the CLI. The user may also configure the router using other methods such as via a Web browser and HTTP, or other network management applications. The network administrator uses the configuration mode to configure the router, that is, enter commands to configure all properties of the router (including interfaces). The user can configure user access, several system and hardware properties, and the routing protocols.

A router configuration is stored as a *hierarchy of configuration statements* and a user can access the JUNOS OS CLI configuration mode hierarchy using either the **configure** or **edit** command [JNCIAGUIDE] [JUNCLIGUID21]:

```
userjames@Router1> configure
Entering configuration mode

[edit]
userjames@Router1#
```

The CLI uses the ">" prompt to visually show that the user is in the operational mode, and the pound character "#" prompt to show that the user is in the configuration mode. The **>** is changed to the **#** prompt when the user moves into the configuration mode, and the current level of the user in the hierarchy is displayed above the router's hostname. The **[edit]** line of the output on **Router1** indicates that the user is at the top of the configuration hierarchy. The user can view the command options at this level with the context-sensitive help system:

```
[edit]
userjames@Router1# ?
Possible completions:
  <[Enter]>     Execute this command
  activate      Remove the inactive tag from a
                statement
  annotate      Annotate the statement with a comment
  commit        Commit current set of changes
  copy          Copy a statement
```

deactivate	Add the inactive tag to a statement
delete	Delete a data element
edit	Edit a sub-element
exit	Exit from this level
help	Provide help information
insert	Insert a new ordered data element
load	Load configuration from an ASCII file
quit	Quit from this level
rename	Rename a statement
rollback	Roll back database to last committed version
run	Run an operational-mode command
save	Save configuration to an ASCII file
set	Set a parameter
show	Show a parameter
status	Display users currently editing the configuration
top	Exit to top level of configuration
up	Exit one level of configuration
update	Update private database

The **status** command shows the users who are currently in the configuration mode, how long each user has been in that mode, and the current configuration hierarchy level of each user. The **run** command is a very useful command in the configuration mode and allows the user to access the operational mode commands from within the configuration mode. This flexibility enables the user to easily verify information on the router (using any operational mode command) while still in the configuration mode. For example, after navigating to the **[edit protocols ospf]** hierarchy directory in configuration mode, the user can examine the current routing table without having to exit the configuration mode and move to the operational mode:

```
[edit protocols ospf]
userjames@Router1# run show route
```

The user may also use the **run** command to list the interface names on the router while in configuration mode:

```
[edit protocols ospf]
userjames@Router1# run show interfaces
```

Just as in operational mode, the configuration mode also supports the pipe variables with each command, output pagination with the (**more**) prompt, Emacs editor strings, and the command-completion function [JNCIAGUIDE] [JUNCLIGUID21].

- **Navigating within the Configuration Mode Hierarchy**: Conceptually, the *configuration mode hierarchy* can be viewed as a vertical top-down directory structure, with the top of the structure at the root of a tree. As is common with a directory system, each branch of the tree below the root forms a subdirectory below; each top-level subdirectory can branch out into its own set of subdirectories. Using the **edit** command, the user can navigate downward through this directory structure to the next lower directory:

```
[edit]
userjames@Router1# edit protocols

[edit protocols]
userjames@Router1#
```

The user can continue into one of the next lower directories as shown below:

```
[edit protocols]
userjames@Router1# edit ospf

[edit protocols ospf]
userjames@Router1#
```

The vertical nature of the configuration mode hierarchy requires the user to only move in an up or down direction. We have already shown above how to move down a directory level. The user can use the **up** command to move up a directory level.

```
[edit protocols ospf]
userjames@Router1# up

[edit protocols]
userjames@Router1# up

[edit]
userjames@Router1#
```

The CLI allows the user to navigate to any lower directory in the hierarchy by entering multiple directories with the **edit** command:

```
[edit]
userjames@Router1# edit protocols ospf

[edit protocols ospf]
userjames@Router1#
```

The **top** command takes the user to the top of the configuration mode hierarchy in a single step:

```
[edit protocols ospf]
userjames@Router1# top

[edit]
userjames@Router1#
```

After reaching the desired directory, the user can use either the **set** or **delete** command to change the current configuration.

- **Altering the Configuration**: Each directory in the configuration mode hierarchy may contain variables that the user can add or remove from the router configuration. The user can enter new information into the configuration with the **set** command. For example, the user can move to the top of the hierarchy and change the hostname to **Router2**. The **host-name** variable is in the **[edit system]** hierarchy directory, and the user can use the **edit** command to move into that directory and then configured the hostname:

```
[edit system]
userjames@Router1# top

[edit]
userjames@Router1# set system host-name Router2
```

Using the **show** command, the user can view the changes made to the configuration. This command displays any configuration in the current directory and all subdirectories below that directory. When this command is issued at the top of the hierarchy, the CLI displays the entire configuration. The user can use the **delete** command to remove variables from the configuration.

- **The Candidate Configuration**: When a user is in the configuration mode, the user is viewing and possibly, changing a file called the *candidate configuration*. This file allows the user to make changes to the configuration without causing operational changes to the *active configuration* (i.e., the current operating configuration). When the user issues the **commit** command, the router will then implement the changes in the candidate configuration (i.e., activates the revised configuration on the router). This abstraction provides flexibility and allows the user to make changes to the router configuration without causing potential problems to the current network operations.
  - o The user can use the **compare** pipe command with the **show** command to compare the current candidate configuration with the active configuration running on the router.
  - o The router displays differences between the two files with either a plus (+) or a minus (−) sign.

○ The (+) sign represents variables in the candidate configuration that are not present in the active configuration; these variables have been newly added to the candidate configuration file.

○ The (−) sign shows the user has deleted variables from the candidate configuration file; the candidate configuration does not contain items found in the active configuration.

- **Saving and Loading Configuration Files**: The current candidate configuration (after being created by the user) can be saved to a file on the router. The candidate configuration is a file that the user can always edit as needed. The user can also load existing candidate configuration files into the router. The candidate configuration file is also very useful when initially configuring a number of similar routers in a network.

  The user may want to have the same configuration information from one router saved, so that the remaining routers can be more easily configured. The user may create a file called **AllRouters** containing the candidate configuration information. The user can then place this configuration information on the other routers using the **load** command.

  ○ Two main options are available for loading the candidate configuration files – **override** and **merge**.

  ○ The **override** option, when issued on a router, completely erases the current candidate configuration on the router and replaces it with the contents of the **AllRouters** file specified. On the router to be configured, the user uses the **load override AllRouters** command to enter the configuration elements.

  ○ The **merge** option when used combines the **AllRouters** file with the current candidate configuration file on the router (**load merge AllRouters**). Elements in the **AllRouters** file that are not in the current candidate configuration are added. Variables in the current candidate configuration file that are not in the merging file **AllRouters** are left unchanged. When an item is in both the **AllRouters** file and the current candidate configuration file, the router will use the value specified in the **AllRouters** file.

- **Using the Commit Command**: Even if a candidate configuration file has been created, no changes are made to the router and become effective until the **commit** command is used. This command can be issued from any level, not necessarily from the top of the configuration hierarchy. The **commit** command has several options that may be used to alter its operation. Each time the user commits a candidate configuration, the router performs several tasks:

  ○ The router examines the candidate configuration for syntax and semantic errors, and if any single problem exists, the candidate configuration is not implemented. One example of a possible error is when the candidate configuration is referencing a routing policy which has not been previously created.

- o The router implements the new candidate configuration if the configuration has no errors, and then proceeds to make changes to the operating environment as needed.
- o Finally, the router saves the existing active configuration for future use. The router displays the **commit complete** message to indicate that the process is successful.
- o The commit process always implements the entire candidate configuration at once; however, any errors encountered during the commit process will result in the router not implementing any portion of the candidate configuration; no changes will be made to the active configuration at all. The router will display an error message informing the user of the problem (**configuration check-out failed** message).

The router always remains in configuration mode, by default, after committing a candidate configuration. The user may exit the configuration mode and move back to the operational mode with the addition of the **and-quit** option (i.e., using **commit and-quit**). When this option is used, the router exits the configuration mode only after it has successfully completed the commit process. If it encounters any errors, it will report the errors and remain in configuration mode.

The user can use the **check** option (i.e., using **commit check**) to instruct the router to verify the validity of the candidate configuration without implementing the changes it contains. A user might need to use this option after making a number of changes to the candidate configuration and wants to make sure that all of the required portions of the configuration are in place. In this case, the router does not implement the changes after running the syntax and semantic checks. The router either notifies the user of a successful check or reports any errors encountered. The router performs syntax and semantic checks to verify only that the information in the candidate configuration will allow the router to implement the file successfully.

If the user has any concerns that configuration changes made will either lock the user out of the router or cause harm to the network operations, the user should use the **confirmed** option (i.e., using **commit confirmed**). This option is designed to allow the router to return to a working configuration automatically, providing the user with a safety net in case of operational problems with the new configuration. After the **commit confirmed** command is issued, the router implements the changes requested and starts a 10-minute timer. If the user is happy with the newly committed configuration, the user must issue a normal **commit** command to stop the timer and end the operation of the **confirmed** option. If the user does not stop the timer and it expires, the router will automatically return to the last operational (active) configuration and implement those changes. The output of the **commit confirmed** command is similar to a normal **commit** operation. The router either displays the **commit complete** message or reports an error. Additionally, the user has the option of changing the timer value used with the **confirmed** option (values range from 1 minute to 65,535 minutes (45 days, 12 hours, and 15 minutes)).

The user can use the **synchronize** option with the **commit** command (i.e., using **commit synchronize**) when the router has two routing processors installed and wants the router to apply the candidate configuration to both route processors. This option is helpful in that, in the event of a route processor failure, the backup route processor will have the latest operational parameters in the network. Most high-end routers support two RPMs on the physical chassis. At any point in time, only one of the route processors is considered the master, and it controls the router's operations. The other route processor serves as the backup and is available to provide failover capability in the event the master ceases to function (see Chapter 1 of this volume).

- **Restoring an Old Configuration**: Anytime a candidate configuration is committed, the router also saves that configuration in a file that is assigned a number in the sequence of committed configurations. This saved file is also the one that the router uses during the **commit confirmed** process. The router saves up to nine previous configuration files that are named **juni-per.conf.1.gz** (i.e., file number 1 representing the most recently saved old configuration) to **juniper.conf.9.gz** (i.e., file number 9 representing the oldest saved configuration) [JNCIAGUIDE]. This old configuration file naming convention starts from 1, and continues with each older file number incremented by 1 until file 9 is reached. The router names the current active configuration **junper.conf** and this represents file number 0.

  The user can place one of the saved old configuration files into the current candidate configuration using the **rollback** command. This command and the **load override** command function exactly the same in that, when issued, the existing candidate configuration is removed and the named old configuration file is used in its place (as the current candidate configuration file). The user must still issue the **commit** command to make this candidate configuration (created from the named old configuration file) the new active configuration. The router does not automatically commit an old configuration file that has been rollbacked. The **commit confirmed** operation is the only exception where the router issues both a **rollback** and a **commit** command.

  For example, a user may have made changes to the properties of the configuration on the router. Then after committing the configuration, the user realizes that the newly committed configuration is not performing as desired. So, the user proceeds to load the most recently saved configuration file and commits that configuration using **rollback 1** (1 representing the old configuration file 1).

3.7.4.6.2 *Cisco IOS CLI Use Modes*

This section describes briefly the three command modes (i.e., *user mode*, *privilege mode*, and *configuration mode*) in the Cisco IOS, each providing access to different command sets [CISCINTRCIOS]:

- **User mode**: This is the first mode through which a user can gain access to the router after logging in. This mode is identified by the > prompt and follows the router name. The user can execute only basic commands in this mode, such as those that display the system's status. The user cannot configure or restart the system from this mode.
- **Privileged mode**: This mode has more capabilities and allows a user to view the system configuration, restart/reload the system, and enter the

configuration mode. This mode also supports all the commands that are available in the user mode. The privileged mode is identified by the # prompt and follows the router name. The user-mode **enable** command informs Cisco IOS that the user wants to enter the privileged mode. If the system is set with an enable password or enable secret password, then the user has to enter the correct password or secret password to be granted access to the privileged mode. Using an enabled secret password provides stronger access protection because it uses stronger encryption when the password is stored in the configuration. The privileged mode should be used with caution because it allows the user to perform all tasks on the router. The user can exit privileged mode by executing the **disable** command.

- **Configuration mode**: This is the mode through which the user can modify the running system configuration. The user enters the configuration mode by entering the **configure terminal** command from privileged mode. The configuration mode supports various sub-modes, starting with the *global configuration mode*, which is identified by the **(config)#** prompt and follows the router name. The configuration mode sub-modes change depending on what is being configured on the system, making the words within the parentheses **()** also change. For example, when the user enters the *interface configuration sub-mode*, the prompt changes to **(config-if)#** which follows the router name. The user can exit the configuration mode by entering **end** or press Ctrl-Z.

In any of these modes, the user can enter the context-sensitive **?** command at any point to show the available commands at that level. The user can also use the **?** character in the middle of a command to show possible completion options. The user can use the **?** within a given command mode command to display the commands available.

3.7.4.7 Microcode

Microcode, sometimes called *firmware*, is a set of hardware-level software instructions (i.e., higher-level machine code instructions) that are processor-specific. A microcode enables and manages the features and functions of a specific processor type. Various hardware devices in routing devices contain their own microcode software images that help to run the various functions in the routing device. For example, network interface processors have their own microcode that can be updated when needed.

Usually, the microcode is a board-specific firmware that resides on a ROM or Flash memory on the route processor, the forwarding engine processors, and on each network interface processor. Typically, a programmable read-only memory (PROM) device on the processor module contains a default microcode boot image that assists the system in locating and loading the microcode image from the router software bundle or Flash memory. Microcode stored on ROM or Flash memory allows new machine instructions to be added when needed without requiring the instructions to be designed into the system electronic circuits. At system startup or reload, the system loads the microcode for each processor type present in the system.

Microcode is very convenient for system design and facilitates the implementation of the following router management and configuration capabilities:

- The microcode (firmware) on the forwarding engine card is a software image that provides card-specific software instructions. Typically, the entire interface processor microcode image is delivered on a Flash memory card (or equivalent portable memory), or is available via download from the website of the router vendor. A router vendor would typically provide downloadable software and microcode for most upgrades. Usually, the downloadable features enable new (upgraded) images to be downloaded remotely, stored in router memory, and to be loaded at system startup without having to physically access the router.
- The router could support downloadable microcode, which enables microcode versions to be upgraded via network download and stored in Flash memory. The system can then be instructed to load an image from Flash memory instead of the default ROM image.
- New microcode is released to enable new features, improve performance, or fix bugs in earlier versions. As new software and hardware features are introduced, the microcode may be updated and maintenance versions released in order to implement new features, improve performance, or fix bugs found in earlier versions.
- A router could support the storage of multiple microcode versions for a specific processor type in Flash memory, and then allow the network administrator to use appropriate router configuration commands to specify which version the system should load at startup. Although the system may store multiple microcode versions for a specific processor module (e.g., network interface) concurrently in Flash memory, only one image can load at startup. Usually, the administrator can use appropriate configuration commands to display the currently loaded and running microcode version for the interface processor.
- In some routers, microcode images for the route processor and all network interface processors are bundled with system software upgrades and load automatically with the new system image. New features and enhancements to the system or interfaces are often implemented in microcode upgrades. Each processor type in the system leaves the factory with the latest available microcode version (in ROM); however, the router vendor periodically distributes updated microcode images with system software images to improve performance, enable new features, or fix bugs in earlier versions. The vendor may bundle the latest available microcode version for each processor type with each new system software maintenance upgrade. Typically, the bundled images are distributed as a single image on a Flash memory card (or compact disk in older systems).
- Online insertion and removal (OIR), also called hot swapping, allows system components such as the RPM and line cards (or network interface processors) to be removed and replaced without causing the system to reboot (see "Understanding Online Insertion and Removal (OIR) or Hot

Swapping" section in Chapter 1 of this volume). However, each time an RPM or line card is removed or replaced, the system reloads the microcode. For example, if a microcode ROM is replaced and the system is already configured to load the microcode for that interface type from ROM, the system automatically loads the new microcode when the interface processor is reinserted in the chassis.

- The system may use the default microcode image that is bundled with the system image or configured to load the microcode for that interface type from a Flash memory file. The default operation may be to load the micro-code from the bundled image. An internal system utility scans for compatibility problems between the bundled microcode images and the installed interface processor types at system startup, then the images are decompressed into running memory (RAM). The bundled microcode images then operate the same as the images loaded from the individual microcode ROMs on the processor modules. The default can be overridden and the system instructed to load a specific microcode image from a Flash memory file or from the microcode ROM with the microcode.

- Typically, the target hardware version will list the minimum hardware revision required to ensure compatibility with the new software and microcode images. When the system is loaded and boots from a new bundled image, the system checks the hardware version of each processor module that it finds installed, and compares the actual version to its target list. Usually, if the target hardware version is different from the actual hardware version, a warning message appears when the router is booting, indicating that there is a disparity between the target hardware and the actual hardware.

3.7.4.8 Inter-Process Communications Services

As discussed in Chapters 2 and 6 of Volume 1, the placement of forwarding functions (i.e., forwarding engines) in a routing device can be done in one of two general ways. From these chapters, the placement of the forwarding engine with respect to the route processor and line cards in routing devices can be done as follows:

- *Central Forwarding*: In this approach, the forwarding table and adjacency tables reside on the RPM, and the route processor performs both routing and forwarding operations.
- *Distributed Forwarding*: In this approach, the line cards maintain identical copies of the forwarding tables and adjacency tables maintained by the route processor. The line cards perform the forwarding operation locally, thereby relieving the route processor engine of involvement in the packet forwarding operation. This is the preferred forwarding method in high-end routers.

Inter-Process Communication (IPC) is a set of mechanisms that an OS provides for the exchange and management of data between two or more threads in one or more processes or programs. The processes may be running on the same processor or on two or more processors connected by a network. The two primary models of IPC are *message passing* and *shared memory*. The IPC method a designer would use depends on the type of data being communicated, and the bandwidth and latency of communication between the threads.

IPC is particularly important in high-end routers with distributed forwarding, where the line cards make the packet forwarding decisions using locally stored copies of the same forwarding tables and adjacency tables as the route processor [CISC12000CEF] [CISCCEFERR]. The databases maintained by the route processor and the line cards must remain synchronized at all times if the line cards are to make the correct forwarding decisions at all times. Any changes to the route processor's tables must be copied to the corresponding tables in the line cards.

A router can use IPC services to maintain synchronization between the route processor's tables and the tables in the line card. IPC becomes an integral part of the communication services required by the router to support distributed packet forwarding. With this, the forwarding table updates that the route processor sends to the line cards are encoded as data information elements within the IPC messages. The IPC service for distributing forwarding table information from the route processor to the line cards is illustrated in Figure 3.28.

As shown in Figure 3.28, IPC messages transport the forwarding and adjacency tables from the route processor to the line cards. The IPC mechanism is used for synchronizing both sets of tables on the route processor and the line cards. The data

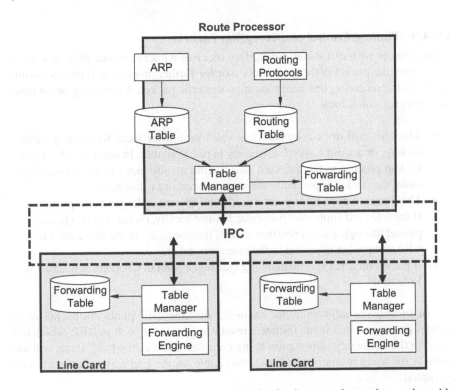

FIGURE 3.28   IPC services to maintain synchronization between the routing engine tables and the line card tables.

structures used by the forwarding engines in the line cards are transported from the route processor to the line cards through the IPC. Also, all relevant statistics collected in the line cards are sent back to the route processor via IPC. Once the forwarding tables in the line cards are synchronized, they can make forwarding decisions using their locally stored (replicated) databases.

As noted above, statistics and forwarding data structures are carried in IPC messages between the route processor and the line cards. Specifically, the IPC messages can carry the following three sets of information [CISCCEFERR]:

- **Control Messages from the Route Processor to Line Cards**: The route processor creates control data in route processor feature sub-blocks in IPC messages to be sent to all the mirroring sub-blocks on the line cards that need to be informed about any changes.
- **Statistics Messages from the Line Cards to the Route Processor**: A line card gathers statistics information from the various local sub-blocks, places the collected information in an IPC buffer, and sends an IPC message to the route processor. The route processor then aggregates these statistics.
- **Asynchronous Event Reporting Messages from the Line Cards to the Route Processor**: Line cards report non-routine events through asynchronous IPC messages that are sent to the route processor as the condition occurs.

3.7.4.9 Punting Control and Exception Packets

When a router with distributed forwarding receives a packet on one of its interfaces, it may punt the packet to the route processor for further processing if the forwarding engine in the receiving line card cannot process the packet. A punt may occur under the following conditions:

- The line card does not produce a valid path after local forwarding table lookup, or a valid Layer 2 adjacency is not available. In other words, if the lookup process in the line card fails to find a valid entry in the forwarding table, the packet is punted to the route processor or dropped.
- The line card does not support a particular feature or Layer 2 encapsulation. If the line card supports a particular feature locally, ownership of a packet is passed through a set of routines in the "feature path" in the line card, otherwise, the packet is punted to the route processor.
- A packet requires special handling (see discussion in Chapters 5, 6, and 7 of Volume 1).

For an *incomplete adjacency*, for example, the line card punts the packet to the route processor so that it can initiate a resolution protocol such as ARP, which may result in the adjacency information being completed sometime later. Punts will also occur if for some reason(s) the forwarding table on the line card is inconsistent or corrupted.

3.7.4.10 Inter-Card Communications Services

Internal communications in the router involve *route processor and line card communications*, and *line card-to-line card communications*. Route processor and line card communications mostly involve the exchange of internal control messages between these two, and also, the forwarding of data such as routing protocol information. Line card-to-line card communications mostly involve the passing of transit traffic through the router from one network interface to another.

In the distributed router architecture, in particular, the route processor communicates with the line cards either *in-band* through the switch fabric, or *out-of-band* through a dedicated bus called the *control and maintenance bus* (CMBus). In some architectures, the route processor communicates with the line cards out-of-band via a dedicated Ethernet link. Generally, the CMBus connects the route processor to other route processors on standby, and also to all line cards. For environmental monitoring and control purposes, the CMBus may connect the route processor to major system components like the power supply, cooling system, on the chassis. The CMBus may also allow the route processor to download a system bootstrap image from another system location, collect or load diagnostic information, and perform general, internal system maintenance operations.

In some architectures without a CMBus, the switch fabric provides the main data path for distributing routing information to the distributed forwarding engines in the line cards, as well as for packets that are sent between the line cards and the route processor, and line card to line card.

3.8 DISTRIBUTED ROUTE PROCESSORS OR ROUTING ENGINES

Some routers rely on a centralized processor to handle tasks like exception packet processing as discussed in Chapters 5, 6, and 7 of Volume 1. If this centralized processor (which also handles all control plane functions) gets overloaded, control plane performance will be affected. In addition, using the control plane processor for normal packet forwarding can make the system more vulnerable to denial-of-service attacks and others.

The extreme bandwidth demands placed on the control CPU (i.e., the control plane processor) for core routers are getting to the point where even the fastest CPU cannot keep pace. In a high-end router, for example, the control CPU must maintain a routing table containing half a million or more entries, using compute-intensive protocols such as OSPF. At the same time, the CPU must also process SNMP packets, perform operations, administration, and management (OAM) functions, support network administrator consoles, download the forwarding table to each network interface card, and handle new network services. With network bandwidth increasing faster than CPU performance (with the deployment of multi-gigabit optical fiber links), more and more router designers are choosing to distribute the control plane workload across multiple CPUs.

The control plane software modules can be distributed to processors that reside in the same or to different cards as shown in Figures 3.29 and 3.30. In both cases, the control plane, running different protocols, is distributed to different modules in order to offload a central control card, as well as to provide more resiliency to the system. Unlike most routers in the market, in these architectures, the line cards are not limited solely to forwarding transit traffic; some control plane functions are also distributed to the line cards. Figure 3.29 shows a distributed processing architecture with centralized control where a central processor keeps track of the overall operation of the system, while Figure 3.30 shows a distributed processing architecture with little or no centralized control plane.

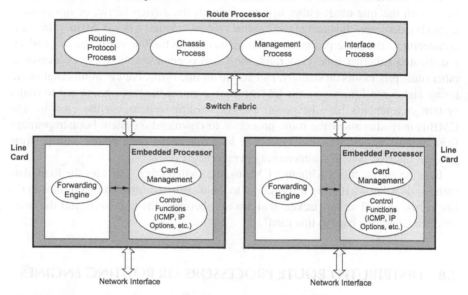

FIGURE 3.29 Partially distributed control plane software modules.

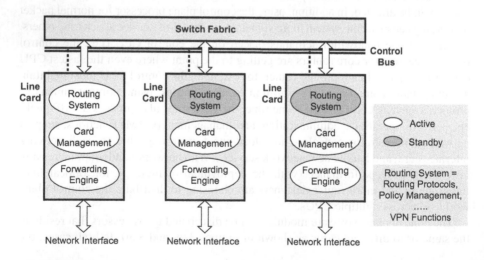

FIGURE 3.30 Fully distributed control plane software modules.

The architecture in Figure 3.29 handles some control plane processing at the card level by combining a network processor with a general-purpose processor on the line card. Through this architecture, the network processor can handle the packet forwarding tasks, while the general-purpose processor handles some specific offloaded control plane tasks (e.g., ICMP, IP header options), thus relieving the centralized processor from handling exception processing. The embedded processor in Figure 3.29 runs the card management software (e.g., for card initialization, checking of status, setting of scheduler weights, etc.) plus control packet processing (e.g., ICMP, IP options, etc.).

In Figure 3.30, the control plane functions are fully distributed across all line cards. The control software modules in one line card exchange information with their peers in other line cards. Typically, the system is designed such that a local software module establishes dynamic communication links when a remote peer is available, and teardowns the links if the remote peer fails. For fault tolerance, software-based heartbeat messages can be used to detect the health status of the remote peer.

In a majority of designs, it is preferable to have just one processor to control and manage the system, thus localizing the complexity of managing the system to a single processor. All the other processors in the system will then handle the packet forwarding functions of the system.

Just as the control and data plane modules can be distributed to different cards, some functions in the switch fabric can be distributed to the line cards. For example, in multiple stage crossbar switch fabrics, some parts of the switch fabrics can also be distributed to the line cards. The extent of distribution, however, will depend on the type of switch fabric used.

Earlier router designs and current low-end routers employed single-processor architectures for all router processing. Improvements over the earlier designs and in the low-end systems produced various forms of distributed processing architectures: *partially centralized* architectures to *fully distributed* architectures. The different router architectures certainly have different levels of performance, depending on the application area of the router in the overall network system (i.e., access, aggregation, or core layers).

The most important characteristics of large Internet backbones are performance, scalability, and reliability. The distributed architecture is the preferred architecture that best satisfies all of these requirements. That is why the high-end routers used in the Internet backbone have distributed forwarding architectures. Chapters 2, 6, and 7 of Volume 1 examine the various designs that fall under this category. These chapters describe the different router architectures in detail including specific control plane and data plane protocols. Hardware processing support is a key consideration in the high-end routers.

3.8.1 Control-Data Plane Separation and the Routing Engine

In Chapters 2, 6, and 7 of Volume 1, we discussed distributed router architectures in which the router OS software runs solely on a route processor, ensuring that control functions are performed without affecting the packet forwarding subsystem. Control operations in the router are performed by the route processor, which runs the router OS software that handles routing protocols, traffic engineering, routing policy, monitoring, and configuration management. Generally, the route processor provides system management, chassis monitoring and control, and control-plane processing functions for the system. The distributed router architecture separates control plane operations from packet forwarding operations, which eliminates or reduces significantly data plane processing and traffic bottlenecks.

Forwarding and packet processing operations in the router may be performed by dedicated microprocessors (i.e., NPUs), or programmable ASICs and other hardware engines that achieve high data forwarding rates that match current fiber-optic capacities. Such designs eliminate processing and traffic bottlenecks, permitting the router to achieve very high performance. Also, by designing the router with further separations between the control plane, forwarding plane, and service plane, the router is able to support multiple services on a single platform. The different services will be supported on different line cards each with its forwarding engine.

REVIEW QUESTIONS

1. What role does the switch fabric play in a network device?
2. What is head-of-line (HOL) blocking and how can it be avoided?
3. What is switch fabric speedup?
4. What role does the backplane play in a network device?
5. What is the main difference between a passive backplane and an active backplane?
6. What is backplane Ethernet?
7. What is auto-negotiation in Ethernet?
8. Explain the main differences between the console, auxiliary, and Ethernet management ports in a router.
9. Which hardware component of a router is responsible for running the CLI?
10. What is the main purpose of the configuration mode in the Juniper Networks JUNOS OS or Cisco IOS CLI?

REFERENCES

[AWEYA1BK18]. James Aweya, *Switch/Router Architectures: Shared-Bus and Shared-Memory Based Systems*, Wiley-IEEE Press, ISBN 9781119486152, 2018.

[AWEYA2BK19]. James Aweya, *Switch/Router Architectures: Systems with Crossbar Switch Fabrics*, CRC Press, Taylor & Francis Group, ISBN 9780367407858, 2019.

[AWEYA2BK21V1]. James Aweya, *IP Routing Protocols: Fundamentals and Distance Vector Routing Protocols*, CRC Press, Taylor & Francis Group, ISBN 9780367710415, 2021.

[AWEYA2BK21V2]. James Aweya, IP Routing Protocols: *Link-State and Path-Vector Routing Protocols*, CRC Press, Taylor & Francis Group, ISBN 9780367710361, 2021.

[CISC12000CEF]. Cisco Systems, Understanding Cisco Express Forwarding, Document ID: 47321, January 17, 2006.

[CISCCEFERR]. Cisco Systems, Troubleshooting CEF-Related Error Messages, Document ID: 7603, June 24, 2008.

[CISCINTRCIOS]. Cisco System, *Internetworking Technologies Handbook*, 4th Edition, Chapter "Introduction to Cisco IOS Software", October 31, 2003.

[EET04.2002]. Chuck Hill, Building Distributed Fabrics with Mesh Backplanes, *EE Times*, April 19, 2002.

[EET07.2002]. Chuck Hill, Distributed Switch Fabric Standards: More Performance, Choices, *EE Times*, July 15, 2002.

[JNCIAGUIDE]. Joseph M. Soricelli, John L. Hammond, Galina Diker Pildush, Thomas E. Van Meter, and Todd M. Warble, *JNCIA Study Guide*, Juniper Networks, 2006.

[JUNCLIGUID21]. Juniper Networks, *Junos® OS – CLI User Guide*, October 22, 2021.

[RFC792]. J. Postel, "Internet Control Message Protocol", IETF RFC 792, September 1981.

[RFC854]. J. Postel, "TELNET Protocol Specification", IETF RFC 854, May 1983.

[RFC1393]. G. Malkin, "Traceroute Using an IP Option", IETF RFC 1393, January 1993.

[RFC2475]. S. Blake, D. Black, M. Carlson, E. Davies, Z. Wang, and W. Weiss, "An Architecture for Differentiated Services", IETF RFC 2475, December 1998.

[RFC2697]. J. Heinanen and R. Guerin, "A Single Rate Three Color Marker", IETF RFC 2697, September 1999.

[RFC2698]. J. Heinanen and R. Guerin, "A Two Rate Three Color Marker", IETF RFC 2698, September 1999.

[RFC4251]. T. Ylonen, and C. Lonvick, Ed., "The Secure Shell (SSH) Protocol Architecture", IETF RFC 4251, January 2006.

[RFC6241]. R. Enns, M. Bjorklund, J. Schoenwaelder, and A. Bierman, Eds., "Network Configuration Protocol (NETCONF)", IETF RFC 6241, June 2011.

[RFC6242]. M. Wasserman, "Using the NETCONF Protocol over Secure Shell (SSH)", IETF RFC 6242, June 2011.

[RFC7589]. M. Badra, A. Luchuk, and J. Schoenwaelder, "Using the NETCONF Protocol over Transport Layer Security (TLS) with Mutual X.509 Authentication", IETF RFC 7589, June 2015.

[XMLW3C1.0]. Extensible Markup Language (XML) 1.0, World Wide Web Consortium (W3C).

4 Case Study
Force10 Networks E-Series Switch/Router Architecture

4.1 INTRODUCTION

To provide a design example that captures the design concepts presented in previous chapters and Volume 1 of this two-part book, this chapter describes the architecture of the Force10 E-Series switch/routers. The goal of Chapters 4 and 5 of this volume is to present practical design examples to illustrate how switch/routers are designed including the steps taken to implement the key features. Although the switch/router design examples presented in both chapters are older designs (that were considered novel and advanced when they were first introduced), many of the design concepts and methods used are still relevant to present-day switch/router designs.

The E-Series is one example of switch/routers, out of many others, that have high capacity switching and support full line-rate Layer 2 and Layer 3 forwarding. They are designed to handle 10 Gigabit Ethernet (GE) ports and line-rate forwarding with access control lists (ACLs) enabled on all ports. Most of today's high-end, high-performance switch/routers are capable of supporting much higher interface speeds. For example, 40 GE, 100 GE, 200 GE, and higher interfaces are not uncommon in the telecom market space (see Chapter 6 of this volume).

The control plane of the E-Series supports millions of routing and forwarding table entries, and tens of thousands of ACLs on every line card. The target network applications of the E-Series switch/routers include server aggregation, cluster/grid computing, next-generation Internet Exchanges (IXs), broadband aggregation, enterprise backbones, and networks supporting metro-Ethernet services.

4.2 E-SERIES SWITCH CHASSIS OVERVIEW

Examples of the E-Series are the E1200, E600, and E300 switch/routers. The E-Series E1200 architecture has switching bandwidth of 1.68 terabits per second (Tb/s) with a throughput rating of 1,000 million packets per second (Mpps) [FOR10RELS06]. The E600 provides a switching capacity of 900 Gigabits per second (Gb/s) and 500 Mpps throughput. The E300 is a compact chassis-based switch/router that scales to 400 Gb/s of switching capacity and up to 196 Mpps throughput. The E1200, E600, and E300 support a total of 56, 28, and 12 10 Gigabit Ethernet ports, respectively, in addition to Gigabit Ethernet and POS/SDH (OC-48/12c/3c) ports.

To enable the E-Series scale to 10 Gigabit port speeds and terabits per second capacity, the design had to address a number of existing switch/router architectural limitations [FOR10RELS06]:

DOI: 10.1201/9781003311256-4

- Backplane design limitations that constrain Ethernet-based designs to 8 Gb/s at the time the E-Series was introduced
- Higher system port densities that prevent a switch from attaining the desired system line rate and offering non-blocking packet forwarding characteristics
- Single CPU-based processing systems that lead to degraded system performance as value-added features such as ACLs, quality-of-service (QoS), bandwidth control, rate limiting, and others are enabled on the system
- Distributed cache-based forwarding architectures that suffer from cache exhaustion caused by traffic patterns that have destinations that are highly uncorrelated and bursty (see Chapters 2, 5, 6, and 7 of Volume 1).

To address these limitations while achieving 10 Gigabit port forwarding speeds, the Force10 E-Series switch/routers designs used new and improved techniques in backplane design, interface module construction, data plane architecture, and control plane processing. The E-Series switch/routers all share a common fully distributed architecture based on a crossbar switch fabric with virtual output queues (VOQs). They also support line cards that process at wire-speed, integrated Layer 2 and Layer 3 forwarding (i.e., Layer 2, IPv4, and IPv6 forwarding). The processing is done entirely in ASIC hardware.

Figure 4.1 shows the main modules of the E-Series switch/routers; distributed router processor modules, distributed forwarding line cards, switch fabric cards, and a passive backplane. In addition to the normal plane control functions, the Layer 2/Layer 3 multiprocessor control plane supports control traffic filtering and rate-limiting capabilities. Control plane functions are performed on the Route Processor Modules (RPMs) and on separate control processors on the line cards.

The distributed forwarding ASICs are designed to distribute packet forwarding, ACL processing, QoS, and buffering to every line card. Compared to other active copper backplanes, the passive copper backplane was determined by the designers to be a more cost-effective and reliable backplane that maximizes system reliability and minimizes cost. Active copper backplanes are more complicated and thus have a non-zero risk of malfunction.

To maximize network uptime, the E-Series architecture supports redundancy, availability, and serviceability features (see "Node Redundancy and Resiliency" section of Chapter 1 of this volume for a detailed discussion on redundancy and high-availability in switch/routers). All key components are redundant, including the RPMs, Switch Fabric Modules (SFMs), power, and cooling components [FOR10ESER05]. The E-Series switch fabric supports 8:1 redundancy, advanced queuing, multicast, and jumbo frame support. The switch fabric is scalable up to 1.68 Tb/s with non-blocking connectivity (in the full chassis E1200). All memory systems are error-correcting code (ECC)/parity protected (see Chapter 3 of this volume for extensive discussion on switch/router memory types).

The E-Series supports system-wide environmental monitoring and persistent configuration synchronization to enable the Force10 Operating Software (FTOS™) to detect, report, and correct faults with minimum system interruption. In addition, the E-Series supports serviceability features such as hot-swappability of all key

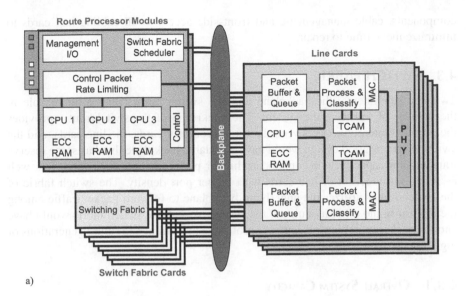

a)

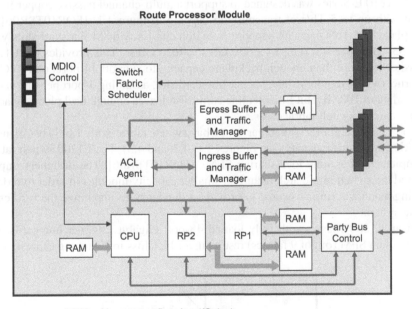

MDIO = Management Data Input/Output

b)

FIGURE 4.1 Force10 E-Series switch/router architecture a) High-level architecture, b) Architecture of the route processor module.

components, cable management, and front-side access to all cabling and cards to minimize mean time to repair.

4.3 SWITCH FABRIC

Two key subsystems, the backplane and the switch fabric, play a major role in the transfer of traffic in the system. The backplane of the switch/router provides the physical interconnect between the route processor cards, the line cards, and the switch fabric. A good design would have a scalable backplane that has some reserve transmission capacity to accommodate higher performance switch fabrics as well as future introduction of line cards with higher port density. The switch fabric of the switch/router utilizes the underlying backplane to forward packet traffic among the line cards and other modules of the system. A scalable switch fabric would have enough packet switching capacity to accommodate current and future generations of higher density line cards.

4.3.1 OVERALL SYSTEM CAPACITY

The Force10 E-Series was designed to support a multi-channel passive copper backplane with up to 5 Tb/s of transmission capacity (Figure 4.1) [FOR10TSR06]. A backplane of 5 Tb/s capacity supports a switch chassis with 14 line card slots with over 330 Gb/s per slot (i.e., 14×330 Gb/s = 4620 Gb/s). This provides the E1200 chassis with more than enough backplane capacity to support 14 ports of 100 Gb/s Ethernet (which requires 200 Gb/s per line card slot, assuming 1 port per line card). Upgrading to 100 Gigabit Ethernet line cards also leaves enough backplane capacity to upgrade the switch fabric cards.

The E1200 supports a modular cross bar switch fabric with 1.68 Tb/s of non-blocking switching capacity (see Figures 4.1, 4.2, and 4.3). The E1200 switch fabric is implemented as nine load-sharing SFMs [FOR10TSR06]. The designers implemented the switch fabric architecture in smaller capacity modules in order to reduce sparing/manufacturing costs and to provide redundancy to maximize the resiliency of the system.

In an E1200 chassis with fully loaded 48-port Gigabit Ethernet line cards, the failure of an SFM (out of 9 fabrics) results in an 11% loss in switching capacity (8:1

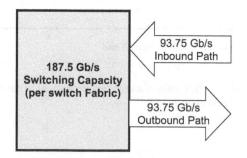

FIGURE 4.2   Individual switch fabric.

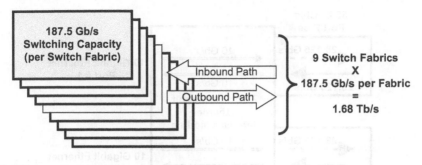

FIGURE 4.3 Switch fabric capacity.

redundancy). In a partially loaded chassis, for example, with 40 or fewer Gigabit Ethernet ports per line card or four 10 Gigabit Ethernet ports per line card, the loss of a single SFM does not result in any reduction in switching capacity (8+1 redundancy) [FOR10TSR06].

4.3.2 System Dimensioning

Traffic travels in and out of each Terabit Switch Fabric module through the backplane as shown in Figure 4.1 [FOR10RELS06]. The Force10 Passive Copper Traces (FPCT) on the E-Series passive backplane physically connects interface module slots to the switching fabric modules. The backplane (which is neither an optical nor an active copper backplane) has 5 Tb/s, 2.7 Tb/s, and 1.2 Tb/s capacity in the E1200, E600, and E300 switch/router chassis, respectively.

Unlike optical backplane interconnect systems or active copper backplanes, the E-Series backplane was designed to have no single points of failure, thus eliminating costly electrical-optical-electrical conversions. At the time of introduction, the designers claimed the E-Series passive copper backplane to be the industry's first high-speed non-optical backplane to achieve 5 Tb/s in a single-rack switch/router chassis [FOR10ESER05].

The passive backplane and the Terabit Switch Fabric work together to provide 187.5 Gb/s total bandwidth per SFM (Figure 4.2). This switching capacity is divided equally between 93.75 Gb/s of incoming and 93.75 Gb/s of outgoing traffic. In the E-Series, the total switching capacity is equal to the aggregate switching capacity of nine load-balanced and active Terabit SFMs working together. Each Terabit SFM contributes up to 187.5 Gb/s of capacity (Figure 4.2), resulting in a total system switching capacity of 1.68 Tb/s (i.e., $= 9 \times 187.5$ Gb/s) for the E-Series as illustrated in Figure 4.3.

In the E-Series architecture, a Port Pipe (Figure 4.4) acts as a backplane communication channel that provides connectivity between slots and the Terabit SFMs (Figure 4.5). The E-Series architecture supports 32 Port Pipes. There are two Port Pipes associated with each interface slot position, and one Port Pipe assigned to each of the RPM slots. While the E-Series architecture contains up to 32 Port Pipes, the E1200 and E600 use only 30 and 16 Port Pipes, respectively.

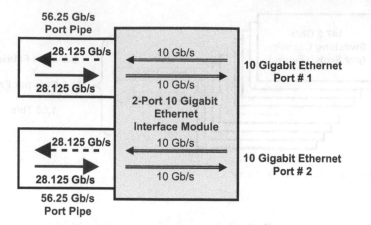

56.25 Gb/s
Port Pipe

28.125 Gb/s

28.125 Gb/s

10 Gb/s

10 Gb/s

2-Port 10 Gigabit
Ethernet
Interface Module

10 Gigabit Ethernet
Port # 1

28.125 Gb/s

28.125 Gb/s

10 Gb/s

10 Gb/s

10 Gigabit Ethernet
Port # 2

56.25 Gb/s
Port Pipe

FIGURE 4.4   Two Port Pipes per interface module.

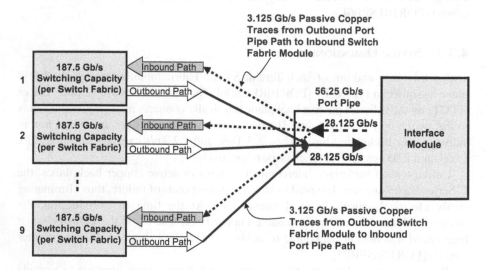

3.125 Gb/s Passive Copper
Traces from Outbound Port
Pipe Path to Inbound Switch
Fabric Module

1   **187.5 Gb/s**
Switching Capacity
(per Switch Fabric)   Inbound Path
Outbound Path

2   **187.5 Gb/s**
Switching Capacity
(per Switch Fabric)   Inbound Path
Outbound Path

9   **187.5 Gb/s**
Switching Capacity
(per Switch Fabric)   Inbound Path
Outbound Path

56.25 Gb/s
Port Pipe

28.125 Gb/s

28.125 Gb/s

Interface
Module

3.125 Gb/s Passive Copper
Traces from Outbound Switch
Fabric Module to Inbound
Port Pipe Path

FIGURE 4.5   Port Pipe and Force10 Passive Copper Traces (FPCT) shown in detail

The relationship between Port Pipes, Switching Capacity, and the Force10 Passive Copper Traces (FPCT) can be explained as follows [FOR10RELS06]:

- Dividing the 1.68 Tb/s total switching capacity by the 30 Port Pipes defined for a fully populated E-Series (E1200) system yields a throughput switching capacity of 56.25 Gb/s in Full Duplex or 28.125 Gb/s per individual Port Pipe (Figure 4.4).
- Given that the E-Series has 9 switch fabrics (see Figures 4.3 and 4.5), a Port Pipe contains 18 FPCTs; 9 FPCTs supporting each 28.125 Gb/s Inbound Port Pipe path and another 9 FPCTs supporting each 28.125 Gb/s Outbound Port Pipe path as shown in Figure 4.5. To attain 28.125 Gb/s, each FPCT

works in combination with the SerDes chipset built into the Terabit Switch Fabric and operates at a throughput rate of 3.125 Gb/s (9 × 3.125 Gb/s = 28.125 Gb/s).

- Within an individual 56.25 Gb/s Port Pipe, 9 Inbound FPCTs (solid blue lines in Figure 4.5) link to the outbound paths on the 9 SFMs, while 9 Outbound FPCTs (dashed black lines in Figure 4.5) link to the inbound paths of the SFMs.
- Every slot contains two Port Pipes with the exception of the RPM slots, where each slot contains only one Port Pipe. Therefore, each slot provides 56.25 Gb/s capacity, and a completely populated E1200 chassis supports 1.68 Tb/s of usable switching capacity [(56.25 Gb/s × 14 slots) + (28.125 Gb/s × 2 slots) = 843.75 Gb/s, i.e., half capacity]. Similarly, a fully populated E600 chassis supports 900 Gb/s of usable switching capacity [(56.25 Gb/s × 7 slots) + (28.125 Gb/s × 2 slots) = 450 Gb/s]. The E-Series E300, similarly, is scaled down to support 400 Gb/s of usable switching capacity [(56.25 Gb/s × 3 slots) + (28.125 Gb/s × 2 slots) = 225 Gb/s].

Figure 4.6 shows the E-Series architecture with the passive copper backplane, providing 56.25 Gb/s between each line card slot. The use of a passive copper backplane is to avoid the cost and complexity of optical or active backplanes. The E-Series provides 56.25 Gb/s of full-duplex bandwidth to each line card slot, ensuring line-rate performance for up to 672 Gigabit Ethernet ports or 56 10 Gigabit Ethernet ports in a single E-Series chassis.

The main advantage of the E-Series architecture (with the FPCT grouped into Port Pipes and working with the switching fabrics) is the ability to support very high throughputs with zero loss (since there is sufficient bandwidth headroom even in a chassis with maximum system port density). The main design goal is to achieve a

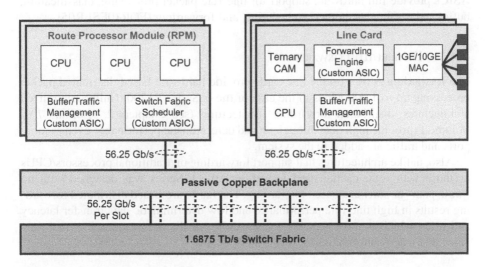

FIGURE 4.6 Route processor module multiprocessor design.

scalable, non-blocking switch fabric that provides low data transfer latency and jitter that is required and critical for streaming media applications.

4.3.3 Non-blocking Features

To prevent head-of-line (HOL) blocking in the switch, the E-Series architecture utilizes 32 x 32 virtual output queues (VQQ) on its crossbar switch fabric. The switch fabric utilizes multiple queues on each line card with global queue monitoring to keep track of the priority and length of each ingress queue in the system. Using sophisticated switch fabric scheduling algorithms, all VOQs are serviced appropriately across the switch fabric. The main design objective is to develop a high-performance switch fabric that eliminates HOL blocking while minimizing cost, fabric complexity, data transfer latency, and jitter. The E-Series switch fabric was designed to forward full Layer 2 (Ethernet) and Layer 3 (IP) packets in hardware, without any need for packet segmentation and reassembly.

The E-Series also integrates QoS into the switch fabric. Both ingress and egress buffering are provided, including backpressure mechanisms that work to prevent the possibility of HOL blocking. The architecture uses separate unicast and multicast queues (with 85 to 200 milliseconds of buffering [FOR10ESER05]) to enable minimal packet loss even in oversubscribed network conditions. The E-Series switch fabric uses a variant of Weighted Fair Queuing (WFQ) to schedule traffic out of the ingress and egress queues, and programmable queue sizes to allow handling of both real-time and bursty traffic patterns.

4.4   DISTRIBUTED LAYER 2/LAYER 3 FORWARDING

The E-Series line cards support distributed forwarding ASICs that are designed for efficient forwarding of Ethernet frames (at Layer 2) and IP packets (at Layer 3). The ASICs provide full hardware support for line-rate packet processing, classification, buffer and traffic management, scheduling, and forwarding [FOR10ESER05].

4.4.1 Distributed Forwarding

The forwarding ASICs on the line cards provide hardware-based distributed packet processing up to 1,000 Mpps in the case of the E-Series E1200. Unlike centralized architectures, distributed forwarding architectures (see Chapters 2, 5, 6, and 7 of Volume 1) provide high data throughput and deterministic performance even as more ports and traffic are added to the system.

Also, unlike architectures that support forwarding on traditional processors/CPUs or those with route-cache-based forwarding, the E-Series was designed with no "slow-path" or software-based forwarding. The hardware-based distributed forwarding results in high line-rate performance and low deterministic data transfer latency and jitter, which is more suitable for streaming media applications like real-time voice and video conversations.

The E-Series forwarding ASICs support the following line-rate features:

- Filtering with standard and extended ACLs
- QoS (DiffServ, IEEE 802.1p)
- Traffic rate policing and shaping
- Layer 2 forwarding features:
  - Source MAC address learning and frame filtering
  - Link Aggregation (defined as *IEEE 802.3ad*) for combining/aggregating multiple network links in parallel in order to increase the traffic carrying capacity beyond what a single link can support
  - VLAN stacking (also called Q-in-Q and defined as *IEEE 802.1ad* – not to be confused with *IEEE 802.3ad*) to allow two VLAN tags to be inserted into a single Ethernet frame (see Chapter 6 of Volume 1)
- Layer 3 routing features:
  - Equal Cost Multi-Path (ECMP) routing
  - Inter-VLAN routing (see Chapters 4 and 6 of Volume 1)
  - IP multicast (see Chapter 1 of Volume 2)
- Statistics collection (see Chapter 5 of Volume 1)

4.4.2 HARDWARE-BASED TABLE LOOKUPS

The packet forwarding engine (which includes packet processing and classification functions) uses hardware lookup engines combined with Ternary Content Addressable Memories (TCAMs). The forwarding engine retrieves the packet-handling instructions required to forward packets from the TCAMs (see Figures 4.1 and 4.6). Based on the source or destination MAC and IP addresses (or other combinations of addresses), Differentiated Services Control Point (DSCP) setting, IP Precedence value, IP protocol number, TCP/UDP port number values, the forwarding engine can retrieve all the necessary information required to forward the packet.

The forwarding ASICs provide line-rate lookups in the Layer 2 and Layer 3 forwarding tables stored in the TCAM. As packets enter the system, the forwarding ASICs are able to support on-the-fly line-rate lookups of ACL entries for network destination, configured policy, and QoS mappings. The simultaneous packet processing and classification allows the E-Series to provide line-rate Layer 2/Layer 3 forwarding performance independent of table lengths, IP address prefix lengths, and packet size – even when all ACLs and QoS features are enabled on the system [FOR10ESER05].

4.4.3 TRAFFIC MANAGEMENT AND QUALITY-OF-SERVICE MECHANISMS

The E-Series QoS features include congestion control with Weighted Random Early Detection (WRED) and Weighted Fair Queuing (WFQ), QoS interworking between Layer 2 and Layer 3, policing with Committed Access Rate (CAR), and detailed statistics collection for accounting and billing (see Chapters 8 and 9 of Volume 1 for detailed discussion on QoS rate management mechanisms). The E-Series implements the traffic management and QoS mechanisms in ASIC-based hardware devices to enable the system to perform all services at line rate [FOR10ESER05].

The ACLs (with thousands of entries), WRED for congestion avoidance, token-bucket-based traffic classification and rate management, WFQ and Strict Priority Queuing (SPQ) for traffic scheduling and policy enforcement, and other QoS features are all implemented in hardware [FOR10ESER05]. This is because, at multi-gigabit line speeds, the effect of turning on each feature can cause performance degradation if they were implemented in software. Implementing all such services directly in the hardware in the packet forwarding path ensures that, for any combination of services, or even when all services are enabled at the same time, users will not experience any variation in system performance, data transfer latency, and jitter [FOR10ESER05].

The E-Series ASICs provide packet QOS marking capabilities, including the ability to re-map packet priorities between IP and Ethernet schemes. Traffic conditioning is based upon the two-rate, three-color token bucket-based metering and marking (see discussion in Chapter 9 of Volume 1). The design supports eight input queues per destination port that map directly to DiffServ and IEEE 802.1p QoS classes. Using WRED, congestion avoidance is provided by configurable packet drop precedence probability curves. The combination of these QoS features (with rate policing and limiting) provides various QoS service processing and offerings.

Using Figure 4.7, QoS within the forwarding ASICs can be explained as follows [FOR10ESER05]. Upon receipt of a packet, the packet classifier ASIC may mark or re-mark the packet based on configured policy including multi-field filtering (i.e.,

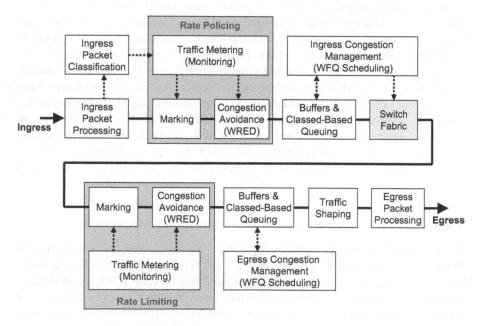

FIGURE 4.7   Packet classification flow chart.

ACL matching on source MAC/IP address, destination MAC/IP address, IP protocol, and TCP/UDP port), VLAN ID, or physical port mapping. After classification, a two-rate, three-color marker mechanism (built into the ASICs) performs ingress traffic shaping. The packet is then placed in the appropriate output queue. Queues are serviced by a WFQ algorithm and sent across the fabric to the egress queues. A second, identical, two-rate, three-color marker mechanism performs traffic shaping on egress.

For control plane protection, the E-Series is able to process, identify, and separate malicious data from control packets (e.g., routing protocol updates and ICMP packets). This control plane feature enables the E-Series to manage and block potentially harmful network traffic from propagating or flooding throughout the network. With the threat of security attacks coming from a large number of sources in today's networks, the ability to apply (potentially) hundreds of ACL criteria to each incoming control packet has become increasingly essential in networks.

To provide control plane security, the control plane of the E-Series architecture was designed to apply its extensive ACLs to incoming control packets (see Figure 4.7) [FOR10RELS06]. With the ASIC-based line-rate ACL processing capability, the E-Series performs filtering of control packets without introducing additional data forwarding latency. The minimization of data transfer latency is crucial for reducing overall route table convergence time across networks.

4.4.4   UNDERSTANDING THE FORCE10'S ENHANCED ACCESS CONTROL LISTS UPDATE

To provide some context to the discussion here, we first start by discussing how port scanning attacks work. Port scanning is often used by an attacker to identify services running on a host (i.e., probe a server or host for open TCP/IP ports) with the intention of compromising it – that is, to exploit the known vulnerability of that service. Since a TCP/IP port is the place where information goes into and out of a host, port scanning identifies open doors to the host. Port scanning has a legitimate use in managing networks, but it can also be used maliciously if an attacker is looking for weak access points to break into a network.

Using this crude yet simple method, attackers probe every port on a network device to find open TCP/UDP ports, and then save that information for malicious attacks. Generally, when an Intrusion Detection System (IDS) identifies the signature of a port scan attack, it adds an ACL entry to the interface receiving the scan with the goal of blocking out the attacker's source IP address in order to protect resources on the network. Even in the case where the attacker spoofs the source IP address (the attacker is not the real owner of the IP address), blocking the source port will prevent the attacker from compromising the network device and resources all the same.

A network administrator may configure network security applications to provide alerts if they detect connection requests across a wide range of ports coming from a single host. As a workaround, attackers typically perform a port scan in strobe or stealth mode. In strobe mode (strobing), the port scan is limited to a smaller target set of ports rather than scanning blanketly all 65536 ports. In stealth mode, the attacker performs scanning but does this slowly. The attacker scans the ports over a much longer period of time, hoping to reduce the chance that the target ports will trigger an alert.

Providing the ability to modify the ACLs in a network device without creating any security holes, and hopefully performance degradation, is essential in thwarting these types of attacks. Comprehensive network security often requires system administrators to update ACL entries of a network device frequently to prevent newly discovered or pending attacks from affecting the network or network equipment.

Most enterprise and service provider networks, under normal regular operations, run a few thousand ACL entries. Usually, the network administrator prunes the ACL entries regularly offline to keep them more manageable in size but the list can grow in size during security attacks. However, given the growing importance of securing networks against network attacks, maintaining forwarding performance with several thousand ACL entries has become a key requirement.

An attack may require increasing the number of ACLs by a few thousand entries which, in turn, could degrade overall system performance. In this scenario, although the end target itself has not yet been factored in, one consequence of the attack (i.e., increase in ACL size) could degrade the entire network service, something that in itself can be viewed as a success from the point of view of the attacker. A more preferred scenario, which is a major requirement for minimizing security threats in the network, is where the modification of ACL entries (in the order of milliseconds) creates no security breaches or black holes.

The security risk to the network or network equipment increases directly in relation to the speed (in packets per second) of the interface where the ACL update occurs. Attackers using sophisticated port scanning technologies can enter the network when those security holes open. An ACL modification or change that takes even a few seconds translates into millions of packets passing unchecked through the network at multi-gigabit line speeds. For these reasons, some equipment vendors have developed sophisticated ways of updating the ACL without creating security holes [FOR10HLACL04].

Traditionally, the process of updating an ACL involves a two-step procedure. Security holes can temporarily be created that leave the network vulnerable to attack when applying a new or modified ACL to an interface. The designers of the E-Series avoid this common security risk by using a distinctive update process, called Hot-Lock, that modifies and applies the ACL in a single pass [FOR10HLACL04].

Hot-Lock, which is built into the Force10 Operating Software (FTOS™), handles ACLs in such a way that avoids removing the ACL from an interface prior to the ACL modification action. The goal is to ensure that no security holes open or disruption of traffic occurs during ACL updates, thus, increasing overall network security, reliability, and availability.

4.4.4.1 Traditional Two-Step Access Control Lists Update Process

As illustrated in Figure 4.8, traditional solutions use a two-pass process to modify or update ACLs. This two-step operation has the potential of opening a security hole because applying a new or modified ACL to the interface first requires the removal of the original ACL. This leaves the network unprotected and vulnerable to port scanning attacks because the interface lacks security policies the existing ACL is

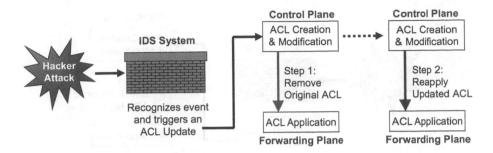

FIGURE 4.8   Traditional 2-step ACL update process.

removed. The problem becomes even acute and poses a greater security threat when this method is used on higher-speed interfaces. This is because the port scanning attack may use a large number of IP addresses almost instantaneously and at the maximum speed of the attacked interface.

The second step in Figure 4.8, which involves the reapplication of the modified ACL can create an additional problem [FOR10HLACL04]. Typically, ACLs are created as an ordered list of policies. For example, if the ACL contains a list of "deny" statements followed by a "permit all", then before the entire ACL has been reapplied, all valid packets that should be allowed to pass are dropped as the last "permit all" rule has not yet been reapplied. Thus, the two-step ACL update process can present decreased security as well as increased traffic disruption during ACL modifications.

ACLs can add a tremendous burden to a switch/router, especially when activating both ingress and egress ACLs for all interfaces and ports. Many traditional solutions suffer from severe performance degradation in terms of reduced throughput when an interface has too many ACL entries. Thus, the designers of the E-Series observed that the preferred solution is one in which the modification of ACL entries is done in the order of milliseconds without creating any security breaches or black holes. Such a solution is a major requirement for minimizing security threats in networks, particularly on high-speed interfaces.

4.4.4.2   Force10 Hot-Lock™ Access Control Lists Update Process

With Hot-Lock, ACL updates are done in a single action (Figure 4.9), thus eliminating the security holes generated during the traditional two-step ACL updates. This process enables new or modified ACLs to be applied directly (almost instantaneously – in the order of milliseconds) to an interface dynamically and directly without disabling them [FOR10HLACL04]. This single action process also eliminates traffic disruptions introduced during ACLs reapplication to the interface.

Hot-Lock organizes ACLs by sequence numbers, which allows the insertion of a new ACL rule at the end, just before the implicit permit- or deny-all statements. This mechanism does not require the ACL to be removed from an interface and then reapplied before ACL modification occurs. The single-pass process locks out the threat of port scanning because it does not remove the ACL rule from the interface at any time during modification, which in turn, does not open any security hole.

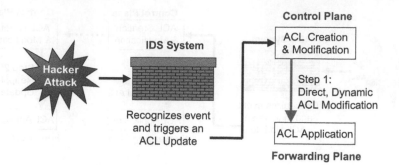

FIGURE 4.9   Force10's Hot-Lock™ ACL update.

With this, Hot-Lock closes potential security holes and eliminates the disruption of traffic that occurs during ACL modification. As noted in [FOR10HLACL04], Force10 E-Series implements ACL rules in hardware thereby allowing line-rate forwarding of packets while filtering them. Line-rate ACL processing allows the switch/router to support several thousand ACL entries per interface without any performance degradations.

4.5   DISTRIBUTED MULTIPROCESSOR CONTROL PLANE

The Force10 Networks switch/routers use a fully distributed architecture with all packet forwarding decisions made by the data plane ASICs, and all control functions performed by the switch/router CPU(s) running the Force10 Operating System (FTOS™) [FOR10FTOS08]. The E-Series supports a multiprocessor control plane that supports line-rate full Layer 2 and Layer 3 protocols and features.

The E-Series has three processors on each Route Processor Module (RPM) and one processor on every line card (see Figures 4.1, 4.6, and 4.10). This allows the

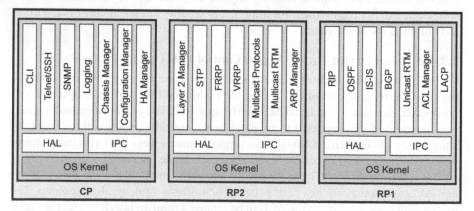

HA = High-Availability
RTM = Routing Table Manager

FIGURE 4.10   Distributed processing control plane with FTOS on the E-Series platform.

control plane of the E-Series to deliver high performance and fault tolerance using the FTOS. In addition to serving as the operating system of the switch/router, the FTOS has a suite of Layer 2 and Layer 3 protocols and features. The Force10 C-Series and S-Series switch/routers have only a single FTOS CPU that performs all control plane and management functions [FOR10FTOS08].

Software processes (to be described below) are distributed among the RPM and line card processors to provide real-time multiprocessing and process isolation with memory protection. Such features have become absolutely necessary for fault tolerance and high-availability in large-scale enterprise and service provider networks (see Chapter 1 of this volume).

The E-series was designed to target the needs of Internet-scale networks where the system control plane supports millions of routing table entries and tens of thousands of ACLs on every line card [FOR10ESER05]. Dedicated 100 Mb/s data paths from the RPMs to every line card facilitate fast and unimpeded forwarding table updates that, when not done quickly, could otherwise jeopardize network stability.

With the ever-increasing amount of Internet traffic in both core and access networks, control plane scalability and security has become mandatory in switch/routers as Internet route tables continue to increase. The next two sections describe the scalability and security features employed in the E-Series architecture.

4.5.1 THE FORCE10 OPERATING SYSTEM

The Force10 Networks Operating System, FTOS, is based on NetBSD, a highly portable, open-source version of the Unix-derivative Berkeley Software Distribution (BSD) operating system that has been optimized for networking applications. NetBSD has a high degree of portability across multiple hardware architectures, allowing a wide selection of control processors. Figure 4.10 shows the general architecture of the FTOS software system that was built based on NetBSD [FOR10FTOS08]. In the E-Series switch/routers, FTOS processing is distributed across processors on each RPM, and on the additional processor on each line card [FOR10FTOS08].

FTOS supports a common set of features across all three E-Series switch/routers (E1200, E600, and E300). It is a real-time operating system with extensive, high-performance Layer 2 and Layer 3 protocols and features. To ensure stability and fault tolerance, FTOS uses process modularity and distribution in a protected multiprocessor environment [FOR10FTOS08].

The FTOS control plane features and functionality work in conjunction with the hardware forwarding ASICs to provide per-packet Layer 2 and Layer 3 forwarding. The critical FTOS features include a robust IP routing control plane, hardware and software fault-tolerance, high granularity traffic management and accounting, industry standard Command Line Interface (CLI), and system diagnostics [FOR10FTOS08].

For Layer 2 operations, the FTOS was designed to support thousands of MAC entries 4096 VLANs, IEEE 802.1Q VLAN tagging, IEEE 802.1ad VLAN stacking, load sharing and failover with IEEE 802.3ad Link Aggregation (LAG), and IEEE 802.1w Rapid Spanning Tree Protocol (RSTP). For Layer 3 operations, FTOS supports routing protocols such as Border Gateway Protocol version 4 (BGP4),

Intermediate System–Intermediate System (IS-IS), Open Shortest Path First version 2 (OSPFv2), and Routing Information Protocol version 2 (RIPv2). The E-Series with FTOS is designed with scalability in mind (millions of routing table entries) [FOR10ESER05].

Virtual Router Redundancy Protocol (VRRP) and Equal Cost Multi-Path routing (ECMP) are implemented in FTOS to enable reliable network design. FTOS also supports IP Multicast and includes protocols such as Internet Group Management Protocol (IGMP), Protocol Independent Multicast-Sparse Mode (PIM-SM), Multiprotocol BGP (MBGP), and Multicast Source Discovery Protocol (MSDP) [FOR10FTOS08]. All of these protocols are essential for enterprise and service provider networks.

User access to the E-Series switch/router is authenticated via Remote Authentication Dial In User Service (RADIUS) and Terminal Access Controller Access-Control System Plus (TACACS+), while secure access methods include Secure Shell (SSH) and Secure Copy (SCP). The architecture has built-in support for SNMP MIBs and a dedicated 100BASE-TX management port that allows seamless integration in any in-band and out-of-band network management environment [FOR10FTOS08]. The E-Series platforms support standards-based SNMP MIBs, as well as MIBs designed to manage each switch/router hardware platform. These MIBs are designed into FTOS to enable easy integration into HP OpenView or other network management software [FOR10FTOS08].

The E-Series was designed to combine the above elements of the architecture with high-availability features. These features include hot-swappability of all key components and system-wide environmental monitoring (see Chapter 3 of this volume). This combination of features enables the E-Series to offer maximum system uptime and serviceability along with high forwarding performance and scalability [FOR10FTOS08].

4.5.2 Control Plane Scalability

The E-Series control plane provides both scalability and security through a multiprocessor-based RPM, control packet filtering and rate-limiting mechanisms, and independent high-speed control data paths to each line card [FOR10FTOS08]. For the control plane, redundant RPMs (as an optional feature) handle all routing and control packet processing while providing protection against Denial-of-Service (DoS) attacks.

The RPMs use a multi-processor architecture with hardware-based ingress traffic rate limiting and filtering. By distributing software processes running among the CPUs, performance can be maximized while minimizing the possibility that a fault in an individual process would take down the entire system.

Three independent CPU subsystems (Figure 4.10) on the Force10 E-Series RPMs run independent software images [FOR10FTOS08]. One CPU handles the local control and management functions (management processor), the second handles Layer 2 processes (Layer 2 processor), and the third handles Layer 3 processes (Layer 3 processor). Each of these CPUs runs an independent software image, maximizing performance and minimizing the possibility that a fault in any one process could lead to catastrophic system failure.

For example, a fault in a management process would not interfere with Layer 2 and Layer 3 forwarding table updates. The modularity of FTOS allows the individual control plane processes to be readily partitioned among the three RPM processors [FOR10FTOS08]. This modularity enables the FTOS to isolate its processes, minimizing the possibility that a fault in any one process could lead to system failure. The operating system's modularity also enables it to share processes across CPUs, raising the computational power available to individual processes.

The line card CPUs perform local control functions including sFlow sample aggregation and reporting [RFC3176] (see discussion on sFlow in Chapter 2 of this volume). The modularity of the FTOS operating system allows for future system redesign and configuration where some central control processes can be distributed across the line card CPUs [FOR10FTOS08]. This distributed multiprocessing can be done to add greater scalability to the processing capacity of the control plane.

The multiprocessor architecture used in the E-Series and C-Series allows for the creation of a manager/agent relationship between the RPM processors and the line card processors [FOR10FTOS08]. Each RPM processor's FTOS compilation includes only the required cross-platform functional modules, while the line card processor's FTOS version includes only the appropriate platform-specific modules plus the cross-platform modules needed to support the manager/agent relationship.

In the E-Series RPM, a CPU running FTOS is dedicated to each of the following three functional areas as stated above: Layer 3 processes, Layer 2 processes, and management/control processes. As discussed above, distribution of control processes greatly increases the processing capacity of the control plane, providing high scalability and processing power. This also significantly improves resiliency because the failure of a process on a CPU will not affect the processes on the other two CPUs (unless they have some data dependencies).

The E-Series supports dedicated memory for each of the three RPM CPUs, thus, providing an additional level of physical memory protection across the functional areas [FOR10FTOS08]. This complements the logical memory protection across processes provided by each CPU's instance of FTOS. In addition, the E-Series and C-Series support the following resiliency features within the control planes [FOR10FTOS08]:

- The E-Series provides separate paths for user traffic and control traffic. Communications within the control plane and communication between the control plane and the line cards are done through out-of-band switched Ethernet channels. A separate control communications path links the RPMs to each line card.

  The E-Series has dedicated 100 Mb/s Ethernet switched paths from the RPMs to every line card. This ensures that configuration and forwarding table updates to the line cards are not impacted by heavily congested user traffic. This eliminates slow forwarding table updates and route convergence which could jeopardize network stability.

- Each line card is equipped with its own CPU dedicated to housekeeping protocol functions such as sFlow or Bidirectional Forwarding Detection (BFD). Local processing reduces messaging between line cards and also preserves bandwidth on the out-of-band switched Ethernet channels.
- The control plane and data plane both use ECC or parity-protected memory to ensure data integrity. ECC protection maintains the integrity of the memory systems in the event of single-bit errors.

4.5.3 HARDWARE ABSTRACTION IN FTOS

To support the portability of FTOS across multiple platforms, Force10 implemented a hardware abstraction layer (HAL) code in FTOS as illustrated in Figures 4.10 and 4.11 [FOR10HAL08]. The HAL is an abstraction layer, implemented in software that decouples the kernel of an operating system from the specific details of the underlying hardware. It hides differences in the underlying hardware from the operating system kernel, so that most of the kernel-mode code does not need to be rewritten to run on systems with different hardware.

The HAL allows porting the FTOS to a new hardware platform to be accomplished without rewriting the FTOS core (i.e., the modular processes and kernel) that comprises the majority of the codebase. With HAL, new code that would require porting is decoupled from the core code (by this layer of abstraction as shown in Figure 4.11) and is restricted to the hardware-specific interfaces of the HAL.

Many operating systems like BSD Unix, Mac OS X, Linux, Windows, and Solaris all implement their own forms of hardware abstraction. Hardware abstraction allows the portability of these operating systems to a variety of processors, with different

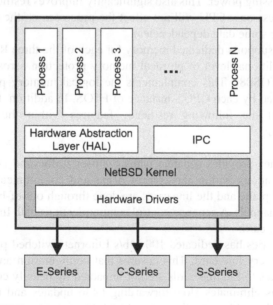

FIGURE 4.11 FTOS software architecture.

memory management unit architectures, range of hardware subsystems (e.g., storage, sound, or video), and a variety of systems with different I/O bus architectures. Hardware abstraction hides the differences in the underlying hardware, thus, isolating the operating system kernel source code from a major rewrite whenever there are changes in the hardware platform.

As illustrated in Figures 4.10 and 4.11, FTOS is based on NetBSD operating system. NetBSD is a highly portable, modular operating system that has been optimized for networking applications and platforms including large-scale server systems, desktop systems, handheld devices, and for the control plane CPUs of numerous network elements, such as switch/routers and firewalls. Designers can port this opensource operating system to new hardware by creating the required driver interfaces to the hardware.

NetBSD is widely noted for its high-quality design, stability, and performance as an operating. It features advanced features including hardware abstraction that allows it to be highly portable and readily adapted to run on a wide range of hardware variants. The hardware abstraction of NetBSD provides a well-defined distinction between chipset drivers and the drivers for peripheral devices, allowing the same device-driver source code to be used across different CPU architectures and bus types.

The HAL allows for the FTOS to be ported from the Force10 E-Series switch/router platform to both the C-Series and S-Series platforms. Also, the HAL enables the FTOS to have a single core base of source code from which all platform-specific images are compiled. Reuse of the same source code for all cross-platform ensures protocol interoperability, a consistent feature set, and unified management interfaces and functionality across all switch/router platforms.

This gives network operators the advantage of running a single operating system across all the switches and routers end to end in the Layer 2/Layer 3 network infrastructure (from the access/aggregation tier to the network core and the data center). By running a single network operating system (FTOS) end to end across the network (i.e., a single switch/router OS that can span the entire network), the complexity of the network operating system environment is reduced and network operators have the flexibility to simplify operation and management of the network.

Figures 4.10 and 4.11 show Force10's FTOS architecture based on the NetBSD operating system. Applications that traditionally execute in the kernel in a monolithic network operating system have been moved to userspace where they run as independent, modular processes. As illustrated in these figures, FTOS creates separate modular processes for each of the following [FOR10FTOS08] [FOR10HAL08]:

- Layer 2 protocols (STP, LACP, etc.)
- Layer 3 routing protocols (IS-IS, OSPF, BGP, static routing)
- Various system services and management functions (SNMP, CLI, etc.)
- Security services and protocols (SCP, SSH, TACACS+/RADIUS, ACLs)

These processes share information via an inter-process communication (IPC) mechanism which is layered on top of the NetBSD kernel (Figures 4.10 and 4.11).

Other features of the FTOS include the following [FOR10FTOS08]:

- FTOS processes are modularized, thus, allowing the control plane functions to be separated from kernel space. The control plane functions can be modified without causing unexpected interactions with other functions.
- The FTOS memory architecture allocates a separate protected address space for each independent process and its associated subsystems. Communications between subsystems within the same process take place directly via sharing of the process' protected memory space.
- Communication between different processes takes place through the IPC message passing mechanism. The IPC allows the processes to communicate while maintaining memory protection between them – memory corruption across process boundaries is virtually eliminated.
- Process modularization and protected memory space ensure that faults in one process are limited to only that process (plus any other processes that require current information from the failing process via IPC). Fully independent processes cannot affect one another or the kernel. Each process is modular and runs in its own protected memory space, thus, preventing a fault in one module from affecting any other modules (software problems are restricted to specific processes preventing them from affecting the entire system). The modular design of FTOS makes it easy to trace software errors to specific processes and facilitates any required remedial action. With this, FTOS eliminates the causes of many potential catastrophic system failures.
- FTOS implements preemptive process scheduling to prevent a single process from hogging the resources of any one of the control plane CPUs. By assigning different priorities to processes depending on their relative importance in the system, preemptive process scheduling allows all processes to have their share of CPU time.

FTOS includes many system management, debugging, and troubleshooting features [FOR10FTOS08]. For system monitoring, the FTOS kernel monitors all processes to ensure operations are within normal limits of resource utilization. FTOS also provides system-wide monitoring for out-of-range environmental conditions and other fault conditions, such as unsynchronized line cards configurations.

Other features include timely fault reporting and automated fault correction to help minimize system interruption [FOR10FTOS08]. In addition, the modularity of the software simplifies tracing of software errors to specific processes and facilitates the remedial action required to return the system to full operation.

4.5.4   PROCESSES RUNNING ON EACH CPU ON THE E-SERIES PLATFORM

As discussed above, the E-Series creates separate processes for each Layer 2 and Layer 3 protocol, as well as management functions, and security services and protocols. FTOS supports a distributed, multiprocessor architecture with separate control CPUs for Layer 2 switching, Layer 3 routing, and management control on the E-Series.

TABLE 4.1

Processes Running on Each CPU on the E-Series Platform

Management Processor	Layer 3 Processor	Layer 2 Processor	Line Card Processor
• Command Line Interface (CLI)	• Routing Information Protocol (RIP)	• Layer 2 Manager	• Interface Agent
• Telnet	• Open Shortest Path First (OSPF)	• Address Resolution Protocol (ARP) Manager	• Layer 2 Agent
• Secure Shell (SSH)			• Layer 3 Agent
• Authentication, Authorization and Accounting (AAA)	• Intermediate System- Intermediate System (IS-IS)	• Spanning Tree Protocols (STP, RSTP, MSTP)	• ACL Agent
• Logging			• sFlow Agent
• Chassis Manager		• Force10 Resilient Ring Protocol (FRRP)	• Bidirectional Forwarding Detection (BFD) Agent
• Configuration Manager	• Border Gateway Protocol (BGP)		• High Availability Agent
• Interface Manager	• Link Aggregation Control Protocol (LACP)	• VRRP	• Line Card Manager
• High Availability Manager		• Internet Group Management Protocol (IGMP)	• Line Card Inservice Diagnostics
• RPM and Data Plane Inservice Diagnostics	• Virtual Router Redundancy Protocol (VRRP)	• Protocol Independent Multicast (PIM)	
	• Unicast Routing Table Manager (RTM)	• Multicast Source Discovery Protocol (MSDP)	
	• Access Control List (ACL) Manager	• Multicast Routing Table Manager	

In addition, FTOS allows the system to offload processes such as sFlow or BFD to line cards on the E-Series and C-Series platforms. This allows the Force10 platforms to run multiple processes at the same time without performance degradation or bottlenecks [FOR10FTOS08]. The FTOS consists of modular processes (as shown in Table 4.1) that in combination target carrier class performance, robustness, and scalability.

4.5.5 CONTROL PLANE PROTECTION AND SECURITY

Control plane security has become very important as DoS attacks directed at the Internet routing infrastructure have become more frequent and sophisticated. Even the most robust network device software and hardware design can still be vulnerable to DoS. DoS attacks are malicious acts designed to bring the operation of a system or network to a halt by flooding it with useless and harmful traffic disguised as specific types of control packets directed at the control plane CPU of the target device.

Distributed DoS (DDoS) attacks amplify the amount of malicious traffic, in some reported cases to many gigabytes per second, by involving hundreds of sources. With multi-gigabit links, the volume of DoS attack traffic that can pass through a single port of a network device can overwhelm even the highest performing control plane CPU.

The E-Series RPMs have been designed to provide high-performance traffic control, rate limiting, and filtering capabilities. These capabilities allow network administrators to suppress harmful DoS attacks, preventing the flooding of unwanted traffic onto the network – an event that places an unnecessary burden on control plane processors. The E-Series architecture addresses this vulnerability by employing a programmable hardware mechanism on the RPM to rate limit traffic to the control plane processors (see Figure 4.1) [FOR10ESER05].

In conjunction with ACLs applied to control packets, the rate-limiting mechanism can be configured to rate limit only specific traffic types. For example, preferential treatment is given to control packet types like routing updates over incidental control traffic such as ICMP echo request. The system can be configured to provide independent rate limits for different control packet types. Rate limiting of control traffic uses the same two-rate, three-color marking scheme and token bucket mechanisms found on the E-Series line cards.

4.5.6 Traffic Statistics Collection

An important tool for statistics collection in the E-Series is sFlow, defined in IETF RFC 3176 [RFC3176] (sFlow is discussed in Chapter 2 of this volume). sFlow is a widely used standards-based sampling technology that is implemented in the forwarding ASICs of E-Series. Assisted by the forwarding ASIC, sFlow provides the ability to continuously monitor Layer 2 to Layer 7 traffic flows at wire speed simultaneously on all ports of the E-Series.

An sFlow Agent, which is a software process that runs on the network management processor of the E-Series, aggregates interface counter values, forwarding table information, and traffic samples into sFlow messages that are forwarded across the network to an sFlow Collector where the statistics are stored and used for analysis and report generation. The sFlow traffic statistics collected can be used by the network operator in a number of ways [FOR10INTEX09]:

- Understand bandwidth consumption by application type (e.g., Peer-to-Peer, Web, FTP, email, etc.)
- Usage in accounting for billing and charge-back
- Input for real-time congestion management
- Provide audit trail analysis to identify unauthorized network activity and trace the sources of DoS attacks
- Route profiling and peering optimization
- Traffic trending and capacity planning

These are just some examples of the application of the traffic statistics collected via sFlow.

4.6 E-SERIES HIGH-AVAILABILITY FEATURES

Like most high-end network devices, the E-Series switch/routers have a number of high-availability features that are required in today's business-critical network infrastructure. These features complement those already discussed above – the multiprocessor-based RPM, ASIC-based filtering and rate-limiting of CPU-bound control traffic, redundancy of critical components, alarms and fault reporting, ECC protected memory, and process isolation. These features are absolutely necessary for fault tolerance and rapid routing convergence in large-scale enterprise and service provider networks.

The hardware resiliency features that exist within the control plane of the E-Series can be summarized as follows [FOR10TSR06]:

- The RPMs are based on a distributed multiprocessor architecture – Each RPM consists of three independent CPU subsystems.
- The system supports hitless failovers between redundant RPMs (see Chapter 1 of this volume). The data plane is designed to be independent of RPM failures which allow each line card to continue forwarding using its local copy of the FIB during an RPM failover or software upgrade. As a result, no traffic is lost and all current traffic flows are protected.
- The system supports control plane ACLs in addition to hardware-based prioritization and rate-limiting of control traffic to provide network stability and protect the control plane from DDoS attacks.
- Control plane communication is done via out-of-band switched 100 Mb/s Ethernet interfaces.
- The system employs protected memory among CPUs to prevent process corruption.
- ECC or parity-protected memory is used in the control plane (as well as data plane) to ensure data integrity.

The high-availability features are discussed in more detail here.

4.6.1 REDUNDANCY OF CRITICAL COMPONENTS AND ENVIRONMENTAL MONITORING

The E-Series provides redundancy and Online Insertion and Removal (OIR) for all critical components thus eliminating single points of failure in the system [FOR10HAVAI08] (see discussion on redundancy and OIR in Chapter 1 of this volume). Redundancy is provided in the following forms: 1+1 redundant RPMs (active/standby); redundant SFMs; redundant DC power modules; redundant AC power supply modules; redundant cooling fan subsystems. In the event of a component failure or removal, the redundant component is brought into service immediately by the switch/router software. The switch/router allows the failure event to be logged in four user-configurable ways: via console trap, via SNMP, via syslog, and via the front

TABLE 4.2
FTOS Actions versus Temperature Levels

Temperature Level	Action
Temperature above maximum 70°C threshold	Major alarm reported and affected component(s) powered down
Temperature above major alarm threshold	Major (red) alarm reported via RPM indicators and alarm contacts, and console/syslog/SNMP messages
Temperature above minor alarm threshold	Minor (yellow) alarm reported via RPM indicators and alarm contacts, and console/syslog/SNMP messages
All temperatures below 70°C threshold	Fan speed increased to maintain temperature below minor alarm threshold

panel major/minor alarm indicators and relay contacts. The E-Series supports OIR of all the above components in addition to the line cards in order to minimize system downtime for repairs and scheduled maintenance.

The redundant power supply supports load sharing and OIR. The system draws power from all working power modules during normal operation. If any single DC or AC power supply module fails or is removed, the system switches to the remaining supplies without interruption of service. Failure of any power module is detected by the FTOS software and reported by any of the four reporting methods mentioned above. In addition, every card in the E-Series (RPMs, line cards, switch fabric cards) incorporates independent power conversion modules to eliminate the possibility that a converter failure could affect other systems beyond the local card [FOR10HAVAI08].

Environmental conditions within the chassis, such as temperature and voltage levels on each card in the system, are continually monitored by FTOS. If temperature levels increase beyond user configurable thresholds, then the actions described in Table 4.2 are taken by FTOS [FOR10HAVAI08]. Similarly, if voltage levels on any card fall below nominal operating ranges, FTOS reports the condition as a minor alarm [FOR10HAVAI08]. If voltage levels drop below the minimum operating voltage, FTOS powers off the affected component and issues a major alarm.

4.6.2 ROUTE PROCESSOR MODULE FAILOVER

Recall that the E-Series has three processors on each route processor module and one processor on every line card. This allows the control plane to deliver high performance and fault tolerance to Layer 2 and Layer 3 protocols and system operations. As described earlier, modular software processes are distributed among the various processors, creating a multiprocessing environment that features process isolation with memory protection [FOR10FTOS08]. By having a modular structure, the E-Series running FTOS restricts faults to specific processes, thus, eliminating many potential catastrophic system failures. With this, software problems are isolated to specific processes and the rest of the system continues to operate.

The E-Series RPM running FTOS has failover capabilities that support hitless forwarding without traffic interruption. The failover capabilities of the RPM are built on the following high availability mechanisms [FOR10HAVAI08]:

- **Automatic Synchronization of Configuration Information**: The E-Series supports automatic synchronization of configuration information between redundant RPMs (see discussion on state synchronization in Chapter 1 of this volume). With this, the E-Series minimizes recovery time in the event of a failure of an RPM. Two levels of synchronization are supported by FTOS: *full synchronize* and *persistent synchronize*.
  - *Full Synchronization*: In full synchronize mode, the E-Series synchronizes the startup configuration, error messages, and task manager status. Synchronizing the task manager minimizes the time needed to bring up a line card after the system fails over to the standby RPM. Synchronization gives the standby RPM knowledge of what line cards are in the system and the tasks that are running on each one, which enables features such as hitless IEEE 802.3ad Link Aggregation Control Protocol (LACP).
  - *Persistent Synchronization*: In persistent synchronize mode, the E-Series copies the startup configuration from the primary to the secondary RPM every time a user saves the running configuration.
- **Hitless Forwarding**: The E-Series with its modular FTOS software supports hitless forwarding in addition to rapid routing convergence. Should any redundant RPM fail, FTOS ensures that all packets are forwarded, preventing application disruption. In systems with redundant (dual) RPMs, hitless forwarding ensures that the system continues to forward traffic in event of an RPM failure. The line cards maintain state during an RPM failure, allowing Layer 2 and Layer 3 traffic to be forwarded without interruption. The E-Series also supports Graceful Restart mechanisms for OSPF, BGP, and PIM, allowing it to support hitless forwarding. Graceful Restart (also known as Non-Stop Forwarding (NSF)) and Non-Stop Routing (NSR) are two different mechanisms used to prevent routing protocol reconvergence during a processor switchover. Graceful Restart and NSR both allow the system to continue forwarding packets along known routes while the routing protocol information is being restored (in the case of Graceful Restart), or refreshed (in the case of NSR) following a processor switchover. System state is maintained for Layer 2 protocols such as STP and LACP, enabling them to continue running during an RPM failover.
- **Hitless Software Upgrades**: With the hitless software upgrade feature, a standby RPM on the E-Series can be loaded with a new version of the FTOS software, and then the system is manually allowed to failover from the primary to the standby RPM without disrupting traffic forwarding. The hitless software upgrade feature is an extension of the hitless forwarding capabilities – these capabilities are used during a software upgrade.

- **Fail-Safe Configuration**: For added system resiliency, FTOS supports advanced version control and configuration rollback mechanisms to protect the network from operator configuration errors. The change control system of FTOS allows network administrators to save multiple device configurations as backup. This makes it easy for a network administrator to view changes from one configuration version to another, thus, preventing fatal configuration errors. For example, a configuration mistake that breaks routing can render a network device unreachable, making it impossible to reconfigure the device without going onsite. With automatic rollback, the router itself can revert to the last known working configuration and restore device reachability.

4.6.3 Pre-Configuration and Persistent Configuration of Line Card Slots

The E-Series with FTOS software supports both pre-configuration and persistent configuration of line card slots. These features reduce system downtime during the replacement of line cards (i.e., the time required for line card swap-outs or for the provisioning of new line card capacity) [FOR10HAVAI08]:

- **Pre-Configuration of Line Card Slots**: With pre-configuration of a line card, the system administrator configures an empty slot as if a line card were present in the slot. Pre-configuration eliminates the need to enter configuration commands after the line card is inserted in the slot and running, thereby, minimizing system downtime. This E-Series feature has significant benefits, in that, it dramatically reduces the time needed to add and provision new ports in the chassis. When a card is inserted in the pre-configured slot, the stored configuration is automatically loaded by FTOS.
- **Persistent Configuration of Line Card Slots**: In persistent configuration of a line card, FTOS stores the line card type, MAC address assignments, and configuration information when a line card is removed. FTOS also automatically reconfigures a replacement card once it is inserted in the slot. When the replacement card is inserted in the slot, FTOS senses the insertion, and automatically gives the new card the stored configuration.

In both configuration options, FTOS can be configured to report an error if an incompatible card type is inserted into the slot [FOR10HAVAI08]. For example, an error would be reported if a 48-port Gigabit Ethernet line card is removed from a slot and a 4-port 10 Gigabit Ethernet line card is inserted in its place.

4.6.4 E-Series Maintenance and Management Features

The E-Series provides resiliency at the manageability/serviceability layer that supports the features that are directed toward enabling system upgrades, re-configuration, fault correction, and component repair. The goal is to allow all of these tasks to be accomplished without taking the switch/router out of service. The E-Series also supports other manageability features that have an impact on resiliency and help to reduce the time to diagnose and repair faults. The E-Series supports the following features that allow for simplified routine maintenance and management:

- OIR of all critical components, e.g., removable air filters
- Hitless software upgrades
- Redundant 10/100 Mb/s Ethernet management ports
- Inline system health checks that monitor control plane and data plane availability. These can be accomplished via proactive monitoring, via service agents, SNMP, and other integrated hardware monitoring
- RPM alarm LEDs
- Alarm reporting to syslog, SNMP, console
- Integrated cable management with front side access to all cabling

These features improve the serviceability of the E-Series and reduce the duration of scheduled and non-scheduled maintenance. Other management features in the E-series include RADIUS and TACACS+ that provide authentication, authorization, and accounting (AAA) services for remote access to the device by network administrators. SSH and SCP are also supported to provide strong encryption, robust authentication, and data integrity for remote system access or file transfers to/from the switch/router.

4.6.5 E-SERIES LINK AND NETWORK AVAILABILITY FEATURES

Link availability and resiliency features allow traffic on a failing link to be re-directed to a parallel redundant link or path. This has to be done in such a way that failover or recovery time is minimized to prevent or limit the disruption of application traffic.

The E-Series support IEEE 802.3ad Link Aggregation (LAG) at Layer 2 and Equal Cost Multi-Path (ECMP) routing at Layer 3. The link availability feature LAG, allows the network manager to aggregate a number of physical Ethernet link (up to 16 Ethernet links) into one logical group and balance traffic across those links. LAG automatically redistributes the traffic to the remaining links (within 100ms) if any link in the group fails [FOR10HAVAI08]. All functioning links are used in the LAG to send traffic, which provides an active/active redundancy scheme.

ECMP in the E-Series allows the network manager to aggregate a number of IP links (up to 16 IP links) in one logical group and balance traffic across those links. If any link in the group fails, ECMP automatically distributes all traffic to the remaining links. Similar to LAG, ECMP is an active/active redundancy scheme since it uses all functioning links to send traffic.

The E-Series ECMP and LAG use the same load-balancing algorithm to distribute the traffic across redundant links [FOR10TSR06]. The algorithm used is based on a unique IP 5-tuple hash (IP source and destination addresses, IP protocol, Layer 4 source, and destination ports) to distribute traffic flows across the links. Since this hash algorithm does not use MAC addresses, LAG load balancing can accommodate either Layer 2 or Layer 3 traffic. As an alternative, the E-Series also supports a LAG load balancing algorithm that is based on MAC addresses, which may be used for non-IP traffic in Layer 2 network segments.

The E-Series like most high-end switch/routers has a number of network availability features in addition to the above link availability features. These network availability features provide a number of benefits to the network operator. To understand these benefits, let us look at the effects of node or link failures on network convergence. It is now well known that when network links fail, or routers fail or require restarts, the network often convergences to a new network topology with temporary sub-optimal routing of traffic and, possibly, packet loss. To address these issues, switch/routers often include a number of resiliency features to minimize re-convergence times and prevent local failures from affecting the overall network.

Some of the network availability mechanisms include Rapid STP (RSTP), Multiple STP (MSTP), Per VLAN STP (PVST+), OSPF and BGP Graceful Restart mechanisms, and VRRP. The newer STP-based protocols allow Layer 2 switched network to have fast (sub-second) convergence beyond that offered by the legacy STP, and also allow active-active load sharing across redundant Layer 2 links and switches.

OSPF and BGP Graceful Restart (like the rest of the Graceful Restart protocols) allow the data plane to continue forwarding packets while the router's control plane software is reloaded or restarted. The main goal of these restart protocols is to allow a router to restart gracefully without causing a routing flap or transient routing loops across the network.

VRRP provides default gateway redundancy and eliminates statically defined default gateways that can create single points of failure in the network. Without VRRP, traffic could be disrupted for a considerable time (seconds or even minutes) while IP configurations are renewed, or the default router is restored.

4.7   E-SERIES DIAGNOSTICS AND DEBUGGING SYSTEM

The purpose of the E-Series diagnostics and debugging system is to improve system uptime and availability by preventing the occurrence of certain types of hard errors (increasing system MTBF) and by improving fault isolation and resolution time (reducing MTTR). Rather than being initiated under operator command, the E-Series diagnostics system consists of autonomous components that can run in the background or can be triggered by events.

The functionality supported include error checking that proactively verifies the correct operation of various hardware and software subsystems, and the capability to take automated actions when an error or exception is detected. The automated actions may take a number of forms, including [FOR10NGDGDB08]:

- Notification of network operations personnel of the detected error or exception.
- Initiation of further system checks and information logging to help isolate the condition to a specific Field Replacement Unit (FRU).
- Notification of network operations personnel of which FRU is causing the problem.
- Saving all relevant information in a core dump file or other crash log file in the event of an unplanned system or subsystem reset.

To accomplish the above, the E-Series implements the following diagnostics and debugging features:

- Reactive diagnostics that are accessible via the CLI using appropriate **show** and **debug** commands to determine the root cause and to isolate the fault to a specific FRU.
- Proactive system health checks that run autonomously whenever the system is in operation. These diagnostics detect and report errors via a syslog message, and also can be configured to take action in real time to minimize the impact of an error.

4.7.1 PROACTIVE DIAGNOSTICS AND DEBUGGING

The following provides an overview of some of the proactive diagnostic and debugging features available in E-Series switch/routers.

4.7.1.1 Runtime System Health Checks

As discussed above, system health checks (or health monitors) run proactively while the system is operational, and are intended to detect errors in data transfers within the system. System health checks are performed using test frames that are interspersed among frames carrying user data and system messages. As a result, system health checks do not disrupt the normal flow of traffic within the device.

The data plane loopback test is one example of a system health check for the E-Series switch/routers (Figure 4.12). This loopback test proactively monitors the data plane for dropped frames. As shown in this figure, the test has two parts, a loopback from the line card through the switch fabric, and a loopback from the RPM through the SFM.

Each RPM and each line card CPU periodically send test frames that loopback through the SFMs. The results of the loopback runtime test reflect the overall health status of the data plane and can be used to identify a single faulty SFM. If three consecutive RPM loopback test frames are dropped when all SFMs are enabled, the system attempts to isolate and then disable the faulty SFM by automatically "walking

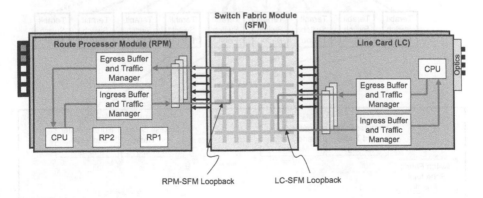

FIGURE 4.12 E-Series data plane loopback test.

through" the SFMs. In this process, each SFM is placed sequentially in an offline state, and then another set of RPM loopbacks is performed until the faulty SFM or SFMs are identified. If the fault is isolated to a single SFM, it is disabled and this event is logged. The system does not perform an SFM walk if one of the installed SFMs is already disabled. When an SFM is disabled, an alarm is sent to notify the operator.

The data plane loopback test offers configuration options that allow the operator to select the action the system will take in the event of a failed test. This allows system behavior to be made consistent with uptime/availability targets and hardware sparing level policies. One option is to allow the system to take the failed SFM offline automatically if a backup SFM is in place and to send an alarm. If desired, the automatic SFM walk (that is launched by default after an RPM-SFM runtime loopback test failure) can be disabled. Also, the automatic shutdown of a single faulty SFM identified by an SFM walk can be disabled. If a line card (LC-SFM) runtime loopback test fails, the system does not launch an SFM walk, but simply logs an error message indicating the failure.

4.7.1.2 Switch Fabric Module Channel Monitoring

SFM channel monitoring is an additional proactive data integrity check that checks for traffic going across the backplane. SFM channel monitoring uses a Per-Channel De-skew FIFO Overflow (PCDFO) polling feature. As with the data plane loopback feature, the PCDFO polling feature is enabled by default. Figure 4.13 gives a view of the E600 and E1200 switch fabric architecture. Each ingress and egress Buffer and Traffic Manager (BTM) ASIC maintains nine channel connections to the Terabit Switch Fabric (TSF) ASICs on the SFMs.

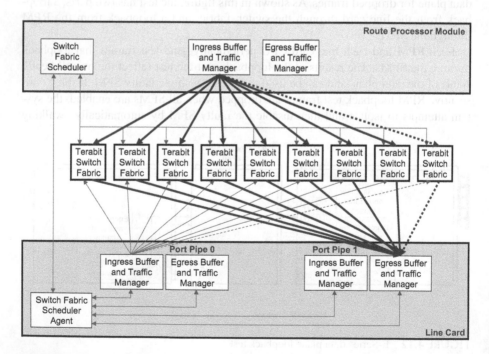

FIGURE 4.13   Switch fabric module channel monitoring on an E1200 or E600 line card.

The PCDFO is designed to detect a bad channel on an SFM, RPM, or line card. It performs this function by breaking a test frame into segments and striping it across all of the SFM channels between the egress BTM and the ingress BTM. The egress BTM ASIC must receive each segment of striped data within a specified time interval (t) in order to consider the segments as having the proper temporal alignment. Small timing skews, less than t, can be tolerated, because the BTMs maintain a small FIFO memory pool to allow the segments to be realigned before forwarding. A PCDFO poll fails when the skew among the segments exceeds t. When this happens, the realignment FIFOs overflow and some of the segments are dropped.

There are two classes of errors that can lead to PCDFO poll failures:

- **Transient Errors**: The errors in this class are transient in nature. Transient errors are considered random events that may occur only one time or may recur only for short time periods. Because such transient events tend to occur very infrequently, they do not have a measurable effect on switch/ router performance or functionality. Since these events are not repeatable, no action is required by network operations staff.
- **Systematic Errors**: This class of errors can cause PCDFO failures and are systematic in nature. Systematic errors are repeatable events because they are generally caused by persistent malfunctions in hardware devices or components.

Events are logged when a PCDFO error first occurs on any SFM and when PCDFO error pattern changes. For transient errors, PCDFO error messages can be expected to disappear after a very short duration without external intervention. The hardware system automatically recovers from the transient error state, and the data plane continues to function properly. For persistent errors, additional reactive diagnostics will need to be run to isolate and resolve the root cause. For example, to confirm that an identified SFM needs to be replaced, a manual data plane loopback test could be executed. If an error persists and cannot be diagnosed, it may be necessary to contact the technical support team.

4.7.1.3 Parity Error Scanning

The RPMs and line cards in the E-Series support multiple parity checking points to ensure data integrity throughout the internal forwarding, lookup, and buffering path. Parity errors can be either transient or persistent. Operating via the FTOS, the E-Series uses an automated memory scanning diagnostic to proactively determine whether a parity error is a transient error or a persistent (hard) error, and if a reset or other action is required. In the case of a persistent error, the FTOS diagnostics would call for the reset or replacement of the line card in question. The traditional, manual approach to differentiating between transient or persistent parity errors, is to reset the card and monitor it for a second occurrence of the error. This approach has the disadvantage of possibly encountering additional data errors until the diagnosis of a persistent error is confirmed.

In the proactive process, all error messages are saved in the syslog regardless of whether the error was diagnosed as transient or persistent. An example application is parity scanning for the Packet Forwarding Engine SRAM, which stores the forwarding table for next-hop lookups. In the case of a persistent error, the diagnostic would identify the exact location of a parity error, and automatically rewrite that location in an attempt to fix the entry. Automatic correction eliminates the need for a customer-reported event. Only extremely rare cases where errors cannot be automatically corrected would call for the reset or replacement of the line card in question. In addition, any non-recoverable parity error events are reported via an SNMP trap in the FORCE10-CHASSIS-MIB.

4.7.1.4   Automatic Information Collection Triggered by Software Exceptions

An additional, proactive diagnostics and debugging feature in the E-Series is the ability to automatically collect information about the system when software exceptions occur. With this feature, the E-Series automatically collects critical fault information when an RPM or line card resets or experiences a failure. This proactively gathered information is saved to flash memory in one of several files, which can be reviewed by Force10 Networks' technical support personnel to help isolate the cause of the error.

The system preserves fault information in the following files:

- *Core Dump*: This is an extensive crash log file, consisting of a dump of the entire memory space on the line card or RPM.
- *Failure Trace Log*: This file preserves the contents of the buffered trace log, which contains messages about internal FTOS software task events. The information captured covers both RPMs and line cards.
- *Sysinfo*: This file captures the status of counters on the Ethernet interfaces of line card, RPM CPUs, and the inter-CPU party bus. Information relevant to FTOS control plane communications among the E-Series CPUs is captured in this file.
- *NVTrace*: This file captures and preserves critical line card status information prior to the occurrence of the last software exception.
- *Command History File*: Command history is a proactive/reactive feature that can be used to check whether a particular sequence of system commands may have led to a software exception. The command history file proactively stores a timestamped log entry that uniquely identifies each command as it is executed. The contents of the file can be used reactively by Force10 Networks' technical support personnel and Force10 Networks' engineering to help isolate the cause of the exception.

4.7.2   REACTIVE DIAGNOSTICS AND DEBUGGING

This section describes the reactive diagnostics and debugging features for the E-Series.

4.7.2.1   Offline Diagnostics

The offline diagnostics test suite available in the E-Series FTOS system image can be useful for reactive fault isolation and debugging of offline line cards. Offline diagnostics are useful in isolating a fault symptom, such as interface errors or unexplained packet loss on a line card. The offline diagnostics tests are grouped into three levels, and generally cover the following functions:

- Verify the existence of an ASIC or other device
- Test the device's internal parts (e.g., registers)
- Perform data-path loopback tests

Offline diagnostics are invoked from the FTOS CLI. While diagnostics are running, the status can be monitored via the CLI. The test results are written to a file in flash memory and can be displayed on screen. Detailed statistics for all tests are collected. Such statistics include last execution time, first and last test pass time, first and last test failure time, total run count, total failure count, consecutive failure count, and error code.

4.7.2.2   Runtime Hardware Monitoring

The E-Series FTOS continuously monitors the status of hardware by polling hundreds of registers on key ASICs and system components on the line cards, RPMs and SFMs, and writing this information to a file. The log is initialized at system startup and continues to dynamically log events and errors as they occur. The hardware monitor feature provides a configurable option for automatic action by the system when certain types of hardware events occur. A number of commands are supported for runtime hardware monitoring like the **show hardware** commands.

The **show hardware** commands comprise a key reactive component of the overall diagnostic/debugging system that allows a detailed look at various counters and statistics gathered at multiple points within the system data and control planes. The **show hardware** commands that apply to the E-Series line cards provide a similar degree of hardware visibility by exposing the contents status registers and drop counters. The **show hardware** commands provide the operator with greatly enhanced visibility into the finer details of the E-Series hardware architecture. For example, an experienced operator can perform detailed fault diagnosis by correlating the results of the **show hardware** command with the proactively gathered information contained in the syslog file. The **show hardware** commands are also an important tool used by Force10 Networks technical support engineers to assist customers in fault diagnosis.

E-Series **show hardware** commands that apply to line cards include the following:

- *Buffer and Traffic Management (BTM) Commands*: BTM commands are used for accessing information on the BTM ASIC on a line card. Command options are available to view or clear various counters, status registers, and queue information.
- *Packet Forwarding Engine (or Flexible Packet Classification) Commands*: The Packet Forwarding Engine ASIC provides line-rate traffic classification for QoS and ACLs. Command options are available for displaying advanced debugging information about the Packet Forwarding Engine. For the packet

forwarding functional area of the Packet Forwarding Engine, it is possible to display the contents of receive and transmit counters, error counters, and status registers. For the forwarding table lookup functional area of the Packet Forwarding Engine, it is possible to display advanced debugging information.

4.8 PACKET WALK-THROUGH

A hardware-only data path in the E-Series processes traffic flows through the system at speeds of more than 10 Gb/s. The E-Series ASICs work together with the following subcomponents to handle all packet forwarding tasks:

- Packet Forwarding Engine
- Buffer and Traffic Manager
- Terabit Switch Fabric
- Switch Fabric Scheduler

Using Figure 4.14 as a guide, the packet walk through the E-Series can be described by the following seven major steps [FOR10RELS06]:

1. Packets enter the system and are processed by the Packet Forwarding Engine. The Packet Forwarding Engine performs the Ethernet MAC level functions.
2. The Packet Forwarding Engine uses a condensed version of the header information to perform simultaneous line-rate TCAM lookups, examining details that pertain to Layer 2 (MAC), Layer 3 (IP), Layer 4 (TCP/UDP), QoS, and ACL-related information. In addition to looking up the destination port, the TCAM returns information that specifies how the packet should be handled to conform to any configured ACLs or QoS policies. At this point, the Packet Forwarding Engine identifies and separates routing, control messages from data packets, thereby, ensuring that heavily congested user traffic cannot slow configuration or forwarding table updates to the line cards.

    When the forwarding table lookups are complete, the TCAM passes information back to the Packet Forwarding Engine, specifying the modifications to be performed on the Ethernet or IP packet header (e.g., to mark the packet with its appropriate traffic class using IEEE 802.1p, DSCP, IP Precedence, or MPLS EXP bits), and populating an internal header that is appended to the packet (that includes the destination egress port and other internal packet handling information).
3. Once the forwarding table lookups are complete, the Packet Forwarding Engine modifies the packet header (if necessary, to remark QoS, etc.) and appends the packet's egress port destination. Once packet modification is complete, the packet is passed to the ingress Buffer and Traffic Manager.
4. The ingress Buffer and Traffic Manager enforces the policies defined by the network administrator and set during packet classification (e.g., packet rate policing, two-rate three color queuing with dual token buckets, RED

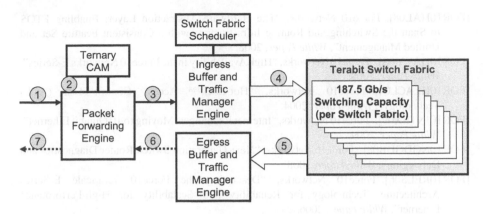

FIGURE 4.14 Diagram describing packet-level walkthrough.

or WRED for congestion avoidance, ingress buffer management to support multiple QoS queues, and queue scheduling using Strict Priority or WFQ as required), and performs queuing for transmission through the Terabit Switch Fabric. After the ingress Buffer and Traffic Manager processing is completed, the packet is passed to the Switch Fabric Scheduler, which supports VOQs at the ingress ports when scheduling traffic to the Terabit Switch Fabric ASICs. The Switch Fabric Scheduler grants access to the ingress Buffer and Traffic Manager to send the packet to the destination line card through the Terabit Switch Fabric ASICs.

The E-Series Gigabit Ethernet and 10 Gigabit Ethernet ports have 55 to 200 milliseconds of buffering. Buffer management provides dynamic access to a pool of buffers shared by all the ports on the line card.

5. On the egress side, the egress Buffer and Traffic Manager receives the packet from the Terabit Switching Fabric and enforces egress policies.
6. The egress Buffer and Traffic Manager inspects and enforces egress policy (token bucket metering (which can be used for rate limiting), congestion avoidance with RED/WRED, egress buffer management, queue scheduling, enforcing ACLs, etc.), and passes the packet to the Packet Forwarding Engine. However, in this case, the policies are being applied to egress traffic aggregated from all the active ingress ports.
7. Finally, the packet is passed to the egress Packet Forwarding Engine. The Packet Forwarding Engine inspects the packet, removes internal header information, modifies the MAC, and directs the packet out the appropriate port.

REFERENCES

[FOR10ESER05]. Force10 Networks, "The Force10 E-Series Architecture", *White Paper*, 2005.
[FOR10FTOS08]. Force10 Networks, "FTOS: A Modular Switch/Router OS Optimized for Resiliency & Scalability", *White Paper*, 2008.

[FOR10HAL08]. Force10 Networks, "The Hardware Abstraction Layer: Enabling FTOS to Span the Switching and Routing Infrastructure with a Consistent Feature Set and Unified Management", *White Paper*, 2008.

[FOR10HAVAI08]. Force10 Networks, "High Availability in the Force10 Networks E-Series", *White Paper*, 2008.

[FOR10HLACL04]. Force10 Networks, "Hot-Lock™ Access Control List (ACL) Technology", *White Paper*, 2004.

[FOR10INTEX09]. Force10 Networks, "Internet Exchanges: Moving to 10 Gigabit Ethernet", *White Paper*, 2009.

[FOR10NGDGDB08]. Force10 Networks, "Next-Generation Switch/Router Diagnostics and Debugging", *White Paper*, 2008.

[FOR10RELS06]. Force10 Networks, "Describing the Force10 TeraScale E-Series Architecture: Technology for Reliability and Scalability for High-Performance Ethernet", *White Paper*, 2006.

[FOR10TSR06]. Force10 Networks, "Next Generation Terabit Switch/Routers: Transforming Network Architectures", *White Paper*, 2006.

[RFC3176]. InMon Corporation's sFlow: A Method for Monitoring Traffic in Switched and Routed Networks, September 2001.

5 Case Study
Foundry Networks BigIron RX Series Switch/Routers Architecture

5.1 INTRODUCTION

The Foundry Networks BigIron RX Series is a family of switch/routers that support Layer 2 (Ethernet) and Layer 3 (IPv4/IPv6) forwarding on high-speed Ethernet interfaces. The BigIron RX Series is designed specifically to address the needs of enterprise and service provider networks, and to support Ethernet interfaces such as 10 Gigabit Ethernet (GE), while allowing growth to accommodate 40 GE to 100 GE interfaces [FOUBIGRX05]. The BigIron RX Series is designed for use in a wide range of customer environments including enterprise network backbones, service provider infrastructures, Internet Exchanges, data centers, and high-performance computing (HPC) clusters (where non-blocking, high-density Ethernet switches are needed.).

For small-size networks, the BigIron RX Series can serve as a "data center-in-a box" platform with high-density 1 Gigabit Ethernet and 10 Gigabit Ethernet interfaces. For mid-size networks, the BigIron RX Series is ideally suited for deployment in the access, aggregation, or core layers. For large-scale networks, the BigIron RX Series is best suited for deployment in the access and aggregation layers.

A key design goal of the BigIron RX Series is to provide high-performance, low-latency, nonstop packet operation (see discussion on redundancy and high-availability features in Chapter 1 of this volume), while meeting the QoS requirements of the respective applications. This chapter describes the architecture of the BigIron RX Series while highlighting the design concepts outlined in the previous chapters and in Volume 1.

5.2 KEY FEATURES OF THE BIGIRON RX SERIES

The BigIron RX Series is available in four different configurations, allowing network designers to select the products that best meets their requirements, ranging from the access to the backbone layer. This family of switch/routers supports the following features [FOUBIGRX05]:

- Non-blocking switching capacity of up to 5.12 Tb/s and routing performance of up to 2284 million packets per second (Mpps).
- Advanced distributed hardware forwarding architecture with fine-grained, wire-speed QoS processing for Layer 2, IPv4, and IPv6 services.

DOI: 10.1201/9781003311256-5

- Distributed packet processing with advanced QoS features implemented at wire-speed rates (see discussion on QoS and traffic rate management in Chapters 8 and 9 of Volume 1).
- Clos-based self-routing, distributed, and non-blocking switch fabric architecture that provides fabric and platform scalability and redundancy.
- High-availability architecture with clear control plane and data plane separation (see discussion on control and data plane decoupling, and high-availability in Chapters 2, 5, 6, and 7 of Volume 1 and Chapter 1 of Volume 2).
- Fully redundant architecture with redundant route processor modules (RPMs), switch fabric modules (SFMs), power supplies, and cooling fan trays to avoid any single point of failure (see discussion on redundancy and high availability in Chapter 1 of this volume). The RPM in the BigIron RX is referred to as the management module.

The BigIron RX Series is available in four different configurations [FOUBIGRX05] [FOUBIGRX08]:

- BigIron RX-4, a 4 RU (*rack unit*), 4 interface-slot system, 960 Gb/s total switch capacity, 286 Mpps packet forwarding capacity per system
- BigIron RX-8, a 7 RU, 8 interface-slot system, 1.92 Tb/s total switch capacity, 571 Mpps packet forwarding capacity per system
- BigIron RX-16, a 14 RU, 16 interface-slot system, 3.84 Tb/s total switch capacity, 1,142 Mpps packet forwarding capacity per system
- BigIron RX-32, a 33 RU, 32 interface-slot system, 5.12 Tb/s total switch capacity, 2,284 Mpps packet forwarding capacity per system

A *rack unit* (abbreviated as U or RU), which is equal to 1.75 inches (44.5mm), indicates the height of a piece of equipment in terms of the amount of rack space it will occupy. RU is most often used to indicate the overall height of rack frames, as well as the height of equipment that is housed in these frames; the height of the frame or equipment is expressed as multiples of RUs. For example, 42 RU is the height of a typical full-size rack, while the typical equipment is 1 RU, 2 RU, 3 RU, or 4 RU high. A 16 RU rack will house in theory 4 pieces of 4 RU gear, 8 pieces of 2 RU gear, or 16 pieces of 1 RU gear.

All modules (RPM and line cards) on the BigIron RX Series are designed to be hot pluggable. The RPM and interface modules can be used interchangeably across the BigIron RX-4, BigIron RX-8, BigIron RX-16, and BigIron RX-32, thereby, providing an effective way for decreasing inventory and maintenance costs for network operators.

The BigIron RX Series is designed with scalability in mind and supports 64 non-blocking 10 GE ports or 768 non-blocking 1 GE on a single RX-16 chassis, for example. The BigIron RX Series can support on a standard 7' rack, up to 192 10 GE ports or 2,304 GE ports. With this, the BigIron RX Series has a density of 55 GE ports per RU and 110 Gb/s of full-duplex switching capacity per RU [FOUBIGRX05].

To future-proof the BigIron RX Series architecture, every interface slot is designed with over 48 Gb/s of full-duplex switching bandwidth available. The physical divider that separates two adjacent interface slots can be removed to convert the half slots into a single full slot. A full slot can handle 100 Gb/s of full-duplex bandwidth from the backplane. This enables the BigIron RX Series chassis to migrate to 40 Gigabit Ethernet or 100 Gigabit Ethernet interfaces when the need arises.

To support migration to IPv6, the BigIron RX Series supports IPv6 over IPv4 tunnels. All network interface modules on the BigIron RX Series have native hardware support for dual-stack IPv4/IPv6 routing with routing table capacity of 200K IPv6 routes. The system supports hardware FIBs on each interface module, allowing wirespeed, full-duplex IPv6 performance even at full chassis capacity. Each interface module also supports IPv6 hardware ACLs. The BigIron RX Series supports the following protocols [BROCBIGRX11] [FOUBIGRX05] [FOUBIGRX08]:

- IPv4 routing protocols: Support for Routing Information Protocol version 2 (RIPv2), Open Shortest Path First version 2 (OSPFv2), Intermediate System-to-Intermediate System (IS-IS), and Border Gateway Protocol version 4 (BGPv4) (see [AWEYA2BK21V1] and [AWEYA2BK21V2] for a comprehensive discussion of IP routing protocols).
- IPv6 routing protocols such as Routing Information Protocol next generation (RIPng), OSPFv3, IS-ISv6, and Multiprotocol BGP (MP-BGP).
- Multicast protocols such as Internet Group Management Protocol (IGMP), Multicast Listener Discovery (MLD), Protocol Independent Multicast–Sparse-Mode (PIM-SM), PIM-Dense Mode (PIM-DM), and Multicast Source Discovery Protocol (MSDP). The support of PIM and IGMP Snooping allows efficient handling of multicast traffic in Layer 2 networks by allowing Layer 2 switches to identify ports that request a multicast stream. This allows multicast traffic to be forwarded only on those ports. PIM and IGMP Snooping dramatically improve the performance of multicast applications, allowing many more multicast streams to be supported on the network.
- Layer 3 redundancy protocols such as Virtual Router Redundancy Protocol (VRRP). VRRP allows a number of routers to operate as backup routers to other network routers. In the event of a router failure, one of the backup routers will automatically and seamlessly perform the tasks of the failed router (see VRRP discussion in Chapter 1 of this volume)
- IEEE 802.3ad Link Aggregation (for up to eight links), and scalable, cross-module trunking, providing resilient high-capacity Layer 2 connections between switches. Trunking allows multiple Layer 2 links of a switch to appear as one logical link.
- Layer 2 redundancy protocols such as Foundry Networks Virtual Switch Redundancy Protocol (VSRP) which provides device redundancy for specific ports in a port-based VLAN (i.e., one or more backups for a device).
  o VSRP combines the Layer 2 Spanning Tree Protocol and Layer 3 VRRP resiliency features – effectively doing the jobs of both protocols at the same time, and significantly improving the failover times provided by either of those protocols. If the active device becomes unavailable, VSRP allows one of the backup devices to take over as the active device and continue forwarding traffic for the network.

- o VSRP provides redundancy in redundant switch configurations while VRRP provides redundancy in redundant router configurations.
  - o VSRP provides sub-second fault detection and failover in mesh topologies with redundant switches in which the redundant switches provide backup operation for one another.
- Layer 2 service protection protocols such as Rapid Spanning Tree Protocol (RSTP) defined in IEEE 802.1w, Multiple Spanning Tree Protocol (MSTP) defined in IEEE 802.1s, Per VLAN Spanning Tree (PVST), Per VLAN Group Spanning Tree (PVGST), Topology Groups, and Super Aggregated VLANs (SAVs):
  - o RSTP, which has replaced STP (in IEEE 802.1D), dramatically improves the convergence time of the spanning tree to sub-second by allowing switches to automatically renegotiate port roles in case of a link failure, without relying on timers.
  - o PVST allows a separate instance of STP/RSTP to be operated for each individual VLAN on a network, and can be used for traffic engineering VLAN traffic (such as load distribution).
  - o PVGST dramatically improves the scalability of VLANs by servicing up to 4096 VLANs with 2 to 16 STPs or RSTP group instances.
  - o Topology Groups dramatically improves the scalability of Layer 2 control protocols by allowing a few instances of STP, RSTP, MRP, or VSRP to control large groups of VLANs.
  - o SAVs allow multiple VLANs to be transparently tunneled through a single backbone VLAN.
- Layer 2 service protection using Brocade's Metro Ring Protocol (MRP) [BRODMRP]. MRP is a protocol for creating Ethernet rings similar to ITU-T Recommendation G.8032, Ethernet Ring Protection Switching (ERPS) [ITUTG8032], which is the industry standard for implementing Ethernet ring networks. MRP provides sub-second fault detection and failover for Ethernet ring topologies.
- Multi-Chassis Trunking (MCT) which allows two BigIron RX Series chassis to be clustered and appear logically as a single device. This allows aggregated links to forward traffic to either chassis, resulting in higher network utilization and sub-second failover in case of a link or node failure.

The device and network management and security features on the BigIron RX Series include the following:

- Centralized network management using the IronView Network Manager which is a Web-based, graphical interface tool that allows network operators to seamlessly control software and configuration updates.
- Command Line Interface (CLI) which supports an industry-standard configuration interface and is consistent and common in all Brocade products (see discussion on CLIs in Chapter 3 of this volume).

- User authentication using RADIUS, TACACS, and TACACS+, IEEE 802.1X, MAC Authentication and MAC Port Security, thus, preventing unauthorized network access (see discussion on user authentication protocols in Chapter 2 of this volume). MAC Port Security controls the MAC addresses allowed per switch port.
- Secure access and management protocols such as Secure Shell (SSH), Secure Copy (SCP), and Simple Network Management. Protocol Version 3 (SNMPv3) (see Chapter 2 of this volume). SNMPv3 is a secured SNMP version that provides authentication and privacy services.
- Embedded sFlow-based Layer 2 to Layer 7 traffic monitoring per port (using hardware-based packet sampling) (see Chapter 2 of this volume).
- Protection against denial-of-service (DoS) attacks, for example, by limiting TCP SYN and ICMP traffic, thus, preventing or minimizing network downtime from malicious users, and protecting against broadcast storms.
- BGP-Guard which complements MD5 security for BGP sessions. BGP-Guard works by restricting the number of hops the BGP session can traverse, thus, protecting against session disruption.

5.3   BIGIRON RX SERIES SWITCH FABRIC ARCHITECTURE

At the core of the BigIron RX Series is a Clos switch fabric architecture that provides high packet forwarding, scalability, and redundancy [FOUBIGRX05]. The Clos switch fabric architecture supports an adaptive self-routing mechanism with virtual output queues (VOQs) at the input ports (see discussion on VOQs in Chapters 1 and 3 of this volume). A key design goal is to optimize the architecture for non-blocking performance, maximum throughput, and low latency for all legal packet sizes.

As shown in Figure 5.1, the architecture supports multiple SFMs with each SFM having multiple switch fabric elements (SFEs). Each SFE has multiple connections to every network interface slot. To ensure that the switch fabric interconnects are optimally utilized at all times, the Clos architecture distributes data across the various interconnects by applying data striping to arriving packets. The striping mechanism distributes arriving traffic equally across all the available links between the input and output interface modules. Arriving packets are segmented into fixed-size cells, which in turn are striped and transported across the switch fabric. The use of fixed-size cells allows the switching architecture to provide predictable performance with very low and deterministic latency and delay variations for all packet sizes. By using multiple switching paths between the input and output interface modules, the system is also able to provide an additional level of redundancy.

The advantages of using a Clos architecture in the BigIron RX Series over the traditional switch fabric architectures are summarized as follows:

- Provides a common switch fabric architecture across the BigIron RX Series as the same SFEs are used on all the four chassis configurations. This provides scalability from the small 4-slot system to the large 32-slot system.

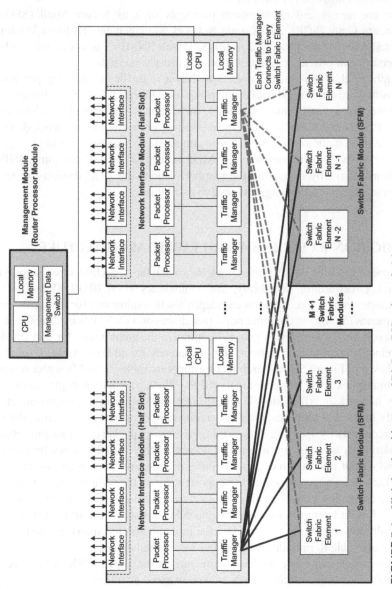

FIGURE 5.1 High-level architecture of BigIron RX series switch/routers.

- Prevents head-of-line (HOL) blocking at any point irrespective of the traffic pattern, type of traffic, or packet size (see discussion on HOL blocking in Chapters 1 and 3 of this volume).
- Allows optimal utilization of the switch fabric resources at all times. The data striping mechanism ensures that the SFEs are optimally utilized at all times without overloading of any single SFE.
- Provides intra-SFM redundancy, where any SFM can withstand the failure of some of its SFEs and still continue to operate with the remaining SFEs. This allows the SFM to provide a very high level of redundancy even within itself.
- Provides enhanced high-availability by supporting SFMs with (N+1) redundancy, allowing the BigIron RX Series to gracefully adapt to the failure of multiple SFEs. Furthermore, because of the presence of multiple SFEs within an SFM, the failure of an SFE does not bring down the entire SFM; provides graceful system degradation in the event of two or more module failures.

5.4 DISTRIBUTED WIRE-SPEED PACKET FORWARDING

The BigIron RX Series supports a distributed forwarding architecture, allowing distributed wire-speed forwarding of any legal packet size. As shown in Figure 5.1, the distributed architecture supports advanced network processors, a high-performance (local) CPU, and high-speed memory on each interface module, providing a scalable, high-performance architecture. The packet processors (Figure 5.1) implement a variety of features such as traffic classification, access control lists (ACLs), policing, and multicast traffic replication. By using fast packet processors on each interface module, the system is able to maintain wire-speed performance, independent of the ACL and QoS features that have been enabled on the system.

The IronWare operating system software of the BigIron RX Series provides several capabilities that facilitate distributed packet forwarding and security (see discussion on distributed forwarding in Chapters 2, 5, 6, and 7 of Volume 1):

- **Distributed maintenance of the Ethernet MAC address table on each network interface module**: The RPM maintains all the learned Ethernet MAC addresses and distributes this information to the MAC address table maintained locally on each network interface module. Each interface module is responsible for locally aging its local MAC address table entries, and updating the RPM in order to keep the MAC address tables consistent across the entire system.
- **Distributed forwarding table (or FIB) on each network interface module**: Each interface module stores the entire forwarding table locally to allow for hardware forwarding of all traffic (up to 512,000 IPv4 routes and 64,000 IPv6 routes per module). The RPM is responsible for downloading the FIB to the hardware-based forwarding engine on each line module.
- **Distributed ACL maintenance on each network interface module**: Each interface module has support for hardware input ACLs on all ports.

As shown in Figure 5.1, the BigIron RX Series has a dedicated out-of-band management link between each network interface module and the RPM, allowing control traffic to be isolated from data traffic. Multiple queues to the RPM allow different types of control traffic to be prioritized. These capabilities, together with secure device access and management (via SSH, SCP and SNMPv3) and ACLs, help in securing the system from potential DoS attacks in the network.

The BigIron RX Series architecture also supports natively spatial multicast forwarding, a critical requirement for offering real-time multicast services in a network. An input network interface module sends only one copy of an incoming multicast packet to the switch fabric, which then replicates the packet internally to multiple output interface modules in the system. The output interface modules in turn replicate the multicast packet locally to the appropriate destination ports and VLANs.

5.5 DISTRIBUTED QUEUING AND FINE-GRAINED QoS CONTROL

The BigIron RX Series architecture uses a distributed queuing scheme that is designed to maximize the utilization of buffers across the whole system during periods of congestion (Figure 5.2). This scheme combines the benefits of input port buffering with VOQs with those of an output-port-driven traffic queuing and scheduling mechanism. The input port VOQs prevent HOL blocking and also ensure that bursty traffic

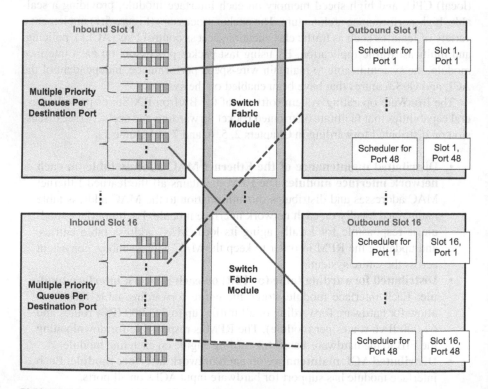

FIGURE 5.2 VOQs with multiple priority queues.

from any one input port does not hog too many buffers on an output port. The output-port-driven queuing and scheduling mechanism ensures that packets are sent to an output port only when that port is ready to transmit packets onto the external network.

Each interface module maintains multiple, distinct priority queues in each VOQ set assigned to each output port on the system. An outbound interface module pulls packets from the VOQs when an output port is ready to send a packet onto the network. The system uses switch fabric messaging to ensure that there is tight coupling between the input queuing and the output queuing and scheduling stages. The input and output stages function as a closed-loop feedback system to ensure that no information is lost between them. The use of VOQs maximizes the efficiency of the system by allowing packets to be stored on the input module until the output port is ready to transmit the packet. The BigIron RX Series chassis supports a total of 512K VOQs that are distributed across the system.

Complementing the large number of input port VOQs is a set of scheduling algorithms that can be applied at each output port. The scheduling mechanisms can be used individually, or in combination to deliver tiered QoS guarantees for different applications on the same output port [FOUBIGRX05]:

- Strict priority
- Enhanced strict priority scheduling
- Weighted Fair Queuing (WFQ) destination-based scheduling
- WFQ source-based scheduling
- Minimum rate-based scheduling
- Maximum rate-based scheduling

For example, using these advanced QoS capabilities, the system can service multiple queues in the order of decreasing priority. The BigIron RX Series handles congestion avoidance by applying Weighted Random Early Discard (WRED) or tail-drop policy to the system queues (see discussion on WRED and tail-drop policies in Chapter 8 of Volume 1).

The BigIron RX Series supports extensive classification and packet marking capabilities that can be configured on the QoS subsystem:

- Prioritization based on Layer 2 (IEEE 802.1p), IP Precedence, or DSCP setting on packets (traffic prioritization mechanisms are described in detail in Chapter 8 of Volume 1)
- Mapping of IP packet and Ethernet frame priority during ingress encapsulation and/or egress encapsulation
- Remarking of the priority of a packet based on the result of the two-rate, three-color policer (rate-limiting mechanisms are described in detail in Chapter 9 of Volume 1)

All network interface modules on the BigIron RX Series support inbound traffic policers in hardware. The system supports both single-rate three-color and two-rate, three-color policers. As discussed in Chapter 9 of Volume 2, the single-rate, three-color marker meters a subscriber traffic flow that exceeds a configured compliant rate, called the committed information rate (CIR), while the two-rate, two-color policers meter a traffic flow by classifying it into a compliant rate (CIR) or a peak information rate (PIR). These capabilities are especially useful when traffic flows with different characteristics are mixed on the same port. All interface modules support input ACLs.

By performing sFlow information collection in hardware, the BigIron RX Series can precisely monitor flows at Layer 2 to Layer 7. This capability facilitates rapid troubleshooting and accurate isolation of faults in a network.

The BigIron RX Series supports Policy-Based Routing (PBR) which provides customizable routing policies using ACLs. Network operators typically use PBR to control network resource usage by controlling the network paths for different traffic flows.

5.6   BIGIRON RX SERIES HIGH-AVAILABILITY FEATURES

The BigIron RX Series supports a hardware and software architecture that is designed to ensure very high Mean Time Between Failures (MTBF) and low Mean Time To Repair (MTTR). By allowing module insertion and cable management on the same side of the chassis, the system provides ease of serviceability when a failed module needs to be replaced or a new module needs to be inserted.

The system is designed to handle the failure of not only an SFM but also the SFEs within the SFM, providing a robust, redundant system suitable for non-stop operation. The overall system redundancy is further enhanced by redundancy in other active system components such as the RPMs, power supplies, and cooling fans. All four BigIron RX Series configurations are designed for non-stop operation, by supporting 1:1 RPM redundancy, N+1 SFM redundancy, M+N power supply module redundancy (for AC and DC power configurations), and N+1 cooling fan redundancy.

The BigIron RX Series chassis uses a passive backplane which increases the reliability of the system (similar to the Force10 E-Series described in Chapter 4 of this volume). Chapter 3 of this volume provides a detailed discussion on the backplane types used in network devices. Using self-adjusting, variable-speed cooling fans, the BigIron RX Series chassis is able to maintain the optimal operating temperature. Modules that exceed a pre-set temperature threshold are powered off by the RPM so as not to completely disrupt system function.

The BigIron RX Series also supports features that allow the system to gracefully shut down an SFM for a scheduled maintenance event with zero packet loss. When the system invokes this facility, the links between the interface modules and the decommissioned SFM will not be used. The IronWare operating system of the BigIron RX Series, which has a modular architecture, has several characteristics that distinguish it from legacy operating systems. Redundant, hot-swappable components allow the BigIron RX Series to provide non-stop service delivery [FOUBIGRX05]:

- Cold system restart time of less than a minute
- Hitless software upgrade
- Hitless Layer 2 and Layer 3 failovers: Stateful failover ensures that the forwarding engines on the network interface modules are not impacted by an RPM failover. This capability enables nonstop packet forwarding in the event of an RPM failover.

- Sub-second detection and switchover to the standby RPM if a communication failure occurs between active and standby RPMs.
- SFE redundancy and a system configured with a redundant SFM supports millisecond failover performance.
- Support for OSPF graceful restart and BGP graceful restart

The hitless software upgrades and graceful restart routing protocol features of the BigIron RX Series allows fast network convergence in the event of an RPM failure. The isolation of control traffic to control plane (i.e., complete separation of the control and data planes) ensures that control packets are processed by the system's RPMs efficiently even during periods of high traffic loads.

The several hardware assist features of BigIron RX Series enable network designers to create robust, scalable, and stable networks. Layer 2 unknown address, broadcasts, and multicast services require support for efficient replication of packets to the entire broadcast domain (VLAN or subnet). Traditional architectures handle Ethernet frames with unknown MAC address by sending them to a centralized processor to replicate the packet to the broadcast domain, a situation that makes the processor vulnerable to potential DoS attacks. The BigIron RX Series handles this scenario by performing such flooding in hardware very efficiently.

The Continuous System Monitoring (CSM) feature in BigIron RX Series is a tool that runs in the background and monitors specific hardware components on all the network interface modules, RPMs, and SFMs. CSM uses both polling and interrupt methods to collect information about the system. If CSM detects a failure, it places an alarm entry in the syslog, allowing network administrators to take appropriate action. Depending on the type of failure, CSM will shut down, reset, or redirect traffic to other modules. CSM increases system availability, allowing failures to be averted before they occur, thus, ensuring maximum network uptime and application availability.

REFERENCES

[AWEYA2BK21V1]. James Aweya, *IP Routing Protocols: Fundamentals and Distance Vector Routing Protocols*, CRC Press, Taylor & Francis Group, ISBN 9780367710415, 2021.

[AWEYA2BK21V2]. James Aweya, *IP Routing Protocols: Link-State and Path-Vector Routing Protocols*, CRC Press, Taylor & Francis Group, ISBN 9780367710361, 2021.

[BROCBIGRX11]. Brocade Communications Systems, *Brocade BigIron RX Series, Modular Layer 2/3 Ethernet Switches*, 2011.

[BRODMRP]. Ruckus FastIron *Layer 2 Switching Configuration Guide*, Metro Ring Protocol, 2018.

[FOUBIGRX05]. Foundry Networks, *BigIron RX Architecture Brief*, Document Version 2.56, 2005.

[FOUBIGRX08]. Foundry Networks, *BIGIRON RX-4, RX-8, RX-16, RX-32: MODULAR LAYER 2/3 ETHERNET SWITCH FAMILY*, Data Sheet, 2008.

[ITUTG8032]. ITU-T Recommendation G.8032, Ethernet Ring Protection Switching, March 2020.

6 The Ethernet Advantage in Networking

6.1 INTRODUCTION

Over the past decades since the 1990s, the performance and cost-effectiveness of Ethernet as a networking technology has been well-understood and recognized, resulting in it becoming the dominant transmission technology both in the MAN and WAN, in addition to its traditional and original application domain, the LAN. The rapid adoption of high-capacity network devices and IP-based services continues to drive the growth and transformation of networks in the LAN, MAN, and WAN. Once considered a networking technology mainly for data services, Ethernet has evolved over the years to be widely used as the network convergence technology for delivering the full array of packet-based data, voice, and video services.

This chapter examines the factors that have contributed to making Ethernet the dominant technology for networking in the LAN, MAN, and WAN. The discussion includes the benefits of using Ethernet, industry trends related to the use of Ethernet, and the role of Ethernet in network infrastructure virtualization.

6.2 ETHERNET PHYSICAL LAYER TYPES

Ethernet networking technology has evolved a lot and gone through several versions to become the ubiquitous technology it is now. The following are some well-known examples of Ethernet Physical Layer types that have been developed over the past three decades (not an exhaustive list):

- **10 Mbps Ethernet**: This variant of Ethernet includes the following:
  - **10BASE-T** (IEEE 802.3i-1990 (Clause 14)) which uses Manchester coding and runs over four copper wires (two twisted pairs) up to 100 m.
  - **10BASE-FL** (IEEE 802.3j-1993 (Clause 15 & 18)) which runs on 850 nm wavelength multi-mode fiber optical fiber cable up to 2000 m.
- **100 Mbps Ethernet (also called Fast Ethernet)**: This variant of Ethernet includes the following:
  - **100BASE-TX** (IEEE 802.3u-1995 (Clause 24 & 25)) which uses 4B5B and MLT-3 coded signaling and runs over Category 5 cable using two copper twisted pair cables.
  - **100BASE-FX** (IEEE 802.3u-1995 (Clause 24 & 26)) which uses 4B5B and NRZI coded signaling and runs over two strands of multi-mode optical fiber up to 2 km.

DOI: 10.1201/9781003311256-6

- **1000 Mbps Ethernet (1 Gb/s or 1 GE)**: This variant of Ethernet includes the following:
  - ○ **1000BASE-T** (IEEE 802.3ab-1999 (Clause 40)) which uses PAM-5 coded signaling, and runs over four copper twisted pairs (at least Category 5 cable, with Category 5e cabling strongly recommended).
  - ○ **1000BASE-SX** (IEEE 802.3z-1998 (Clause 38)) which uses 8B10B and NRZ coded signaling and runs over 850 nm carrier, short-range multi-mode fiber up to 550 m.
  - ○ **1000BASE-LX** (IEEE 802.3z-1998 (Clause 38)) which uses 8B10B and NRZ coded signaling on 1310 nm carrier, multi-mode fiber up to 550 m or single-mode fiber up to 5 km.
- **2.5 and 5 Gb/s Ethernet (2.5 GE and 5 GE)**: These variants of Ethernet (IEEE 802.3bz-2016 (Clause 126)) include **2.5GBASE-T** and **5GBASE-T** which are scaled-down variants of 10GBASE-T.
  - ○ 2.5GBASE-T runs over Category 5e twisted pair copper cabling up to 100 m, while 5GBASE-T runs over Category 6 cabling up to 100 m.
- **10 Gb/s Ethernet (10 GE)**: This variant of Ethernet includes the following:
  - ○ **10GBASE-T** (IEEE 802.3an-2006 (Clause 55)) which uses PAM-16 with DSQ128 line coding and runs over Category 6A twisted-pair wiring, four lanes at 800 Mega-Bauds (MBd) each.
  - ○ **10GBASE-SR** (IEEE 802.3ae-2002 (Clause 49 & 52)) which uses 64B/66B encoding and runs over multi-mode fiber cabling using 850 nm wavelength. It has a range of between 26 m and 400 m depending on optical fiber cable type.
  - ○ **10GBASE-LX4** (IEEE 802.3ae-2002 (Clause 48 & 53)) which uses 8B/10B encoding and NRZ over four lanes with wavelength division multiplexing (WDM) (1275, 1300, 1325, and 1350 nm). It runs over multi-mode cabling to support ranges of between 240 m and 300 m, and over single-mode fiber up to 10 km.
  - ○ **10GBASE-LR** (IEEE 802.3ae-2002 (Clause 49 & 52)) which uses 64B/66B encoding and runs over single-mode fiber using 1,310 nm wavelength up to 10 km.
  - ○ **10GBASE-ER** (IEEE 802.3ae-2002 (Clause 49 & 52)) which uses 64B/66B encoding and runs over single-mode fiber using 1,550 nm wavelength up to 30 km.
  - ○ **10GBASE-SW** (IEEE 802.3ae-2002 (Clause 50 & 52)) which has 9.95328 Gb/s line rate (WAN-PHY) and is designed to be mapped directly to carry OC-192/STM-64 SONET/SDH streams (850 nm wavelength).
  - ○ **10GBASE-LW** (IEEE 802.3ae-2002 (Clause 50 & 52)) which has 9.95328 Gb/s line rate (WAN-PHY) and is designed to be mapped directly to carry OC-192/STM-64 SONET/SDH streams (1,310 nm wavelength).
  - ○ **10GBASE-EW** (IEEE 802.3ae-2002 (Clause 50 & 52)) which has 9.95328 Gb/s line rate and is designed to be mapped directly to carry OC-192/STM-64 SONET/SDH streams (1,550 nm wavelength).

- o **10GBASE-LRM** (IEEE 802.3aq-2006 (Clause 49 & 68)) which uses 64B/66B encoding and runs over multi-mode fiber (1,310 nm wavelength) up to 220 m.
- **25 Gb/s Ethernet (25 GE):** This variant of Ethernet includes the following:
  - o **25GBASE-T** (IEEE 802.3bq-2016 (Clause 113)) which is a scaled-down version of 40GBASE-T and runs over Category 8 up to 30 m.
  - o **25GBASE-SR** (IEEE 802.3by-2016 (Clause 112)) which runs over 850 nm multi-mode fiber with 70 m (OM3) or 100 m (OM4) reach.
  - o **25GBASE-LR** (IEEE 802.3cc-2017 (Clause 114) which runs over 1310 nm single-mode fiber with 10 km reach.
  - o **25GBASE-ER** (IEEE 802.3cc-2017 (Clause 114)) which runs over 1550 nm single-mode fiber with 30 km 4 km reach.
- **40 Gb/s Ethernet (40 GE):** This class of Ethernet includes the following:
  - o **40GBASE-T** (IEEE 802.3bq-2016 (Clause 113)) which runs over Category 8 cabling up to 30 m.
  - o **40GBASE-SR4** (IEEE 802.3ba-2010 (Clause 86)) which runs over 2000 MHz·km multi-mode fiber (OM3) with at least 150 m reach, or over 4700 MHz·km multi-mode fiber (OM4) with at least 100 m reach.
  - o **40GBASE-LR4** (IEEE 802.3ba-2010 (Clause 87)) which runs over single-mode fiber, Coarse WDM (CWDM) with 4 lanes using 1270, 1290, 1310, and 1330 nm wavelength and with at least 10 km least reach.
  - o **40GBASE-ER4** (IEEE 802.3ba-2010 (Clause 87)) which runs over single-mode fiber, CWDM with 4 lanes using 1270, 1290, 1310, and 1330 nm wavelength and with at least 30 km to 40 km reach.
  - o **40GBASE-FR** (IEEE 802.3bg-2011 (Clause 89)) which runs over 1550 nm wavelength, single lane, single-mode fiber, up to 2 km.
  - o **40GBASE-CR4** (IEEE 802.3ba-2010 (Clause 85)) which runs over twinaxial (twinax) copper cable assembly (4 lanes, 10 Gb/s each) up to 7 m.
- **50 Gb/s Ethernet (50 GE):** This class of Ethernet includes the following:
  - o **50GBASE-SR** (IEEE 802.3cd-2018 (Clause 138)) which uses PAM-4 encoding and runs over OM4 multi-mode fiber with 100 m reach, or over OM3 multi-mode fiber with 70 m reach.
  - o **50GBASE-FR** (IEEE 802.3cd-2018 (Clause 139)) which uses PAM-4 encoding and runs over single-mode fiber with 2 km reach.
  - o **50GBASE-LR** (IEEE 802.3cd-2018 (Clause 139)) which uses PAM-4 encoding and runs over single-mode fiber with 10 km reach.
  - o **50GBASE-ER** (IEEE 802.3cd-2018 (Clause 139)) which uses PAM-4 encoding and runs over single-mode fiber with 30 km to 40 km reach.
  - o **50GBASE-CR** (IEEE 802.3cd-2018 (Clause 136)) which runs over twinax cable with 3 m reach.
- **100 Gb/s Ethernet (100 GE):** This variant of Ethernet includes the following:
  - o **100GBASE-SR10** (IEEE 802.3ba-2010 (Clause 86)) which runs over 2000 MHz·km multi-mode fiber (OM3) with at least 100 m reach, or over 4700 MHz·km multi-mode fiber (OM4) with at least 150 m reach.

- ○ **100GBASE-SR4** (IEEE 802.3bm-2015 (Clause 95)) which runs over 4 lanes, 2000 MHz·km multi-mode fiber (OM3) with at least 70 m reach, or over 4700 MHz·km multi-mode fiber (OM4) with at least 100 m reach.
- ○ **100GBASE-SR2** (IEEE 802.3cd-2018 (Clause 138)) which uses PAM-4 encoding and runs over two 50 Gb/s lanes over OM4 multi-mode fiber with 100 m reach, or over OM3 with 70 m reach.
- ○ **100GBASE-LR4** (IEEE 802.3ba-2010 (Clause 88)) which runs over single-mode fiber, DWDM with 4 lanes using 1296, 1300, 1305, and 1310 nm wavelength and with at least 10 km reach.
- ○ **100GBASE-ER4** (IEEE 802.3ba-2010 (Clause 88)) which runs over single-mode fiber, DWDM with 4 lanes using 1296, 1300, 1305, and 1310 nm wavelength and with at least 30 km reach.
- ○ **100GBASE-DR** (IEEE 802.3cu-2021 (Clause 140)) which runs over single-mode fiber using a single lane with at least 500 m reach.
- ○ **100GBASE-FR** (IEEE 802.3cu-2021 (Clause 140)) which runs over single-mode fiber using a single lane with at least 2 km reach.
- ○ **100GBASE-LR** (IEEE 802.3cu-2021 (Clause 140)) which runs over single-mode fiber with at least 10 km reach.
- ○ **100GBASE-ZR** (IEEE 802.3ct (Clause 153 & 154) which runs over single-mode fiber using a single wavelength over a DWDM system with at least 80 km reach. This variant also forms the base for 200GBASE-ZR and 400GBASE-ZR.
- ○ **100GBASE-CR10** (IEEE 802.3ba-2010 (Clause 85)) which runs over twinax copper cable assembly (10 lanes, 10 Gb/s each) with up to 7 m reach.
- ○ **100GBASE-CR4** (IEEE 802.3bj-2014 (Clause 92)) which runs over twinax copper cable assembly (4 lanes, 25 Gb/s each) with up to 5 m reach.
- ○ **100GBASE-CR2** (IEEE 802.3cd-2018 (Clause 136)) which runs over twinax cable (two 50 Gb/s lanes) with 3 m reach.
- ○ **100GBASE-CR** (IEEE 802.3ck) which runs over single-lane, twin-axial copper with at least 2 m reach.
- **200 Gb/s Ethernet (200 GE):** This variant of Ethernet includes the following:
  - ○ **200GBASE-DR4** (IEEE 802.3bs-2017 (Clause 121)) which runs over four PAM-4 (26.5625 GBd) lanes using individual strands of single-mode fiber (1310 nm) with 500 m reach.
  - ○ **200GBASE-FR4** (IEEE 802.3bs-2017 (Clause 122)) which runs over four PAM-4 (26.5625 GBd) lanes using four wavelengths (CWDM) over single-mode fiber (1270/1290/1310/1330 nm) with 2 km reach.
  - ○ **200GBASE-LR4** (IEEE 802.3bs-2017 (Clause 122)) which runs over four PAM-4 (26.5625 GBd) lanes using four wavelengths (DWDM, 1296/1300/1305/1309 nm) over single-mode fiber with 10 km reach.
  - ○ **200GBASE-SR4** (IEEE 802.3cd-2018 (Clause 138)) which runs over four PAM-4 lanes over OM3 multi-mode fiber with 70 m reach, or OM4 multi-mode fiber with 100 m reach.
  - ○ **200GBASE-ER4** (IEEE 802.3cn-2019 (Clause 122)) which runs over four-lane using four wavelengths (DWDM, 1296/1300/1305/1309 nm) over single-mode fiber with 30 km to 40 km reach.

- o **200GBASE-CR4** (IEEE 802.3cd-2018 (Clause 136)) which runs over four-lane over twinax cable with 3 m reach.
- o **200GBASE-CR2** (IEEE 802.3ck) which runs over two-lane, twinax copper with at least 2 m reach.
- **400 Gb/s Ethernet (400 GE):** This variant of Ethernet includes the following:
  - o **400GBASE-SR16** (IEEE 802.3bs-2017 (Clause 123)) which runs over 16 (26.5625 Gb/s) lanes using individual strands of over OM3 multi-mode fiber with 70 m reach, or OM4/OM5 multi-mode fiber with 100 m reach.
  - o **400GBASE-DR4** (IEEE 802.3bs-2017 (Clause 124)0 which runs over four PAM-4 (53.125 GBd) lanes using individual strands of single-mode fiber (1310 nm) with 500 m reach.
  - o **400GBASE-FR8** (IEEE 802.3bs-2017 (Clause 122) which runs over eight PAM-4 (26.5625 GBd) lanes using eight wavelengths (CWDM) over single-mode fiber with 2 km reach.
  - o **400GBASE-LR8** (IEEE 802.3bs-2017 (Clause 122)) which runs over eight PAM-4 (26.5625 GBd) lanes using eight wavelengths (DWDM) over single-mode fiber with 10 km reach.
  - o **400GBASE-FR4** (IEEE 802.3cu-2021) which runs over four lanes/ wavelengths (CWDM, 1271/1291/1311/1331 nm) over single-mode fiber with 2 km reach.
  - o **400GBASE-LR4** (IEEE 802.3cu-2021) which runs over four lanes over single-mode fiber with 10 km reach.
  - o **400GBASE-SR8** (IEEE 802.3cm-2020 (Clause 138)) which runs over eight-lane using individual strands of multi-mode fiber with 100 m reach.
  - o **400GBASE-SR8** (IEEE 802.3cm-2020 (Clause 138)) which runs over eight-lane using individual strands of multi-mode fiber with 100 m reach.
  - o **400GBASE-SR4.2** (IEEE 802.3cm-2020 (Clause 150)) which runs over four-lane using individual strands of multi-mode fiber with 100 m reach.
  - o **400GBASE-ER8** (IEEE 802.3cn-2019 (Clause 122)) which runs over eight-lane using eight wavelengths over single-mode fiber with 40 km reach.
  - o **400GBASE-ZR** (IEEE 802.3ct (Clause 155 & 156)) which runs over single-mode fiber using a single wavelength with 16 QAM over a DWDM system and with at least 80 km reach.
  - o **400GBASE-CR4** (IEEE 802.3ck) which runs over four-lane, twinax copper with at least 2 m reach.
- **800 Gb/s Ethernet (800 GE):** This variant of Ethernet includes 800GBASE-R (being defined in IEEE 802.3ck) which runs over eight lanes of 100 Gb/s each, and connects with the transceiver module through a C2M or C2C interface.

Other types of Ethernet Physical Layers have also been developed for multi-mode and single-mode fiber, automotive and industrial applications, point-to-multipoint topologies, electrical backplanes, and Ethernet first mile networking (for providing Internet access service to homes and small businesses directly from service providers). At the time of writing, the networking industry is also looking at developing 1 Terabit per second (Tbps), that is, 1000 Gb/s Ethernet as a further extension of Ethernet.

The technology evolution of Ethernet has made it possible to greatly simplify network infrastructures using one network technology type end to end, from the residential network up to the service provider networks and their point-of-presences (POPs). For example, deploying 10, 25, 40, 50, or 100 Gigabit Ethernet within the POP and for WAN/MAN links makes it possible to leverage the simplicity and cost-effectiveness of switch/routers in both the core and access tiers of the enterprise and service provider networks. Switch/routers typically use Ethernet at the Layers 1 and 2, and IP at Layer 3.

6.3 BENEFITS OF ETHERNET

The shift to Ethernet as the preferred technology for enterprise and service provider networks including data centers and cluster interconnects is dictated by the number of advantages that Ethernet has over other networking technologies and more specialized switching fabrics like those based on Infiniband and Fibre Channel.

6.3.1 LOWER COST

Advances in manufacturing technologies and the highly competitive global market have created an environment for very high production volumes for electronic products. This has ensured that Ethernet will continue to offer the lowest cost network devices and host adapters. These advances and developments have enabled network operators to deploy lower-cost network infrastructures. Ethernet is used in high-performance network devices and also supports port densities with small footprint (high-density switching). This has led to Ethernet being able to conserve space in the POP and to minimize the number of devices that need to be managed.

Ethernet economics have made multi-gigabit switch ports and adapters very low cost and affordable, thereby, lowering total cost of ownership (TCO). Multi-gigabit Ethernet has matured significantly, with 40 Gb/s, 100 Gb/s, and 200 Gb/s gaining much attention. Traditionally, service providers have turned to Synchronous Optical Networking/Synchronous Digital Hierarchy (SONET/SDH) to build their networks, favoring its resiliency features over Ethernet. However, 10 Gigabit Ethernet interfaces, for example, cost as much as 10 times less than OC-192/STM-64 ports. Also, the growing volume of 10 Gigabit Ethernet ports in the market continues to drive the overall port cost, as well as component costs even lower, while OC-192/STM-64 ports (with prices that have seen only modest declines over the past years) are seen as legacy technology and no more used in new network deployments.

In the face of falling profit margins and rising traffic volumes, service providers are finding it difficult to justify the expense of their legacy SONET/SDH networks, particularly, given the traffic mix (of data, voice, and video) seen in today's networks. With traffic growing exponentially, Ethernet offers a higher density, more scalable, cost-effective alternative to SONET/SDH. In addition, basic Ethernet core switches are significantly less expensive than comparable SONET/SDH switches.

In addition to the cost advantage over SONET/SDH, the combination of multi-gigabit Ethernet and basic IP routing in the core also has significantly reduced network operations overhead. A simple best-effort Ethernet network is easier to manage than a comparable SONET/SDH network. And setting up SONET/SDH services can be time-consuming and complex, particularly for a large network. This operational overhead becomes prohibitive when the traffic does not warrant such handling.

Ethernet has the ability to scale bandwidth gracefully from 10 Mbps to 800 Gb/s (and even higher as discussed above) to meet demand with minimal disruption of the existing network infrastructure. Switches are able to support high-density multi-gigabit Ethernet and capable of scaling gracefully to incorporate next-generation Ethernet interfaces (100 and higher Gigabit Ethernet). The high-end switches, in particular, are able to support multiple upgrades through simple replacement of line cards and modules. Recent developments have also allowed Ethernet to support carrier-grade device "zero packet loss" reliability and Network Layer redundancy features at a reasonable cost.

6.3.2 UBIQUITOUS CONNECTIVITY

Ethernet is the most widely deployed technology in today's networks. Practically, almost every host system (end devices, servers, etc.) shipped today comes with an Ethernet interface built in. Also, a large percentage of business data begins or ends as Ethernet frames, and with voice and video being transmitted as IP traffic over Ethernet. An increasingly popular option for servers is to incorporate the Ethernet interface on the motherboard, with higher performance servers now commonly supporting on-board 10 Gigabit Ethernet intelligent network interface cards (NICs).

One of the advantages that Ethernet offers is its support for a wide range of physical media and cable lengths as discussed above. For example, 10 Gigabit Ethernet supports a range of distances and is compatible with OC-192c/STM-64c SONET/SDH links. Interest in 10 Gigabit Ethernet as a WAN transport technology led to the introduction of a WAN PHY for 10 Gigabit Ethernet. The WAN PHY (10GBASE-SW, 10GBASE-LW, 10GBASE-EW) allows Ethernet packets to be encapsulated in SONET OC-192c frames, and operates at a slightly slower data rate (9.95328 Gb/s) than the LAN PHY. As a result, Ethernet delivers the flexibility needed to work with the most cost-effective media available, whether it is SONET/SDH, DWDM, or direct fiber or copper connection. Ethernet delivers the needed flexibility to extend the LAN, and even storage area network (SAN), across the metro and wide area.

Today, even multi-national enterprises are building "all-Ethernet" networks around the world in order to extend LANs over a global Ethernet WAN. The corporate move to all-Ethernet networks mirrors carriers' desire to migrate to IP-optimized networks. As a result, "IP-friendly" Ethernet has become the ubiquitous service delivery technology that can be supported over any transport network. Other global Ethernet applications include traditional data center connectivity as well as video broadcasting, IP voice, and mobile network backhauling.

6.3.2.1 Ethernet in the LAN

The biggest market for 10 GE and 25 GE in the LAN environment is in the enterprise network backbone. The cost of 1-Gigabit and 10-Gigabit Ethernet interfaces and ports has dropped significantly, which has triggered the demand for 10 GE, 25 GE, and 40 GE services and products in the LAN.

In the enterprise network, Ethernet is used for the following:

- Switch-to-switch links to provide high-speed connectivity within the same wiring closet, between different buildings, or to/from data centers.
- Aggregation of multiple lower speed copper and fiber network segments onto higher speed uplinks. Presently, nearly all new laptops and desktops are shipped with 10/100/1000BASE-T ports, and data servers and storage devices with 10-Gigabit interfaces.
- High-capacity Ethernet interfaces play a key role in interconnecting high-performance servers with the network devices used for transporting a large amount of data between server farms. High-speed Ethernet networks can provide server interconnectivity for clusters of servers as discussed in Chapter 7 of this volume.

Driven by 10 GE to 100 GE deployment in enterprise and service provider networks, and by the proliferation of 1 GE to desktops/laptops as a standard interface and 10 GE as interfaces in data servers and storage devices, the strongest growth in the Ethernet switch/router market has come from demand for 10 GE to 200 GE chassis ports.

6.3.2.2   Ethernet in the MAN

Ethernet has been defined to support a wide range of interface capacities and distances, and now find applications in the residential network, the edge and MAN, and the WAN at large. At the MAN level, service providers continue to face immense pressures to expand capacity for local broadband access, access to high-speed WANs, access to data farms and storage, and bandwidth connectivity for creating campus and enterprise grids. High-speed Ethernet interfaces enable service providers to extend their networks into the metro edge using optical fiber links, especially, when 10 GE, 25 GE, 40 GE, and 50 GE are combined with metro area optical fiber networks based on WDM.

Ethernet-based MANs can be deployed in either a star or ring topology. Resilient Packet Ring (RPR), also known as IEEE 802.17, is an IEEE standard for implementing optical fiber ring networks. Unlike RPR networks, Ethernet ring networks use the standard Ethernet MAC protocol. Ethernet ring networks are designed to provide network reliability similar to networks based on SONET/SDH rings.

The industry standard for implementing Ethernet ring networks is ITU-T Recommendation G.8032, Ethernet Ring Protection Switching (ERPS) [ITUTG8032]. Some vendor-based (proprietary) techniques for creating Ethernet rings include Cisco's Resilient Ethernet Protocol (REP) [CISCDOC116384], [CISCREPGUID], [CISCREPWP07], Extreme Networks' Ethernet Automatic Protection Switching (EAPS) [RFC3619], 3Com and Huawei (H3C) Rapid Ring Protection Protocol (RRPP) [H3CRRPP], Allied Telesis' Ethernet Protection Switching Ring (EPSRing™) [ALLIEDTEPSR], Brocade's Metro Ring Protocol (MRP) [BRODMRP], and Force10 Redundant Ring Protocol (FRRP) [DELLFRRP].

6.3.2.3   Ethernet in the WAN

The flexibility, acceptance, and universal availability of high-capacity Ethernet interfaces have fueled the demand for multigigabit Ethernet deployment in network segments, from corporate networks, to data centers, and service provider networks. The continuous demand for bandwidth and performance has driven equipment vendors to deliver high-performing switches, switch/routers, and routers that meet service providers' stringent requirements – requirements that ensure high network availability and meet stringent service level agreements (SLAs).

Ethernet interfaces with rates from 10 Gb/s and beyond are being deployed in today's WAN. 10 GE was designed to have variants (10GBASE-SW, 10GBASE-LW, and 10GBASE-EW) that leverage existing SONET/SDH networks rather than new infrastructure. As discussed above, the 10 GE types that were designed to be compatible with SONET/SDH are referred to as WAN PHYs. SONET/SDH, for some time, was the dominant transport protocol in the WAN. At the time the WAN PHYs were ready for deployment, enterprise and service provider network backbones, and most MANs were based on SONET/SDH.

The 10 GE WAN PHY interfaces are compatible with SONET/SDH and have the payload rate of OC-192c/SDH VC-4-64c. The WAN PHY interfaces facilitate the transport of native Ethernet packets across the SONET/SDH WAN transport networks, with no need for protocol conversion. Other than improving the performance of the transport network, this makes it simpler and less costly to operate and manage. With SONET/SDH, networks are being phased out or replaced with pure Ethernet-based networks, today's WANs are generally based on pure Ethernet interfaces (10 GE, 25GE, 40 GE, 50 GE, 100 GE, and higher), following the same trend in the LAN and MAN.

6.3.3 PROTECT EXISTING NETWORK INVESTMENTS

The Ethernet standards define interfaces that provide a significant increase in bandwidth while maintaining backward compatibility with the installed base of IEEE 802.3 standard-based interfaces. This backward compatibility protects existing investments in Ethernet technology. For example, the 10-Gigabit Ethernet standard (IEEE 802.3ae) defines both LAN and WAN Physical Layers, the latter being compatible with existing SONET/SDH infrastructure.

Ethernet is not only the dominant technology in the LAN, it has also taken hold in the MAN. It has extended into the WAN arena as both its capacity and distance have increased as discussed in the "Ethernet Physical Layer Types" section above. Ethernet at 10-Gigabit to 100-Gigabit speeds continues to attract increasing interest from carriers operating in the MAN and WAN market segments.

6.3.4 PROVEN INTEROPERABILITY

Ethernet guarantees successful interoperation of standards-compliant products, regardless of manufacturer. Ethernet has been well tested over the years and all protocol versions and multiple vendor equipment offerings have been proven to interoperate very well. From the onset, Ethernet networks have relied on interoperability between the products of multiple vendors of NICs, hubs, switches, and routers. Proven multi-vendor interoperability has continued to be a major strength of Ethernet through successive generations of higher link speeds, including 10 Gb/s 40 Gb/s, 100 Gb/s, and higher.

Ethernet interoperates with a good number of networking technologies (e.g., MPLS), and there are continuous efforts in the industry to make it interoperable with newer technologies. While service providers have accepted Ethernet as the enabling convergence layer technology, today's global carrier infrastructure is a diverse mix of technologies. While delivering Ethernet services within a local geographical area may involve just one kind of network, a global carrier delivering an end-to-end Ethernet service will more likely need to traverse a mixture of Ethernet, IP/MPLS, SONET/SDH, CWDM, and DWDM local and long-haul networks. The good news is Ethernet is interoperable with all these technologies and is able to cross geographic and technology boundaries.

6.3.5 EASE OF MANAGEMENT

Another important characteristic of Ethernet as a networking technology is, it is easy to understand, implement, manage, and maintain. Ethernet also provides extensive topological flexibility for network installation. Ethernet is a familiar technology with a mature set of management and debugging tools. There is now an abundance of well-developed management tools for Ethernet networks. Ethernet-based cluster and storage interconnect can be readily assimilated into the existing Ethernet network management environment without requiring the additional management tools or training needed for special purpose fabric technologies and protocols like Infiniband, Fibre Channel, and SONET/SDH.

Ethernet now supports a lot of security and QoS features, allowing robust segmentation of networks, and isolating and controlling traffic flows. There are tools for extensive traffic statistics collection, providing data that can be used to manage bandwidth consumption and plan for capacity expansion and SLAs.

6.3.6 ADVANCED NETWORKING FEATURES

Optimizing a LAN or MAN for the delivery of Ethernet services is challenging enough, but the challenges increase further for carriers looking to offer Ethernet services to large enterprises and to other carriers on a wholesale basis. Because capacity is a finite and valuable resource on their networks, carriers must find ways to maximize bandwidth efficiency. In addition, carriers may interconnect with multiple other carriers and offer a range of Ethernet connectivity services requiring bandwidth scalability, advanced QoS, and SLAs. As a result, many carriers continue to transition their older legacy platforms to newer generation Ethernet platforms supporting far more advanced service delivery features.

Fortunately, Ethernet has a number of advanced features that address these networking concerns. These features allow enterprises and service providers to minimize the costs and complexity associated with delivering advanced Ethernet services. The features are robust and flexible enough, and are supported in many Ethernet service delivery platforms:

- Ethernet has flexible QoS and traffic policing features to support simplified provisioning of tiered services in the service provider networks (see Chapter 9 of Volume 1 of this two-part book). A network device may support dynamic committed information rate (CIR)-like services by policing and/or shaping bandwidth based on traffic loading demands and then bill according to usage. Ethernet switches now have the means to aggregate and manage advanced Ethernet services and provide specific bandwidth (CIR) and class of service (CoS) guarantees.

- Ethernet supports high service granularity and capacity, allowing the carrier to rate-limiting mechanisms to deliver Ethernet services at any bandwidth level desired by the customer ranging from 1 Mb/s up to multi-gigabit rate leased lines.
- Ethernet networks can be designed with advanced QoS and performance metrics (based on packet loss, latency, and delay variation) to support Ethernet SLAs. Carriers now have ways to provide advanced QoS to support Ethernet-based SLAs while maximizing bandwidth efficiency.
- Ethernet devices can be designed with features to support carrier-grade hardware and software availability (see Chapter 1 of this volume). The devices can be designed to support redundancy in order to minimize operational costs and to provide SLA guarantees to customers.
- Security and traffic control features are available to protect the Ethernet network infrastructure and customers from disruption by accidental and malicious traffic.
- Ethernet can support a wide array of service offerings, including MEF (Metro Ethernet Forum) compliant Ethernet Line or E-Line (point-to-point), Ethernet LAN or E-LAN (multipoint-to-multipoint), and Ethernet Tree or E-Tree (point-to-multipoint). Ethernet services ranging from speeds of 1 Mb/s to 10 Gb/s can be provisioned across diverse transport networks.
- Ethernet switches can support diverse optical interconnect options including SONET OC-n, SDH STM-n, and DWDM/CWDM wavelengths.
- Ethernet switches can support TDM aggregation and service switching (STS-1/, VC- /4, VT1.5/VC-1) functionality.
- The switches can support intelligent Generalized Multi-Protocol Label Switching (GMPLS)-based control plane to simplify end-to-end Ethernet provisioning and monitoring.
- Ethernet has scalable Layer 2 aggregation and service switching functionality like IEEE 802.1ad (also known as Provider Bridging, Stacked VLANs, or simply QinQ or Q-in-Q). Ethernet is able to support Ethernet-related technologies such as IEEE 80.1ah (Provider Backbone Bridging (PBB), also known as Mac-in-Mac), IEEE 802.1Qay (Provider Backbone Bridge Traffic Engineering (PBB-TE)), Ethernet NNIs (Network-to-Network Interfaces), PWE (Pseudowire Emulation), and others.

As carriers move to deliver next-generation Ethernet services and enter new markets for enterprise and wholesale services, they will need multiservice carrier Ethernet products that unify advanced switching and transport capabilities (like those cited above) in a single system. Integrating a mix of advanced Ethernet/IP and MPLS services in a single platform with optical interconnect/transport flexibility simplifies the network, and dramatically reduces costs by eliminating the need for separate switches.

Furthermore, by deploying multiservice carrier Ethernet switching platforms, carriers can simplify their networks, increase the range and scalability of their Ethernet service offerings and have a smooth transition to accommodate new technologies and features as they evolve. A multiservice carrier Ethernet switching platform that supports the above features provides a future-proof solution that enables the transition to be seamless.

6.4   INDUSTRY TRENDS

The advantages cited above have also earned Ethernet a well-deserved place in data center networking. There were a number of technology trends that had the potential to significantly increase the complexity of building data center network infrastructures, but Ethernet was able to standup to those challenges. These challenges were driven by the following technology transitions: IP/Ethernet storage, high-performance computing (HPC) clusters, multi-core processing and cluster computing for enterprise applications, grid computing, and virtualization. Ethernet network infrastructures have been able to adapt to changes in systems such as server farms and storage, and HPC clusters, and are still changing to meet the needs of emerging businesses.

6.4.1   IP/ETHERNET STORAGE

IP/Ethernet storage refers to storage networks that allow users to access data resources over IP/Ethernet rather than direct attached storage (DAS) devices. IP/Ethernet storage has become an alternative to special-purpose SANs based on storage-specific networking protocols, such as the Fibre Channel and Infiniband.

6.4.1.1   IP/Ethernet Storage Categories
The three main categories of IP/Ethernet storage are as follows:

- *IP/Ethernet SANs*: In this category, TCP/IP and Ethernet infrastructure is used to provide the Transport Layer, Network Layer, and Data Link Layers for an SAN solution. The primary IP/Ethernet SAN protocol is iSCSI (Internet Small Computer System Interface) [RFC7143], which layers the SCSI block-level protocol over TCP/IP. This is in contrast to the Fibre Channel standard which defines a completely different suite of layered protocols.
- *Storage-Specific Networking Protocols over IP/Ethernet*: Storage-specific gateways can be used to provide connectivity between remote Fibre Channel SANs across the IP MAN or WAN. The gateways can provide interconnection from the Fibre Channel SAN to the IP network by using protocol stacks that layer the Fibre Channel block-level protocols (Fibre Channel Protocol (FCP)) over IP using either FCIP (Fibre Channel over IP) [RFC3821], or iFCP (Internet Fibre Channel Protocol) [RFC4172].
- *NAS over IP/Ethernet*: Network Attached Storage (NAS) uses file-access protocols such as Network File System (NFS) [RFC1813] and Common Internet File System (CIFS) to provide file-level access to file servers or file storage appliances on the IP/Ethernet network.

6.4.1.2 IP/Ethernet Storage Devices

A number of iSCSI and NAS appliances are available that support IP/Ethernet network connections, including the following:

- **Hybrid iSCSI/NAS Storage Device**: This storage device provides native support for both file-level and block-based I/O into the disk array attached to the back end. Using a combination of iSCSI and NAS over a common IP/Ethernet LAN, applications requiring file-level access and block-level data access can share a common storage infrastructure. iSCSI SANs and NAS solutions based on the Gigabit Ethernet LAN interconnects have been widely deployed in smaller, more homogeneous data centers that are typically found at the third tier of large enterprise data infrastructures or in the data centers of small- to medium-sized enterprises (SMEs). 10 GE and higher capacity Ethernet provide the high bandwidth connectivity that enables consolidation of file and block storage within an iSCSI/NAS device.
- **NAS Gateway**: The NAS Gateway acts as an interface between a file-level request and the block-level storage. The device allows end systems and servers to access files over the LAN, with the files stored on iSCSI SAN disk arrays connected via an internal or external 1 GE, 10 GE, 25 GE, and higher gigabit Ethernet interfaces. The NAS appliance, on the other hand, provides only file access via NFS and CIFS over the multi-gigabit Ethernet LAN, with the files stored on devices directly attached to the appliance.

The emergence of higher multi-gigabit Ethernet has generated demand for devices that support both iSCSI SANs and NAS, as users seek to meet both block storage requirements (e.g., database, messaging applications) and file storage requirements on a single system. This consolidation simplifies IT management, increases storage capacity utilization, and reduces data center space, power, and cooling costs.

6.4.2 HIGH-PERFORMANCE COMPUTING CLUSTERS

Clustering has emerged as a cost-effective architecture for building HPC facilities. A cluster can be loosely defined as a group of homogeneous, whole computer systems running cooperatively to process in a divide and conquer fashion, a computing task. The cluster is used as a unified computing resource. Clusters can include specially built HPC nodes, or off-the-shelf PCs or servers. Clusters in HPC frequently use Ethernet as the cluster interconnect technology.

The role of Ethernet in HPC clusters, multi-core processing and cluster computing for enterprise applications, and virtualization is discussed in more detail in subsequent sections of this chapter and in Chapter 7 of this volume.

- The performance of multi-threaded enterprise applications is highly dependent on having low end-to-end latency and high bandwidth between the sub-processes running on different processor cores or servers. Ethernet is fast becoming the preferred interconnect for these computing applications.

- Modern data centers rely heavily on higher performance networking and resource virtualization. Effective and flexible resource virtualization allows entire pools of computing resources to be shared by end systems at a very high speed.

6.4.3 OTHER INDUSTRY TRENDS

Innovations in multi-gigabit switched Ethernet (10 GE and higher) have made it an excellent choice for the above applications. Ethernet switching has the advantages of being a highly scalable, very low-cost alternative for implementing the switching fabric and interconnects of data centers and cluster computing systems. The cost of multi-gigabit Ethernet switch ports and high-performance server adapters continues to drop as more enterprise LANs and desktops migrate to multi-gigabit Ethernet interfaces.

Most enterprise and service provider networks are now based on Ethernet and IP; Ethernet/IP is used as the standard network transport. Furthermore, older generation Ethernet switches are being replaced with newer generation Ethernet switches that support high-density, multi-gigabit Ethernet interfaces. Particularly, high-density Ethernet switches can simplify the network infrastructure by replacing numerous older generation switches.

High-performance Ethernet switches in conjunction with IP can also provide the high-speed interconnect required for virtualization, inter-process communication (IPC) channels for computational clusters and Grids, and SANs with iSCSI. iSCSI SANs and NAS solutions based on the multi-gigabit Ethernet LAN interconnects are now widely deployed from smaller to large data centers. Ethernet-based storage can be readily assimilated into the existing Ethernet network management environment without requiring additional management tools, or the training necessary for using special-purpose protocols, such as Infiniband and Fibre Channel.

IP/Ethernet storage is now firmly established as a viable alternative to Fibre Channel SANs, and the iSCSI/NAS market has dramatically expanded as IP/Ethernet storage devices leverage the performance and cost-effectiveness inherent in newer generation multi-gigabit Ethernet enterprise switches, routers, and switch/routers as well as multi-gigabit Ethernet adapters. The capabilities of multi-gigabit Ethernet network infrastructures and adapters have greatly accelerated TCP/IP and iSCSI protocol processing by providing increased data transfer throughput and I/O performance lower end-to-end latency, and reduced CPU utilization.

6.5 ROLE OF ETHERNET IN NETWORK INFRASTRUCTURE VIRTUALIZATION

Today's data centers are continuously evolving toward a virtual data center model where enterprise applications have the flexibility to draw on pools of shared computing, storage, and networking resources rather than being rigidly constrained to dedicated physical resources. Virtualization not only greatly increases the flexibility of the data center infrastructure to accommodate changing workload requirements, it also improves resource utilization and power efficiency, thereby, contributing to reduction in network TCO.

Maximizing the effectiveness and flexibility of resource virtualization implies that entire pools of computing resources can be shared at a very high speed. This degree of resource sharing requires not only a high-speed LAN for server interconnect, but also a high-speed, low-latency SAN for shared access to virtual machine images and data. Consolidation of network and data center resources also offers an excellent opportunity for overall network architectural improvement based on the use of scalable, high-density, high-availability technology solutions, such as high port density switches and switch/routers.

6.5.1 DATA CENTER INTERCONNECT AND SWITCHING FABRIC UNIFICATION

Well before the benefits of Ethernet were well appreciated by the industry, data center and cluster computing systems managers were facing the prospect of deploying three distinct switching fabrics to support the LAN, the SAN, and to provide low-latency cluster interconnects. This created the possibility that data center servers would be attached to three distinct switching fabrics: a LAN for connecting users and for general networking (generally, using Ethernet and IP), a SAN for shared access to storage/file resources, and a cluster fabric for IPC between cluster nodes (i.e., a fabric for low latency message passing between compute cluster applications).

Although each fabric might be well optimized for its function, deploying three separate fabrics, each with its own cabling, host adapters, and management systems, results in high cost and complexity for most enterprise IT departments to accept. In addition to being expensive to purchase and install, this mix of disparate network elements is difficult and costly to maintain and monitor. Because most enterprises and service providers have specialized provisioning and management systems for each type of network, it may take weeks to coordinate the various systems and teams to provide a service for just one customer.

It became clear for most enterprise data center applications that, having multiple switching fabrics could prove excessively expensive in terms of both capital expenditures or CAPEX (from deploying multiple switching fabrics and host adapters per server) and operational expenditures or OPEX (from using a much broader range of required operational and management expertise). Deploying multiple switching fabrics/technology certainly has some disadvantages, some of which are as follows:

- Each technology requires its own network administration and management.
- Each technology requires its own set of maintenance and support staff with the right skills.
- Each technology requires its own set of spare parts, inventory, etc.
- Provisioning redundancy in the network results in higher cost of network design and deployment.
- Overall footprint of the networks can be very large – taking up large space. This can be a problem when space is premium (normally is for most organizations).
- Additional interconnect complexity can prove to be a hindrance to achieving a fully virtualized data center infrastructure, and in developing the level of automated management of virtual resources required to implement service-oriented architectures (SOA) for enterprise applications.

A more viable alternative to using multiple switching fabrics is to standardize on a single (unified) data center switching fabric that can simultaneously serve as a LAN interconnect, a SAN interconnect, and a low-latency cluster computing interconnect. The industry now sees Ethernet as the best option for data center switching fabric unification. Ethernet advantages in bandwidth, scalability, manageability, and its ability to provide parity for cost and latency are driving the vast majority of clusters to using Ethernet as the cluster interconnects as well as a LAN and SAN interconnect.

A unified switching fabric leverages Ethernet's TCO advantages, including massive economies of scale, widespread administrative expertise, and familiar management tools. Ethernet NICs, cables, switches, and wiring practices are well known, and installing and debugging an all-Ethernet cluster is well established and understood. The advantages of connecting everything using a single Ethernet network are compelling. The TCO is dramatically improved – there is only one network to operate and maintain.

6.5.2 ETHERNET AS THE DATA CENTER INTERCONNECT AND SWITCHING FABRIC

Ethernet has become the de facto technology for the general-purpose LAN, and more so, has found increasing use in other networking applications areas (MAN and WAN) where other technologies were traditionally favored. Continuous innovation in Ethernet is resulting in continuously improving performance, higher speeds, new capabilities, rapidly falling price points, and rapid evolution to meet new requirements.

Up until the development of 10 Gigabit Ethernet, (Gigabit) Ethernet was considered a sub-optimal switching fabric for very high-performance cluster computing and storage networking. Ethernet was widely considered as a sub-optimal switching fabric for very high-performance cluster interconnect for IPC (e.g., for those parallel applications that require extremely low end-to-end latency). This view stemmed from the performance limitations of Gigabit Ethernet compared to InfiniBand and Fibre Channel which were developed with cluster interconnect and storage networking in mind. Gigabit Ethernet has end-to-end message passing latency in the range of 50–70 microseconds which is significantly higher than the less than 10 microseconds for the more specialized cluster interconnects like InfiniBand and Fibre Channel. As a block storage fabric, Gigabit Ethernet with iSCSI offered a good medium performance solution for access to networked block storage.

The performance limitations are due primarily to the fact that Gigabit Ethernet has lower bandwidth than InfiniBand and Fibre Channel, and typically exhibits significantly higher end-to-end latency and host CPU utilization. The situation has dramatically changed today due to recent developments in low-latency 10 and higher Gigabit Ethernet switching and intelligent Ethernet NICs that offload protocol processing from the host processor.

These recent switching and NIC technological advancements have allowed server end systems to fully exploit multi-gigabit Ethernet line rate, while reducing the one-hop end-to-end latency and CPU utilization for line rate transfers. These improvements have resulted in 10 and higher Gigabit Ethernet end-to-end performance now being comparable to that of more specialized data center interconnects like InfiniBand

and Fibre Channel, thus, eliminating performance as the main factor holding back the adoption of an Ethernet unified data center switching fabric.

As a result of the advances and enhancements in IP storage, enterprise data center managers now have the option of adopting multi-gigabit Ethernet as a unified switching fabric for cluster interconnectivity, server LAN connectivity, and server interconnectivity to networked storage resources. A unified switching fabric is an appealing concept because it avoids the inherent cost and complexity of deploying multiple switching fabrics, each with its own cabling, host adapters, and management systems.

A unified switching fabric also greatly simplifies the implementation and management of a virtual data center that spans several networks, I/Os, computing and storage resources. These new developments allow network managers to minimize the complexity of the data center. By using multi-gigabit Ethernet as the converged switching technology, the network is better positioned to meet the highest performance requirements of each type of data center traffic.

6.5.3 Advances in Ethernet Switching and NIC Design – Enablers for Data Center Networking

Advances in Ethernet switching and NIC design have pushed Ethernet from being a predominantly a LAN technology to applications such as data center switching fabric. These technological enhancements to Ethernet are discussed here.

6.5.3.1 Low Latency Switching

Factoring out latency due to network traffic congestion and propagation (wire) latency, end-to-end latency in the LAN comprises two basic components:

- **Host latency**: This is the delay incurred in the end systems and NICs when data is being transferred between the application buffers and the network. Host latency is incurred in both the sending and receiving end systems.
- **Network transmission latency**: This is the delay involved in serializing the data unit on the transmission medium and switching it through network nodes to its destination.

The round trip send/receive message latency includes the switch latency (on the order of 10 microseconds in each direction in some switches) and the receive/send latency of the remote host due to TCP/IP processing and buffer-to-buffer transfers (both of which typically involve software processes). Most of the end-to-end latency incurred with Ethernet data transfer is due to software processing on the hosts.

The two main switching approaches employed by Ethernet switches are as follows:

- **Cut-through switching**: In this method, the switch starts forwarding a frame before the whole frame has been received, normally, as soon as the destination MAC address is processed. The downside is, corrupted frames are potentially forwarded because the Ethernet Frame Check Sequence

(FCS) appears at the end of the frame. The switch cannot identify and discard all corrupted packets because the packets are forwarded before the FCS field is received and, thus, no CRC calculation can be performed.
- **Store-and-forward switching**: In this method, the switch makes a forwarding decision on a frame after it has received the whole frame and checked its integrity. After packet processing is complete, the switch has to re-serialize the packet on the transmission medium to deliver it to its destination. In this mode, corrupted frames are dropped.

Typically, low latency Ethernet switches employ cut-through switching to reduce the network transmission latency component of end-to-end latency. Cut-through switching involves the switch delaying the packet only long enough to read the Layer 2 packet header, allowing it to make a forwarding decision based on the destination MAC address and other header fields (e.g., VLAN tag and IEEE 802.1p priority field).

In cut-through switching, the switch begins forwarding the packet to the destination system as soon as the destination address is mapped to the appropriate output port. The resulting effect is switching latency is reduced because packet processing is restricted to the header itself rather than the entire packet, and the packet does not have to be fully received and stored before forwarding can begin.

The network latency for the cut-through switch is lower because of the simplicity of cut-through switch, and also because, only one instance of serialization delay is encountered, which can be a significant factor for larger frame sizes. According to [FOR10LOWL10G07], the typical switch latency for a 10 Gigabit Ethernet store-and-forward switch is in the range of 5-35 microseconds, while the switch latency for a 10 Gigabit Ethernet cut-through switch is typically only 300-500 nanoseconds.

Low latency cut-through switching also reduces serialization time in multi-hop networks because the packet is serialized only one time rather than twice per hop as in a network of store-and-forward switches. As the diameter of the interconnect network increases, the advantage of cut-through switching becomes more significant.

With store-and-forward Ethernet switching, the network latencies would be typically higher due to higher switching latency and the additional serialization delay at every hop across the network. Cut-through switching is currently the preferred switching mode for specialized multi-gigabit Ethernet data center interconnects – for those applications that are highly sensitive to switch latencies in the microsecond range.

6.5.3.2   Off-Loading Protocol Processing
Traditionally, TCP/IP protocol processing in an end system has been performed in software by the end system's CPU. With a software-based protocol stack, the host CPU is shared between the various applications and the network. A drawback of processing network I/O in software is that it results in fairly high CPU utilization during intense network activity. The load on the CPU increases linearly as a function of the rate of packets processed by the end system.

A simple rule of thumb for assessing the load on the end system is that each bit per second of network traffic (TCP/IP bandwidth) consumes about a Hz of CPU clock (e.g., 1 Gb/s of network traffic consumes about 1 GHz of CPU). Thus, a sustained Ethernet transfer at 800 Mb/s would involve approximately 80% CPU utilization of a 1 GHz CPU. Which means, as more of the host CPU is consumed by the network load, both CPU utilization and host send/receive latency become significant issues. The traditional approach of network I/O processing in software results in the CPU itself becoming the bottleneck, which limits throughput and adds significantly to the end-to-end latency.

The obvious implication of the 1 Gb/s per GHz rule is that, 10 Gigabit Ethernet NICs will have to be designed to eliminate host network processing in software in order to achieve line-rate data transfers. Vendors of intelligent Ethernet NICs, together with a number of industry consortiums and standards groups (e.g., the RDMA Consortium and the IETF), have defined specifications for hardware-accelerated TCP/IP protocol stacks that can support the continuously increasing performance demands of general-purpose networking, cluster IPC, and storage interconnectivity over Gigabit Ethernet and 10 Gigabit Ethernet.

A TCP/IP Offload Engine (TOE) is a hardware device that processes the TCP/IP protocol stack on a module outside of the main CPU. A dedicated TOE is incorporated in the NIC, and the TOE offloads essentially all the TCP/IP processing from the host CPU. This is done by moving the processing to the TOE which resides on the "network side" of the internal PCI bus (which is slow and is a performance bottleneck). This greatly reduces CPU utilization and also reduces latency, because the protocols are executed in the TOE hardware rather than in host CPU software. This provides a faster response to network traffic while reducing the workload for the CPU.

NICs that offload the host CPU are suitable for Gigabit Ethernet end-to-end performance. Off-loading protocol processing from the central CPU to the intelligent NIC (with ASIC-based processors) can also improve the power efficiency of end stations. This is because off-load ASIC processors in NICs are generally more power-efficient in executing protocol workloads. Intelligent Ethernet NICs offload protocol processing from the host CPU, thereby, eliminating software performance bottlenecks, minimizing CPU utilization, and greatly reducing the host component of the end-to-end latency.

Remote Direct Memory Access (RDMA) is a direct memory access technique that enables two hosts on a network to exchange (read or write) data between their main memories without relying on either host's CPU or operating system. For example, RDMA allows the copying of data from the memory of one host into the memory of another host without involving either host's CPU or operating system. To use RDMA, the network adapters on the hosts need to support RDMA capability; an RDMA-enabled NIC must be installed on each host that participates in RDMA-based data exchange. RDMA provides high-throughput, low-latency communication, which is especially useful in data center networks and massively parallel HPC clusters.

RDMA also allows the copying of data directly from a network adapter to the host CPU's memory without involving the CPU. With RDMA, data is deposited directly into the host CPU's memory. RDMA data transfers bypass the operating system networking stack in both hosts, improving data transfer throughput. RDMA conserves memory bandwidth and reduces latency by eliminating kernel interrupts for copying data between the network adapter buffer pool and host application buffers. RDMA is, particularly, useful for the movement of large blocks of data, such as that required for storage interconnects and cluster computing.

RDMA NICs (RNICs) also greatly reduce CPU utilization with the addition of TOE, which executes TCP/IP processing in NIC hardware/firmware rather than in kernel software. By implementing a transport protocol in the NIC of each communicating device (e.g., a server), RDMA facilitates more direct and efficient data movement into and out of each host. This permits high-throughput, low-latency networking, which is especially useful in HPC clusters. By removing CPU intervention in data copying, RDMA removes another performance limiter encountered when using conventional NICs.

For example, two hosts on a network can each be configured with an RNIC that supports the RDMA over Converged Ethernet (RoCE) protocol, enabling the hosts to participate in RoCE-based communications. Typically, RNICs support one or more of the following three network protocols: RoCE (also known as InfiniBand over Ethernet (IBoE)) which is a network protocol that enables RDMA communications over an Ethernet network; Internet Wide Area RDMA Protocol (iWARP) which leverages TCP or Stream Control Transmission Protocol (SCTP) for data transmission; InfiniBand which is the standard protocol for high-speed InfiniBand network connections and provides native support for RDMA.

RNICs can also support iSCSI and socket upper-layer interfaces, enhancing the role that Ethernet plays as a single "converged" fabric, satisfying the needs of cluster communications, IPCs, and storage interconnectivity. A converged fabric, potentially, allows even very high-performance clusters to be based on a single switching fabric versus the more complex and costly approach of using multiple switching fabrics.

Vendors of intelligent Ethernet NICs, together with the RDMA Consortium and the IETF, have defined specifications for hardware-accelerated TCP/IP protocol stacks. This is to support the ever-increasing performance demands of general-purpose networking, cluster IPC and storage interconnectivity over multi-gigabit Ethernet. Although InfiniBand and Fibre Channel in the past offered better end-to-end performance benchmarks for SANs and IPC cluster interconnectivity, Ethernet/TCP/IP has evolved to provide very flexible, cost-effective, and comparable solutions in both of these areas.

As the next generation of intelligent multi-gigabit Ethernet NICs become available, and the prices of multi-gigabit Ethernet NICs and switch ports continue to fall, the end-to-end performance gap of Ethernet will either disappear completely (for SANs) or be reduced to relative insignificance (for IPC).

6.6 VIRTUALIZATION AS A VEHICLE FOR NETWORK AND COMPUTING RESOURCE CONSOLIDATION

As today's enterprise and service provider networks continue to converge, there is an increasing need for the logical separation of applications, and/or user groups within the network. Virtualization abstracts a physical resource made of several parts (e.g., hardware compute platforms, storage devices, and networking resources), creating one or more virtual (logical) instances of the resource running on (and representing that) single physical resource. Virtualization can take several forms, the most prominent of which are infrastructure virtualization, cluster computing, Grid computing, and Layer 3 virtual switching (the virtual router).

The underlying principle of all these virtualization concepts is that, applications and users should be able to tap a common pool of computing resources (networking, servers, and storage) by making them appear collectively as a large virtual or logical resource and as if they were a single large physical system. Virtualization allows for the allocation of applications and users in the system to logical resource pools, thereby, simplifying the alignment of physical resources to application and user needs.

6.6.1 Virtualization Technologies

The main virtualization concepts are outlined below.

6.6.1.1 Infrastructure Virtualization

This virtualization approach allows applications and software services to be decoupled from the underlying physical infrastructure by providing a logical view of the infrastructure's compute, storage, and networking resources. Thus, infrastructure virtualization, in general, consists of compute (e.g., server), storage, and network virtualization. This decoupling allows infrastructure hardware management (adds, removals, changes, etc.) to be completely separated from application software and user management. Infrastructure hardware can be allocated and re-allocated to various software services on the fly via a management console that provides the interface for the required resource allocation, load balancing, and monitoring of resource utilization.

6.6.1.2 Cluster Computing

Generally, cluster computing can be implemented as one of two types, computational cluster or load-balancing cluster. Both types of clusters are based on software techniques that make the array of homogeneous servers appear as a single virtual system. In a computational cluster, a tightly coupled cluster of computers provides supercomputer-like power for executing applications specifically programmed to run on parallel computers.

Computational clusters generally comprise highly homogeneous systems built with commodity hardware and software. Applications including clustered databases typically run on small clusters with less than 40 processors, while scientific/ engineering applications involving complex simulations generally run on much larger clusters (with as many as hundreds or even thousands of cluster nodes).

Cluster middleware provides the single system image (SSI) virtualization needed to make a multiplicity of networked computer systems appear to both the user and application as a single parallel computing system. A key aspect of the computational cluster middleware is a message-passing system for IPC between the servers in the cluster. The speed of IPC is critical for a class of parallel applications known as tightly coupled synchronous applications. These applications require communication among all of the cluster processors (global synchronization) before the computation can be completed.

Load balancing clusters are mainly used to provide high availability and scalable performance by distributing client connections among the application servers in the cluster. Various load balancing algorithms may be used to distribute the client sessions to the servers to achieve relatively equal distribution of the workload. Each client session is independent of the others; as a result, the servers can operate in parallel – there is no requirement for IPC among the servers or modification of existing applications to run on the cluster.

6.6.1.3 Grid Computing

There is a considerable amount of debate as to whether cluster computing (i.e., a local computational cluster of computers) should be classified as a Grid. Some view cluster computing and Grid computing as entirely different things as explained here. With Grid computing, computer systems and other resources on an Intranet or the Internet are not necessarily constrained to be dedicated to individual users or applications, but are made available for dynamic pooling or sharing in highly granular fashion to meet the changing computing needs of an organization.

To some, Grid Computing is a general-purpose distributed computing model where heterogeneous systems on an Intranet or the Internet are pooled together to offer services (such as computational services, data services, or other types of service) that can be accessed and utilized by other systems participating in the Grid. In this model, the group of heterogeneous computers (forming the grid) donate spare cycles to accomplish a computing task. The Global Grid Forum (GGF) and Enterprise Grid Alliance view Grid computing from different perspectives:

- **GGF:** The Grid concept from the GGF is modeled on the idea that ubiquitous shared computing services can turn the entire Internet into a single large virtual computer. The GGF middleware is a standards-based initiative that promotes Internet-wide resource sharing and collaborative problem solving by multi-institutional "virtual organizations".
- **EGA:** The EGA approaches Grid computing from the perspective of pooling the computing resources within the enterprise network for sharing and collaborative problem-solving. The EGA is more focused on business applications of the Grid rather than technical or scientific supercomputing applications. Particularly, interest is in static grids located within the confines of the enterprise data center, and the goal is to build as much as possible on the existing infrastructure of applications, servers, SANs, and networks in the enterprise.

The USA NSF-funded TeraGrid project is another Grid computing initiative that allows for the harnessing and use of distributed computers, data storage systems, networks, and other resources, as if they were a single massive system, regardless of physical location, thereby, creating "virtual supercomputers".

6.6.1.4 Layer 3 Virtual Switching (Virtual Routing and Forwarding)

Layer 3 Virtual Switching (also called Virtual Routing and Forwarding (VRF)) allows the partitioning of a single switch/router or router into many virtual routers (VRs). VRF is an IP technology that allows multiple instances of a routing table to coexist on the same router at the same time. Because the routing instances are independent, the same or overlapping IP addresses can be used without conflict.

A VR has the same capabilities and properties as a physical router. It inherits all the same routing mechanisms for configuration, operation, and troubleshooting. As a result, each virtual switch domain can be separately managed and isolated for security safety measures

Recently, enterprise and service provider networks have been demanding more flexibility, partitioning, and security in the design and operations of their networks.

VLAN technology alone has limitations that do not allow it to meet these emerging needs. VRF is designed to complement VLANs by providing additional layers of separation, control, and security for network designs and operations.

As a result, VRF delivers higher levels of availability and security than what is possible with VLAN technology alone. VRF enables network managers to deal with rapid changes in their business requirements such as accommodating a new organizational structure and changes in their network requirements.

Virtualization of system software and hardware resources has been applied to various IT infrastructures and applications related to data center computing and storage. VRF implements a number of VRs on the same physical switch. A VR has the same capabilities and properties as a physical router does and inherits all the same routing mechanisms for configuration, operation, and troubleshooting. In a nutshell, a VRF implementation has several instances of a router and protocol code running on it, each running independently from the other and serving a different purpose.

Each VR in a routing device that supports VRF technology maintains a separate logical forwarding table, which allows the different VRs to have overlapping IP address spaces per VLAN. Since each VR maintains its own separate routing information, and switch ports can belong to only one VR, packets arriving at a port on one VR can never be switched to ports on another VR. Protocols such as PIM, RIP, OSPF, or BGP routing protocols can be used on each VR.

VRF has become a valuable tool for partitioning a network as well as consolidating multiple Layer 3 routing domains onto a single routing device. Its benefits are significant since it reduces capital and operations costs for equipment, space, power, and management.

****

All of the above virtualization technologies are highly dependent on having a high-performance network that provides flexible interconnection among the physical devices that may belong to a variety of dynamic logical groupings. The network (which serves as the data interconnect) provides shared access to data and files and the high-speed transport medium that allows migration of image files and processes among systems.

When Ethernet switches are used, VLAN functionality can be used to partition and isolate resources into multiple groupings and allowing bandwidth to be adequately allocated to ensure predictable performance for specific critical applications. The switches and other physical network devices can perform a range of networking and security functions, such as web load balancing, firewall filtering, and intrusion detection.

6.6.2 Virtualization Middleware

Depending on the application, the resources that are virtualized might be located in a single data center or distributed across a number of data centers, or even the entire enterprise network. For all of the virtualization technologies discussed above, the single system view (i.e., consolidation into a single logical resource) is accomplished by interposing a virtualization middleware layer between the application and the resource pool (Figure 6.1). The primary function of the virtualization middleware is to provide an abstraction layer or virtualization interface between the application and the resource pool.

6.6.2.1 Primary Functions of the Virtualization Middleware

Computational clusters are conceptually similar to Grids. Both are dependent on middleware software (Figure 6.1) to provide the virtualization needed to make a multiplicity of networked computer systems appear to the user as a single system. Therefore, the middleware for clusters and Grids addresses the same basic issues, including message passing for parallel applications.

For example, a Grid middleware provides the location transparency that allows the applications to run over a virtualized layer of networked resources. The key aspect of middleware is that it gives the Grid the semblance of a single computer system, providing coordination among all the computing resources that comprise the Grid. These functions usually include tools for handling resource discovery and monitoring, resource allocation and management, message passing system, security, and performance monitoring and accounting [FOR10HPCROE06].

6.6.2.1.1 Resource Discovery and Monitoring

The end systems participating in the Grid must have the means to discover what resources or services are available to them and also to monitor their status. After resources discovery, the systems must be able to access system configuration and status information in order to define the Grid's topology for the job in question. Typically, well-known tools, such as Lightweight Directory Access Protocol (LDAP), Domain Name Service (DNS), network management protocols, and indexing services are used for Grid resource and services discovery and monitoring.

6.6.2.1.2 Resource Allocation and Management

The end system will need to support resource allocation and management software to dynamically match application requirements to the available computing resources in the discovered Grid topology and to create the remote jobs as required. The software will be responsible for scheduling the jobs and monitoring their progress to ensure optimum load balancing and resource utilization in the Grid. This software is responsible for configuring resources in the network to support a given service.

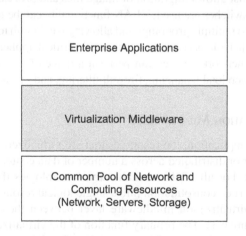

FIGURE 6.1 Network and computing resource virtualization.

6.6.2.1.3   Performance Monitoring and Accounting

A computational cluster or Grid needs a process(es) (within the distributed system of applications and resources) for collecting and storing state data. The monitoring system can gather performance monitoring data for the application itself, and data for the system being monitored. It deals with keeping track of resources in the network and how they are assigned. The resource allocation and management system can also gather the necessary data from its operations and make them available for the purposes of accounting, billing, and reporting.

6.6.2.1.4   Message Passing System

A key functionality of a Grid is one that supports efficient communication between processes running on the different computing systems on the Grid. Typically, communications among processes that model a parallel program running on a distributed memory system, are done over a message passing system. This is also the dominant communication model used in HPC today. Compute-intensive applications developed for parallel supercomputers frequently use a message-passing model for IPC.

The Message Passing Interface (MPI) or the Parallel Virtual Machine (PVM) programming models are the two most prominent message-passing systems. These parallel programming models have also been adapted for the Grid, an example, being MPICH-G which is a Grid-enabled version of MPI that supports the migration of MPI programs from parallel super-computers to a Grid based on Globus services.

6.6.2.1.5   Security

As with most distributed network applications, security is a key requirement in a Grid. Security is a critical issue because the resources and services in the Grid being shared among end-users may include sensitive information like company trade secrets or intellectual property that need to be properly safeguarded. Security concerns such as these often call for the middleware to support secure communications, authenticate user identities, and restrict user activities based on authorization policies. The Grid typically consists of numerous resources spread over a wide area, thus, making a single sign-on authority the best option. Existing software security standards like Kerberos, Secure Socket Layer (SSL), Transport Layer Security (TLS), and X.509 certificates are available for implementing this functionality.

6.6.2.2   Integration of Virtualization Middleware

While the virtualization middleware can take various forms, the potential benefits include the following:

- Decoupling and shielding the enterprise applications from the details of the physical resources in the pool
- Providing flexibility for static or dynamic changes to the infrastructure with minimal disruption of applications
- Providing a conduit to simplified centralized system management and providing the basis for automating management functions

There are several ways of integrating the middleware in the system:

- The middleware may be embedded within the application
- It may be layered on the operating systems of the servers and other intelligent devices in the resource pool
- It may be installed on specialized servers/appliances

10 Gigabit Ethernet adapters designed for virtualization applications have I/O virtualization functionality designed into their core. As a result, they deliver the high levels of 10 Gigabit Ethernet performance and server utilization that enable IT managers to virtualize I/O intensive applications. Such 10 Gigabit Ethernet adapters also support iSCSI line-rate throughput between the initiator and target systems in an IP storage environment, which lowers latency, improves access to storage resources, and drastically reduces backup times.

The 10 Gigabit Ethernet adapter has features that deliver the bandwidth needed to enable 10 Gigabit Ethernet IP storage. These features can include a multi-channel device model with independent hardware-based transmit and receive paths that carry traffic that may be prioritized for real-time QoS support. This model may implement independent interrupt schemes by application or processor core, and maybe reset independently to enable the virtualization of more I/O intensive applications per physical server. Other features may include a comprehensive set of hardware offloads that process a significant amount of TCP/IP traffic to maximize performance while reducing CPU utilization and lowering latency.

REFERENCES

[ALLIEDTEPSR]. Allied Telesis, EPSRing™ (Ethernet Protection Switched Ring), *White Paper*, 2006.

[BRODMRP]. Ruckus FastIron Layer 2 Switching Configuration Guide, *Metro Ring Protocol*, 2018.

[CISCDOC116384]. Cisco Systems, "Resilient Ethernet Protocol Overview", *Document ID: 116384*, July 22, 2016.

[CISCREPGUID]. Cisco Systems, "LAN Switching Configuration Guide, Chapter: Resilient Ethernet Protocol (REP)", August 5, 2019.

[CISCREPWP07]. Cisco Systems, "Cisco Resilient Ethernet Protocol", *White Paper*, September 2007.

[DELLFRRP]. Dell Force10 S4810P Configuration Manual, Force10 Resilient Ring Protocol (FRRP); Protocol Overview.

[FOR10HPCROE06]. Force10 Networks, "Building Scalable, High Performance Cluster and Grid Networks: The Role of Ethernet", *White Paper*, 2006.

[FOR10LOWL10G07]. Force10 Networks, "Low Latency 10 GbE Switching for Data Center, Cluster and Storage Interconnect", *White Paper*, 2007.

[H3CRRPP]. Huawei, "Configuration Guide - Ethernet Switching, Understanding RRPP", July 12, 2020.

[ITUTG8032]. ITU-T Recommendation G.8032, 'Ethernet Ring Protection Switching", March 2020.

[RFC1813]. B. Callaghan, B. Pawlowski, and P. Staubach, "NFS Version 3 Protocol Specification", *IETF RFC 1813*, June 1995.

[RFC3619]. S. Shah and M. Yip, "Extreme Networks' Ethernet Automatic Protection Switching (EAPS) Version 1", *IETF RFC 3619*, October 2003.

[RFC3821]. M. Rajagopal, E. Rodriguez and R. Weber, "Fibre Channel Over TCP/IP (FCIP)", *IETF RFC 3821*, July 2004.

[RFC4172]. C. Monia, R. Mullendore, F. Travostino, W. Jeong, and M. Edwards, "iFCP - A Protocol for Internet Fibre Channel Storage Networking", *IETF RFC 4172*, September 2005.

[RFC7143]. M. Chadalapaka, J. Satran, K. Meth, and D. Black, "Internet Small Computer System Interface (iSCSI) Protocol (Consolidated)", *IETF RFC 7143*, April 2014.

7 Applications of Switch/Router

7.1 INTRODUCTION

As discussed in the previous chapters, the switch/router combines a wide range of unique features to help organizations overcome the most challenging business requirements. This chapter discusses the typical networking applications of switch/routers, particularly, in enterprise and Internet Service Provider (ISP) networks. A switch/router uses Ethernet at Layer 1 and 2, and IP at Layer 3. Thus, given that Ethernet has evolved to become the dominant and widely adopted technology for building LANs, MANs, and WANs, this evolution has also influenced how switch/routers are used in enterprise and ISP networks. The switch/router provides a high-performance, cost-effective solution for many applications such as in data centers, midsize to large branch offices, high-speed server connectivity, multi-gigabit Ethernet aggregation, iSCSI storage, and High-Performance Computing (HPC) environments.

7.2 DATA CENTER AND APPLICATION HOSTING

Data centers are at the core of many business operations and require high-density, high-performance, high-security, and low-latency switching to ensure service connectivity to mission-critical applications. The increasing value of the data center to business operations necessitates that data and network integrity, confidentiality, and security must be maintained without impacting performance.

7.2.1 SWITCH/ROUTERS IN DATA CENTERS

As discussed in Chapter 6 of this volume, Ethernet has become the premier technology for implementing converged (or unified) data center switching fabrics that integrate general purpose LAN connectivity with storage and cluster interconnect capabilities. Recent improvements in the throughput, latency, and CPU utilization characteristics of intelligent Ethernet NICs allows data center managers to take advantage of the simplicity and cost-effectiveness of Ethernet to reduce or limit the use of specialty switching fabrics such as InfiniBand and Fibre Channel.

Particularly, multi-gigabit Ethernet technologies have reached maturity and now provide competitive network interconnectivity solutions for many networking applications. Solutions based on 10 Gigabit and higher capacity Ethernet are now widely deployed in data centers. The ubiquity and reliability of Ethernet make it a desirable solution for high-performance data center networks, storage, and compute fabric applications. The mass-market availability of 10 Gigabit Ethernet adapters and switches has enabled Ethernet-based networks to deliver high bandwidth, high

DOI: 10.1201/9781003311256-7

throughput, and low latency solutions for the rigorous demands of data center applications.

As discussed in Chapter 6, the combination of low latency 10 Gigabit Ethernet switching and intelligent NICs gave data centers the full benefits of 10 Gigabit Ethernet interconnectivity without the drawbacks of high CPU utilization and high end-to-end latency (that were issues with traditional Gigabit Ethernet). This opened the door for data center managers to implement unified switching fabric strategies that leverage the many strengths of TCP/IP and Ethernet as pervasive networking standards supported by huge ecosystems of hardware, system, and software vendors.

The virtualized data center provided the foundation for service-oriented architectures (SOAs). There are two aspects of SOA – the application and infrastructure aspects. These two aspects are highly complementary and are enablers for resource virtualization and data centers. From an application perspective, SOA is a virtual application architecture where the application comprises a set of component services (e.g., implemented with web services) that may be distributed throughout the data center or across multiple data centers. From an infrastructure perspective, SOA is a resource architecture where applications and services draw on a shared pool of resources rather than having physical resources rigidly dedicated to specific applications.

The industry has accepted that Ethernet leverages the broadest choice of application and service-aware appliances required by both legacy networked applications and SOA applications. These appliances include not only the usual load balancers and firewalls, but also intrusion prevention systems, SSL (Secure Sockets Layer) offload devices, and XML (Extensible Markup Language)-aware firewalls, accelerators, and gateways.

One important step to prepare for SOA is to develop a service-oriented infrastructure that provides the flexibility and adaptability required for highly dynamic application environments. Multi-gigabit Ethernet has become the essential networking technology for next-generation data centers. It provides optimal flexibility and investment protection in the evolution of the data center architecture regardless of how slow or how fast the application environment is transitioned toward SOA.

Particularly, multi-gigabit Ethernet also provides the performance overhead to deal with the less structured traffic patterns implicit with SOA, where new services are continually defined and demand for services can change significantly in a short time. SOA now accounts for a significant percentage of distributed applications, which means more network bandwidth is required to deal with the markup overhead of XML-based messaging. XML messages can consume more than 10 times the bandwidth of traditional binary messaging protocols.

The switch/router addresses these needs by acting as the gateway and switch fabric of the data center. The integration of Layer 2 and Layer 3 forwarding and the high port density potential of the switch/router allows for growth from the smallest to the largest data center. Also, port aggregation via, for example, IEEE 802.3ad Link Aggregation allows the switch/router to provide high-performance, high-capacity interconnectivity, increasing the availability of the server farm.

Security is a critical requirement in today's data centers and enterprises, and the switch/router provides robust security through a wide range of advanced features.

Organizations can use both regular and extended ACLs to control access to and through data center networks. Organizations can use control policies that permit or deny traffic based on a wide variety of packet identification characteristics – such as source/destination MAC addresses, source/destination IP addresses, and TCP/UDP ports – further protecting and restricting network access. In addition, for maximum security the switch/router can also leverage IEEE 802.1X security [IEEE 802.1X] (see Chapter 2 of this volume), MAC authentication, port MAC security, and MAC filter enhancements.

The switch/router can be provisioned with security features to protect the server farm against DoS attacks as well as provide security for maintaining network integrity. The sFlow functionality [RFC3176] (see Chapter 2 of this volume), for example, can provide the network access information required to track who has accessed to which server on the network, providing network usage audit trails. By utilizing wire-speed switching and filtering to screen and direct traffic to the appropriate server with minimal latency (while blocking undesired traffic), organizations can ensure the optimal operation, security, and integrity of their networks and data centers.

In data center environments where most servers are at least 1 Gigabit Ethernet-capable, the switch/router could serve, for example, as a compact and cost-effective multi-gigabit Ethernet aggregation switch. It connects to the data center core through wire-speed 10 Gigabit Ethernet while other devices (switches or routers) with 1 Gigabit Ethernet interfaces connect the 1 Gigabit Ethernet servers through 10 Gigabit Ethernet uplinks to the switch/router (Figure 7.1). This example is just for illustrative purposes – present-day servers can support directly 10 gigabit and higher Ethernet interfaces.

The switch/router could be designed to fit in server racks, and consume only one rack unit. To simplify cabling, the 10 Gigabit Ethernet interface cards in the servers connect to the switch/router's 10 Gigabit Ethernet ports using SFP+ (enhanced Small Form-Factor Pluggable) or direct attached SFP+ copper (Figure 7.2).

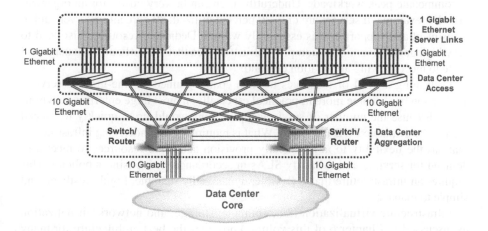

FIGURE 7.1 Switch/router provides data center aggregation with 1 Gigabit Ethernet interfaces to data centers.

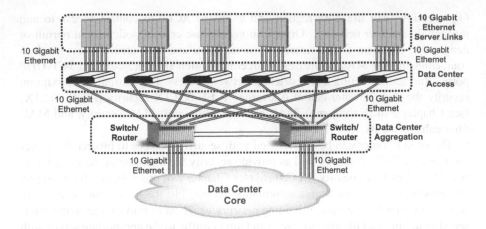

FIGURE 7.2 Switch/router provides data center aggregation with 10 Gigabit Ethernet interfaces to data centers.

7.2.2 VIRTUALIZATION OF DATA CENTERS

Legacy data center architectures are generally based on a modular approach in which each application is assigned dedicated resources in a point of delivery (POD) module consisting of servers, storage, switches, and other network devices. The physical and logical architecture within each POD has typically evolved to suit the needs of a particular application. For example, many enterprise applications, such as ERP (Enterprise Resource Planning), have PODs that are based on three tiers of servers: web servers, application servers, and database servers.

Dedicating physical resources to each application results in PODs with resources that are underutilized most of the time because their capacity must be scaled to accommodate peak workloads. Underutilization can be very costly for an organization. For example, if servers are only 20 percent utilized, 80 percent of server capital investment and support cost is essentially wasted. Dedicated resources also lead to very high device counts and complex infrastructures that are difficult to manage and scale to meet growing application demand.

Dedicating a POD of resources to each coarse-grained or fine-grained service/application has become more impractical especially in most large enterprise environments because the number of services/applications would grow quickly to exceed greatly the available POD resources. What is required is an underlying infrastructure that has the flexibility to automatically provision and scale resources to meet user demand for services and to satisfy SLAs in accordance with business policies. This requires an infrastructure that is flexible, dynamically scalable, highly resilient, and simple to manage.

Infrastructure virtualization (i.e., compute, storage, and network virtualization as discussed in Chapter 6 of this volume) provides the best architecture for many services in a cost-effective manner. Virtualization provides applications with technology-neutral logical interfaces that are abstractions of the underlying hardware and software resources. This is quite analogous to the logical interfaces that support the virtualization of the infrastructure's underlying compute, storage, and networking resources.

Ethernet has become the preferred transport technology for data centers and offers the best support for infrastructure virtualization within the PODs of the virtual data center. Popular virtual machine software, such as VMware, provides native support of Ethernet networking for both LAN and storage networking, allowing the infrastructure to be fully virtualized using Ethernet as the unified fabric for all information flows required to support virtualization.

Figures 7.1 and 7.2 show some examples of virtual data center POD employing aggregation/access switches (i.e., switch/routers). Multiple servers running the same application/service can be placed in the same application VLAN together with Ethernet-attached storage resources and appropriate appliances providing packet and application-level services (load balancing and security services). This means, physical and virtual servers dedicated to a specific application are placed in a VLAN reserved for that application.

Enterprise applications, such as ERP, that are based on distinct, segregated sets of web, application, and database servers can be implemented within a single tier of scalable Layer 2/Layer 3 switching using server clustering and distinct VLANs for segregation of web servers, application servers, and database servers. Alternatively, where greater scalability is required, the application could be distributed across a web server POD, an application server POD, and a database POD.

Where multiple applications or services are supported by a single POD, inter-switch links supporting IEEE 802.1Q VLAN trunking can be used for Layer 2 traffic. Where an application or service spans multiple physical PODs, the VLAN can be extended across the data center core. In general, the links between the aggregation/access switches and the core switches carry a combination of trunked intra-VLAN traffic switched at Layer 2 and inter-VLAN traffic switched at Layer 3.

7.3   HIGH-PERFORMANCE COMPUTING

The main barrier to performance improvements in current day microprocessors is on-chip power consumption. The rise in on-chip power consumption using current semiconductor technology is preventing major increases in clock speed. As a result, performance improvements in newer microprocessors come primarily from increased aggregate performance of multiple-cores per chip, coupled with only modest increases in clock rate.

This shift in microprocessor technology has resulted in compute-intensive enterprise applications transitioning to multi-threaded programming models that support efficient, parallelized program execution on multi-core servers and clusters of multi-core servers. However, performance for multi-threaded enterprise applications is highly dependent on low end-to-end latency and high bandwidth between sub-processes running on different processor cores or servers. As more enterprise applications transition to multi-threaded programming models, low latency cluster interconnects (based on higher-speed Ethernet technologies) have become the main-stream requirement for enterprise data centers.

Enterprises are increasingly deploying clusters of commodity servers that have been configured into highly cost-effective forms of supercomputers. As hardware and software technologies have matured and costs have declined, enterprises across a wide range of industries are leveraging HPC for product design and simulation, data analysis, and other highly compute-intensive applications that were previously beyond the reach of IT budgets.

Other applications are scientific/engineering applications involving complex simulations of discrete event systems, particle systems, and lumped/continuous variable systems. It is now not uncommon to see off-the-shelf clusters frequently use 10 Gigabit Ethernet as the cluster interconnect technology, but a number of cluster vendors are exploiting faster cluster interconnect fabrics (based on, for example, 25, 40, 50, 100, and higher Gigabit Ethernet) that feature very low message-passing latency.

7.3.1 SWITCH/ROUTERS IN HPC

HPC entered the mainstream marketplace with Ethernet switching and interconnects as the technology of choice. Ultra-low latency and high-density Ethernet switching are required for successful Ethernet-based HPC deployment. High-capacity, high-density switch/routers are ideal for this environment. For example, such a compact size switch/router could offer low latency through the device with high port densities of 1, 10, 25, 40, and higher Gigabit Ethernet.

This high-performance architecture offers terabits per second of data switching capacity to meet the needs of the most demanding HPC environment. The combination of performance, density, and reliability makes such a switch/router an excellent choice for enterprise HPC environments.

The above design features of switch/routers have become key for the most demanding HPC environments. In environments where high-speed inter-cluster connectivity is required over distance, organizations can use Ethernet ring technologies (e.g., ITU-T Recommendation G.8032, Ethernet Ring Protection Switching (ERPS) [ITUTG8032]) to provide dual-ring, fault-tolerant connectivity (Figure 7.3). Figure 7.3 shows HPC computing as a Grid of specialized clusters interconnected by a high-performance MAN, although such an arrangement can be implemented over a LAN or WAN.

The most important decision in the design process is whether to deploy an application on a local cluster or on a Grid of clusters dispersed within a department, campus, or enterprise. This decision will typically be based on performance goals in conjunction with a number of other considerations, including system management, security, and resource sharing policies within the organization.

From a performance perspective, the decision on whether to use a local cluster or Grid of clusters should be based on the analysis and implications of a number of factors. These factors include the degree of application parallelism, number of processors, types of processors (speed, memory size, etc.), operating systems, IPC mechanism, storage architecture and storage interconnection characteristics, and processor interconnect bandwidth, latency, and CPU utilization.

A Grid of clusters generally relies on the enterprise IP LAN/MAN/WAN network infrastructure for both internodal data transfers and IPC. For a campus LAN Grid of

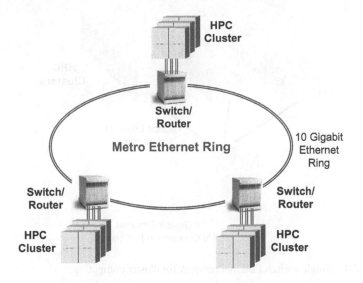

FIGURE 7.3 Example application where a switch/router provides HPC cluster inter-connectivity.

clusters, the IP network is typically implemented as a multi-tier switched Ethernet network based on multi-gigabit Ethernet interfaces. The network needs to have sufficient capacity for both Grid and non-Grid traffic coupled with QoS functionality to provide predictable bandwidth and latency for Grid IPC communications.

The network design should control subscription ratios in order to minimize the possibility of congestion that could result in considerable additional latency or even packet loss to the higher priority Grid IPC traffic. End-to-end latency for file transfers and message interchange to remote Grid sites may also be minimized by extending the QoS-enabled Ethernet network beyond the campus into the MAN and WAN.

The higher priority critical cluster/Grid traffics, such as Message Passing Interface (MPI) or other IPC traffic, are sensitive to throughput and latency in the network; thus, it is desirable to provide such traffic with the necessary QoS. QoS may be based on packet identification/classification using TCP port number or end system packet marking using the IEEE 802.1p or IP DiffServ specifications (see Chapters 8 and 9 of Volume 1). QoS may be enforced by providing strict priority to cluster/Grid traffic, in conjunction with enforcing policies for rate-limiting (policing and shaping) other traffic sources.

7.3.2 SWITCHED ETHERNET NETWORKS FOR HPC

Switched multi-gigabit Ethernet interconnects are excellent technologies for HPC cluster configurations such as the one shown Figures 7.4–7.7 [FOR10HPCROE06]. As discussed above, Ethernet switching has the advantage of being a highly scalable, very-low-cost alternative for IPC switching fabrics. The cost for switch ports and server adapters continues to drop as more enterprise desktops continue to migrate to Gigabit and 10 Gigabit Ethernet.

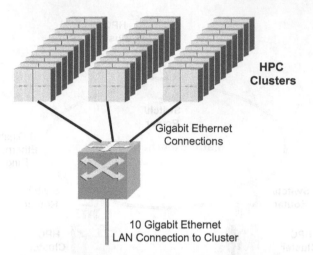

Gigabit Ethernet
Connections

10 Gigabit Ethernet
LAN Connection to Cluster

FIGURE 7.4   Single-switch Ethernet network for cluster computing.

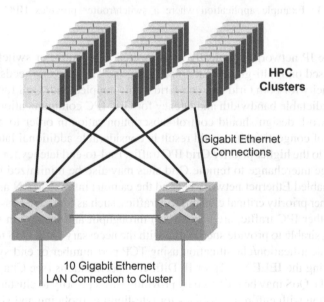

Gigabit Ethernet
Connections

10 Gigabit Ethernet
LAN Connection to Cluster

FIGURE 7.5   Dual-switch Ethernet network for cluster computing.

7.3.2.1   Single Switch Ethernet Network for HPC

A high availability, high-capacity switch/router can be used as a single cluster interconnect for a small HPC cluster as shown in Figure 7.4 [FOR10HPCROE06]. This
is the simplest type of cluster network, but requires a switch/router with high-performance features: non-blocking architecture, high-availability, high scalability, high
port density, etc. The maximum size of a single-switch cluster is determined by the
non-blocking port capacity of the switch. A possible configuration for this simple
setup could be as follows:

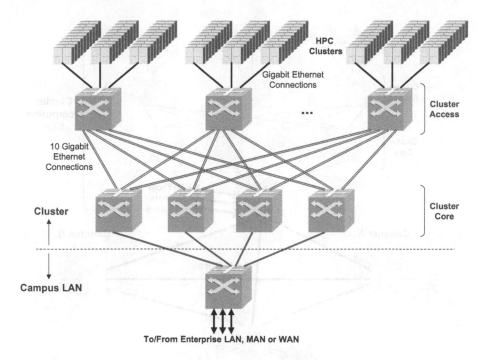

FIGURE 7.6   Meshed Ethernet switches for large clusters.

- All the server ports can be configured to be in the same port-based VLAN/ subnet with all server-to-server traffic switched at Layer 2.
- The uplinks can be configured for Layer 3 routing for connectivity to the site's general-purpose campus LAN. This is to allow control of the traffic sent between the cluster and the rest of the network using available Layer 3 routing and traffic control mechanisms. Traffic control mechanisms such as Layer 3 extended ACLs, policing, and shaping can be enabled on the Layer 3 interface of the switch/router to further control access to the cluster.

With cut-through switching (see discussion in Chapter 6 of this volume), the stand-alone switch/router in Figure 7.4 can be viewed as a single-tier cut-through switching fabric and provides the advantages of cut-through switching for smaller clusters and storage networks. Cut-through switching requires only one instance of serialization delay per hop, which can be a significant factor for larger frame sizes, and also is inherently simpler than store-and-forward switching.

The design in Figure 7.4 has some noticeable limitations. For a given size HPC cluster, the complexity of the network aggregating the cluster connections directly relates to the number of line-rate ports supported in the single chassis. The most cost-effective network is built with the highest density switch/routers. However, when the size of the cluster grows beyond the capacity of a single switch, retaining non-blocking connectivity between servers becomes a very big problem: the size of the aggregation network increases non-linearly and its cost rises dramatically. Also, with these increases come reliability concerns and additional management complexity.

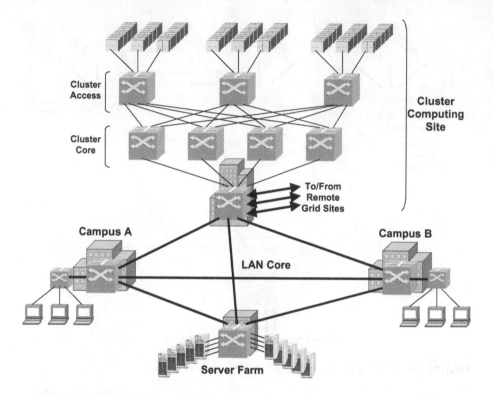

FIGURE 7.7 Switched Ethernet network for a campus grid of clusters.

A single switch/router may be very limited in its ability to meet future networking demands in a single chassis. This is because it lacks strong capacity growth features, the ability to support higher number of line rate 1, 10 and higher Gigabit Ethernet ports, and advanced high-availability features. With a blocking network architecture, the network manager needs to consider a number of variables in relation to the levels of service required in the network. The network manager can use traffic prioritization and QoS capabilities to fine-tune the network to avoid chokepoints and incorporate ACLs for network security and control.

It is not always easy and practical to find single switches that have the right features that can be used to build highly available clusters. Here, the impact of any downtime, scheduled, or not, is magnified. Aside from software-related high-availability features such as redundant protocols (e.g., VRRP as discussed in Chapter 1 of this volume), switch/routers deployed in Figure 7.4 will also be required to deliver network-availability features that include the following (see discussion in Chapter 1 of this volume):

- Redundancy of all critical elements (redundant Route Processor Modules (RPM), redundant Switch Fabric Modules (SFM), redundant power and cooling systems)
- Stateful failover of RPMs

- In-service software upgrades and maintenance
- Protected memory systems (all memory systems ECC/parity protected)
- Hot-swap and Online Insertion and Removal (OIR) of all components
- Clean separation of control and data planes
- System-wide environmental monitoring
- Persistent configuration synchronization between RPMs
- Cable management and front-side serviceability

The ultimate questions that need to be addressed when designing the HPC network in Figure 7.4 are as follows.

1. What is the ultimate capacity of the switch/router? Is there enough internal capacity in the chassis and chassis slots today to accommodate future changing requirements? Can the switch support both the LAN and WAN ports for 10 and higher Gigabit Ethernet interfaces and will they operate at full line rate?
2. Does the system offer line-rate performance when implementing all the QoS control and security features?
3. Can the line cards actually support all the bandwidth introduced by the ports or are they limited to local switching capacities only?
4. What happens if a critical system element fails? Can it be hot swapped? Does the chassis reboot? How long does it take the system to restore to full 100% operation?

7.3.2.2 Dual-Switch Ethernet Network for HPC

In Figure 7.5, the HPC cluster is equipped with dual redundant switch/routers configured for Layer 2 switching between nodes in the cluster, and Layer 3 switching (IP routing) between the cluster and the remainder of the enterprise LAN/WAN. Servers can be connected to the switch/routers using Gigabit Ethernet, Gigabit Ethernet trunks, or 10 Gigabit Ethernet depending on the performance level required.

In Figure 7.5, the switch/routers provide redundant routed connections to the Layer 3 core of the site LAN. One of the switch/routers is configured as the IEEE 802.1D primary root bridge for the Layer 2 connectivity among the servers, and as the VRRP primary router for connectivity to the core. The second switch would be configured to act as the secondary root bridge and the secondary VRRP router. This way, each switch/router plays a consistent role at both Layer 2 and Layer 3.

VRRP creates a virtual router interface that can be shared by two or more routers. The interface consists of a virtual MAC address and a virtual IP address. Each server is configured with the VRRP virtual interface as the default gateway. Each of the routers participating in the virtual interface is assigned a priority that determines which is the primary router. The VRRP routers multicast periodic "hello" messages that include the priority of the sending router. In the event the primary VRRP router fails, the secondary router detects the missing hello, determines that it has the highest remaining priority, and begins processing the traffic addressed to the virtual interface.

7.3.2.3   Meshed Ethernet Networks for HPC

Larger HPC clusters can be built using a meshed network of high-performance switch/routers as shown in Figure 7.6. The meshed network can be configured, for example, with following features:

- The servers on each access switch of the cluster can be configured to belong to a single port-based VLAN with Layer 2 switching enabled to provide connectivity among directly attached servers.
- The design offers fully redundant connections between the distribution and core tiers. Uplinks from the access layer to the distribution layer can be based on 1, 10, or higher Gigabit Ethernet interfaces. For higher bandwidths, IEEE 802.3ad Link Aggregation trunks based on multiple Gigabit Ethernet or even 10 Gigabit Ethernet links could be deployed.
- The uplink ports that provide connectivity to the core, and the core switches themselves, can be configured for Layer 3 routing. Routing is required because the core of the cluster is characterized by numerous parallel paths that need to load share effectively in order to provide non-blocking connectivity to each node switch. Layer 3 routing supporting Equal Cost Multi-Path (ECMP) routing provides superior load sharing and path recovery capabilities compared to Layer 2 protocols such as Spanning Tree Protocol (STP) and its newer variant Rapid Spanning Tree Protocol (RSTP).

The meshed network in Figure 7.6 can be built as a Layer 2 interconnect fabric using cut-through switches in both the aggregation and the access tiers of the network. All the inter-switch links in the access and core tiers are configured as IEEE 802.1Q trunks for all the access VLANs. For resiliency, two (or more, four shown in Figure 7.6) redundant aggregation switches can be used, so that, for any particular access VLAN, one of the aggregation switches is the primary switch (primary root) while the other switch plays a secondary (secondary root) or backup role. For a particular VLAN, only the Link Aggregation Group (LAG) to the primary aggregation switch would carry traffic under normal operating conditions, and the second uplink LAG would be blocked as demanded by the RSTP.

Furthermore, by exploiting IEEE 802.1Q VLAN tagging and the IEEE Multiple Spanning Tree Protocol (MSTP), the traffic load between the distribution and core tiers can be shared between the primary and secondary forwarding paths (assuming Figure 7.6 has two aggregation switches). This approach avoids the overhead costs of inactive redundant networking elements that stand by in a passive mode until a failure occurs. When a failure does occur, RSTP allows traffic to rapidly fail over from the primary to the secondary paths without operator intervention. With RSTP, failover periods can be as short as milliseconds to hundreds of milliseconds compared to as much as 30 seconds for the original IEEE 802.1D STP.

In a two-aggregation switch configuration, for example, access VLANs can be designated as either green or blue. One core switch is designated as the primary root for the green VLANs and another as the secondary root for the blue VLANs. The second core switch plays the complementary primary/secondary role. In normal operation, traffic for each color of VLAN flows through its primary root bridge. The green uplinks between the distribution and core normally block blue VLAN traffic and vice versa for the blue uplinks.

If the uplink carrying green VLAN traffic fails, the green VLAN traffic will fail over to the alternate (blue) uplink, which will now be carrying traffic for both green and blue VLANs. In this failure scenario, the green VLAN traffic would then traverse the core 802.1Q trunk to its primary root bridge, and from there, be forwarded to its destination. If the green VLAN primary root bridge (core switch) fails, the green VLAN traffic will be diverted to its secondary root bridge (the blue primary root). In this scenario, the remaining core switch will be forwarding traffic for both green and blue VLANs. This simple example shows how load sharing in a Layer 2 switched network can be based on having at least two classes or "colors" of VLAN and multiple instances of the Spanning Tree Protocol.

The number of access switches can be varied in proportion to the aggregation switches, depending on the degree of over-subscription that can be tolerated in the specific application. As described below, further expansion of the cluster network can be achieved by Layer 3 meshing of the switches, and by adding load sharing and the path recovery capabilities of ECMP routing. Incorporating multiple levels of high-availability in the network to drastically reduce the likelihood of "single component" catastrophic failures provides better protection.

7.3.2.4 Switched Ethernet Networks for Grid of Clusters

Figure 7.7 shows an example of a multi-tiered switched Ethernet network for a campus Grid of clusters. This network can be built with high-performance switch/routers supporting high-density 1, 10 and higher Gigabit Ethernet interfaces. This example design further highlights the advantages of Ethernet by showing how a single, homogeneous Ethernet switching solution supports cluster interconnection and also allows the cluster to participate in a broader campus or enterprise Grid structure. The main features of such a network are as follows:

- The core of the campus LAN is based on a meshed Layer 3 routing over 10 or higher Gigabit Ethernet links or trunks.
- Server farms outside of the cluster are connected via Gigabit Ethernet, Gigabit Ethernet trunks (using Link Aggregation), or 10 Gigabit Ethernet links to Layer 2/Layer 3 (multilayer) server site switches.
- End-user computers and workstations are connected to the campus Grid infrastructure via Gigabit Ethernet connections to Layer 2/Layer 3 (multilayer) access/distribution switches at each campus site.
- Connectivity to remote sites participating in the Grid may also be provided over the MAN and WAN by 10 or higher Gigabit Ethernet over dark fiber, multiple 10 Gigabit Ethernet links over Coarse Wavelength Division Multiplexing (CWDM) or Dense Wavelength Division Multiplexing (DWDM) over dark fiber, or by other telecommunications services that support multi-gigabit Ethernet interfaces.

The network in Figure 7.7 will have to be designed to maximize the predictability of the service level delivered to Grid traffic. The first consideration would be to ensure that the network has sufficient capacity for the incremental traffic expected due to the Grid applications. Once this has been done, the network designer can exploit the QoS capabilities of the switch/router to prioritize critical Grid traffic. The design goal is to give high priority to general Grid traffic, while ensuring that IPC traffic and other more critical Grid traffic does not incur any added latency by queuing behind less critical traffic.

7.4 ENTERPRISE INFRASTRUCTURE

Ethernet's high performance, simplicity, and cost-effectiveness has made it the networking technology of choice in the enterprise. Enterprise architectures are now handling increasingly diverse application traffic. They are also supporting more Gigabit Ethernet-attached servers and storage devices. Packetized media traffic, streaming content, and VoIP are increasingly being supported and demand more stringent latency and delay variation performance from the network. In addition, data storage is being centralized in SANs with iSCSI connectivity. As a result, the enterprise is required to accommodate and secure the increased volumes of storage data.

10 and higher Gigabit Ethernet interconnects are fast becoming the backbone technology for enterprise and service provider networks (Figure 7.8). High-performance multi-gigabit Ethernet interfaces give enterprises the functionality and scalability they need to build their next-generation backbones cost-effectively. For example, 10 Gigabit Ethernet has matured to the point that it provides cost-effective network connectivity with high bandwidth and low latency, in addition to all the available network management capabilities that come with IP and Ethernet

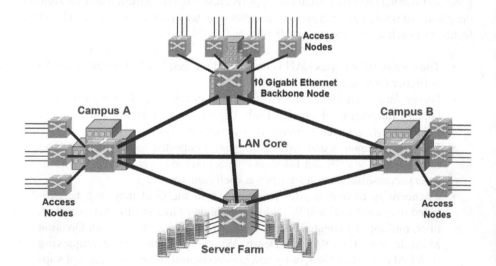

FIGURE 7.8   Enterprise network with 10 Gigabit Ethernet backbone.

networking. The wide-scale adoption of multi-gigabit Ethernet is mainly driven by the following factors:

- There are now switches with 10, 25, 40, 100, and higher Gigabit Ethernet interfaces in the market that provide the line-rate ACLs, QoS, and queuing features needed for network security, performance, and availability.
- 10 Gigabit Ethernet, for instance, is competitive to InfiniBand and Fibre Channel, and delivers the bandwidth and low latency required to support storage networking. The bandwidth and latency characteristics of 10 Gigabit Ethernet provide a common transport mechanism for supporting data and storage traffic across the entire Enterprise.
- Multi-gigabit Ethernet interconnects simplify backbone engineering by enabling remote buildings, data centers, and storage facilities to appear as simple extensions of the LAN.
- Multi-gigabit Ethernet interconnects reduce the need to support other parallel MAN and WAN technologies and connections by providing transport over multiple wide-area media: SONET/SDH, DWDM, CWDM, and dark fiber. Ethernet is a flexible technology supporting short distances across the LAN, intermediate distances across the MAN, and long haul across the WAN with DWDM, CWDM, and SONET/SDH compatibility.

Today's enterprise networks are critical to the operations of organizations. Network administrators are concerned about zero downtime on the network, securing the network from attacks (DoS attacks, cyber-spying, malicious users), and maintaining data integrity and confidentiality, without adding excessive cost or impacting performance. All of these require a structure that allows for graceful growth as the enterprise grows. To build a network that will continue to provide non-blocking access between users and resources as it grows, the enterprise must deploy a backbone that can scale with meshed nodes and redundant paths. This topology requires a switch/router that can support multiple multi-gigabit Ethernet trunks and forward traffic at a line rate between them.

The switch/router in this case must incorporate high resiliency, security, and scalability and provide an architecture that can possibly scale from the edge to the core in order to minimize the total cost of ownership (TCO). As traffic begins to travel at multi-gigabit speeds across the backbone, carrier-class redundancy and availability become essential (see Chapter 1 of this volume). The resilient design features include redundant RPMs, SFM, cooling fans, and power supplies. The hardware resiliency can be enhanced with software resiliency including hitless system failover, graceful restart, and VRRP for Layer 3 resiliency.

As enterprise networks move to high-performance Ethernet and IP technologies, the amount of mission-critical data flowing across the enterprise backbone continues to increase dramatically. From remote servers to centralized storage systems, each network link is expected to carry traffic to/from hundreds of users and applications. Protecting the surging amount of mission-critical data is imperative. This situation also clearly demands enhanced network security and higher availability.

Wire-speed security is required in the switch/routers to lock out unauthorized users, filter DoS and unauthorized traffic with ACLs, and monitor traffic flows with mechanisms such as sFlow, NetFlow, and RMON (see Chapter 2 of this volume). At the same time, high-priority real-time traffic that travel through the switch/router must utilize desirably high-performance hardware-based QoS mechanisms.

Today, more enterprise users are participating in shared sessions such as audio/video events, seminars, lectures, distributed meetings, workgroup collaboration, tutorials, and training. As multicast traffic grows, backbone switch/routers must support hardware-based multicasting to keep the backbone traffic running at line rate. Multicast support, as it existed in previous generation switch/routers, was implemented in the "slow-path" packet forwarding path (see Chapters 2, 5, and 6 of Volume 1). Networks with this software-based form of multicast can support only a small percentage of multicast traffic before their overall performance begins to decline. Leveraging ASIC-based multicast forwarding, the switch/router is better positioned to deliver full line-rate multicasting with packet replication across the switch fabric.

7.5 FEATURES OF 10 GIGABIT ETHERNET SWITCH/ROUTERS

In addition to offering a highly cost-effective network interconnect, the current generation of market-leading multi-gigabit Ethernet enterprise switches has been designed from the outset to meet the most demanding requirements for increased levels of aggregated traffic in the data center, as well as the core and access/aggregation tiers of the network. Switch/routers with these attributes provide an enabling infrastructure for modern data center consolidation and virtualization of networking, compute, and storage resources.

The typical features of switch/routers that support multi-gigabit Ethernet interfaces (for consolidation and virtualization of storage resources) include high port density, non-blocking performance, resiliency/high availability, low latency, cut-through switching, and WAN extensibility. These features are described below:

- **High Port Density**: This simplifies the design of data center networks and allows a large number of servers to readily share IP storage resources.
- **Non-Blocking Performance**: This feature assures that line-rate performance can be delivered simultaneously on all switch ports, maximizing storage access performance.
- **Resiliency/High Availability**: This includes a high degree of software resiliency and hardware redundancy among critical switch/router subsystems (e.g., RPMs, SFMs, cooling fans, and power supplies) to provide the reliability required for mission-critical storage access.
- **Low Latency, Cut-Through Switching**: This mode of switching is required for reducing network latency within the IP storage portion of the Ethernet switching fabric. Layer 2 cut-through switches support switching latencies in the order of 300 nanoseconds compared to the latencies of several microseconds in typical store-and-forward switch.
- **WAN Extensibility**: WAN extensibility of the Ethernet network allows geographically dispersed IP storage to be virtualized into a logical

Ethernet-based storage resource, simplifying data access and data management (including disk-to-disk backups and disaster recovery facilities). Alternately, IP storage requests can be routed over any existing IP MAN or WAN to reach remote storage facilities without requiring additional gateway devices.

REFERENCES

[IEEE 802.1X]. IEEE Std 802.1X-2004- Port-Based Network Access Control.

[FOR10HPCROE06]. Force10 Networks, "Building Scalable, High Performance Cluster and Grid Networks: The Role of Ethernet", *White Paper*, 2006.

[ITUTG8032]. ITU-T Recommendation G.8032, Ethernet Ring Protection Switching, March 2020.

[RFC3176]. InMon Corporation's sFlow, "A Method for Monitoring Traffic in Switched and Routed Networks", *IETF RFC 3176*, September 2001.

Index